GROWTH HORMONE SECRETAGOGUES
Basic Findings and Clinical Implications

Growth Hormone Secretagogues

Basic Findings and Clinical Implications

Edited by

E. Ghigo
M. Boghen
F.F. Casanueva
C. Dieguez

University of Torino, Italy
and
University of Santiago de Compostela, Spain

1999
ELSEVIER
AMSTERDAM – LAUSANNE – NEW YORK – OXFORD – SHANNON –SINGAPORE – TOKYO

ELSEVIER SCIENCE B.V.
Sara Burgerhartstraat 25
P.O. Box 211, 1000 AE Amsterdam, The Netherlands

First edition 1999

Library of Congress Cataloging in Publication Data:
A catalog record from the Library of Congress has been applied for.

ISBN: 0-444-82933-4

∞ The paper used in this publication meets the requirements of ANSI/NISO Z39.48-1992 (Permanence of Paper).

Transferred to digital printing 2006.

Prologue

With the recent cloning of the growth hormone secretagogue (GHS)-receptor, the early vision of Cyril Bowers in which GH-releasing hexapeptides were considered to represent a new physiological system implicated in the regulation of GH secretion has been confirmed. GHSs, administered alone or in combination with GHRH, are effective probes for the diagnosis of GH deficiency in both children and adults. Currently, most pharmaceutical companies involved in the GH field, and several independent groups, have developed and tested different GHS compounds that are active by the oral route, and have improved potency and bioavailability, giving rise to exciting therapeutic possibilities.

The volume and diversity of recent research on GHS makes it almost impossible to present a comprehensive and coherent review of the whole subject. However, in undertaking the challenge of gathering, in one multiauthor volume, the existing data covering the entire field of GHS, we were lucky in obtaining an enthusiastic response from the leading experts in this area. Each contributor has advanced the field of knowledge, and was asked to emphasise the practical aspects of their work, reviewing the subject in the light of their own experience. Therefore, the theme of this monographic book is a practical one. It deals with all aspects of GHS that are relevant to the field, from the chemical structure of the different analogues, to the cloning and expression of the GHS-receptor and the role of these compounds in the physiological control of GH secretion. Also discussed at length in several chapters, are the most recent advances in relation to the possible role of these compounds in the diagnostic therapeutic settings in different clinical situations, either in children, adults or the elderly.

We are pleased that this volume meets the requirements of covering most, if not all, of the advances in the field. It will therefore enable scientists and clinicians to keep abreast of the rapidly evolving knowledge that we have witnessed in recent years, and should prove useful as a review for those interested in the topics discussed.

We would like to thank all the authors for helping to make this book a reality. We would also like to acknowledge the financial support given by Pharmacia through an educational grant.

The Editors

Contents

Contributing Authors

Erik F. Adams
Pharmaceutical Sciences Institute, Aston University, Birmingham, UK

Emanuela Arvat
Division of Endocrinology, University of Turin, 10126 Turin, Italy

Ariel L. Barkan
Professor of Medicine (Endocrinology and Metabolism) and Surgery (Neurosurgery), University of Michigan Medical Center, Division of Endocrinology and Metabolism, 3920 Taubman Center, Ann Arbor, MI 48109-0354, USA

Bengt-Åke Bengtsson
Research Centre for Endocrinology and Metabolism, Gröna Stråket 8, Sahlgrenska University Hospital, S-413 45 Göteborg, Sweden

Barry B. Bercu
Professor of Pediatrics, Pharmacology and Therapeutics, All Children's Hospital, Department 6900, 801 Sixth Street South, St. Petersburg, FL 33701, USA

Ferruccio Berti
Department of Pharmacology, Chemiotherapy & Medical Toxicology, University of Milan, Via Vanvitelli, 32, 20129 Milan, Italy

Cyril Y. Bowers
Tulane University Medical School, 1430 Tulane Avenue, New Orleans, LA 70112, USA

Nathalie Briard
Laboratoire de Neuroendocrinologie Expérimentale, INSERM U 501, Faculté de Médecine, Boulevard Pierre Dramard, 13916 Marseille Cedex 20, France

Fabio Broglio
Division of Endocrinology, University of Turin, 10126 Turin, Italy

Franco Camanni
Division of Endocrinology, University of Turin, 10126 Turin, Italy

Felipe F. Casanueva
Endocrinology Section, Dept. of Medicine, Santiago de Compostela University, P.O. Box 563, 15780 Santiago de Compostela, Spain

Chen Chen
Prince Henry's Institute of Medical Research, P. O. Box 5152, Clayton, Victoria 3168, Australia

Jens Sandahl Christiansen
Institute of Experimental Clinical Research, University of Aarhus, and Medical Dep M (Endocrinology & Diabetes), Aarhus Kommunehospital, DK-8000 Aarhus, Denmark

Ross Clark
Research Centre for Developmental Medicine & Biology, School of Medicine and Health Sciences, University of Auckland, Auckland, New Zealand

Iain J. Clarke
Prince Henry's Institute of Medical Research, P. O. Box 5152, Clayton, Victoria 3168, Australia

Frédéric Dadoun
Service d'Endocrinologie, Maladies Métaboliques et de la Nutrition, Hôpital Nord, Chemin des Bourrely, 13915 Marseille Cedex 20, France, and Laboratoire de Neuroendocrinologie Expérimentale, INSERM U 501, Faculté de Médecine, Boulevard Pierre Dramard, 13916 Marseille Cedex 20, France

Vito De Gennaro Colonna
Dept. of Pharmacology, Chemiotherapy & Medical Toxicology, University of Milan, Via Vanvitelli, 32, 20129 Milan, Italy

Romano Deghenghi
Europeptides, Bt. Aristote, 9, Avenue du Marais, 95108 Argenteuil Cedex, France

Suzanne L Dickson
Department of Physiology, University of Cambridge, Downing Street, Cambridge CB2 3EG, UK

Carlos Dieguez
Department of Physiology, Faculty of Medicine, University of Santiago de Compostela, Spain

Eleni V. Dimarki
Research Fellow, Division of Endocrinology and Metabolism, Department of Internal Medicine, University of Michigan, Ann Arbor, MI 48109-0354, USA

Anne Dutour
Service d'Endocrinologie, Maladies Métaboliques et de la Nutrition, Hôpital Nord, Chemin des Bourrely, 13915 Marseille Cedex 20, France, and Laboratoire de Neuroendocrinologie Expérimentale, INSERM U 501, Faculté de Médecine, Boulevard Pierre Dramard, 13916 Marseille Cedex 20, France

Scott D. Feighner
Dept. of Metabolic Disorders, Merck Research Laboratories, Building RY-80-265, P.O. Box 2000, Rahway, NJ 07065, USA

Lawrence A. Frohman
Department of Medicine, University of Illinois at Chicago, 840 S. Wood Street (MC 787), Chicago, IL 60612, USA

Ricardo V. Garcia-Mayor
Internal Medicine, Hospital Xeral Cies, Vigo, Spain

Ezio Ghigo
Division of Endocrinology, University of Turin, Ospedale Molinette, C. Dogliotti 14, 10126 Turin, Italy

Roberta Giordano
Division of Endocrinology, University of Turin, 10126 Turin, Italy

Michel Grino
Laboratoire de Neuroendocrinologie Expérimentale, INSERM U 501, Faculté de Médecine, Boulevard Pierre Dramard, 13916 Marseille Cedex 20, France

Ashley B. Grossman
Department of Endocrinology, St Bartholomew's Hospital, West Smithfield, London EC1A 7BE, UK

Viviane Guillaume
Service d'Endocrinologie, Maladies Métaboliques et de la Nutrition, Hôpital Nord, Chemin des Bourrely, 13915 Marseille Cedex 20, France, and
Laboratoire de Neuroendocrinologie Expérimentale, INSERM U 501, Faculté de Médecine, Boulevard Pierre Dramard, 13916 Marseille Cedex 20, France

Andrew D. Howard
Dept. of Metabolic Disorders, Merck Research Laboratories, Building RY-80-265, P.O. Box 2000, Rahway, NJ 07065, USA

Donna L. Hreniuk
Dept. of Metabolic Disorders, Merck Research Laboratories, Building RY-80-265, P.O. Box 2000, Rahway, NJ 07065, USA

John-Olov Jansson
Research Centre for Endocrinology and Metabolism, Gröna Stråket 8, Sahlgrenska University Hospital, S-413 45 Göteborg, Sweden

Richard C. Jenkins
Department of Medicine, University of Sheffield, Clinical Sciences Centre, Northern General Hospital, Herries Road, Sheffield S5 7AU, UK

Rhonda D. Kineman
Department of Medicine, University of Illinois at Chicago, Chicago, IL 60612, USA

Márta Korbonits
Department of Endocrinology, St Bartholomew's Hospital, West Smithfield, London EC1A 7BE, UK

Steven W.J. Lamberts
Professor of Medicine, Department of Medicine, University Hospital Dijkzigt, 40 Dr. Molewaterplein, 3015 GD Rotterdam, The Netherlands

Zvi Laron
Director, Endocrine & Diabetes Research Unit, Schneider Children's Medical Center, 14 Kaplan Street, 49202 Petah Tikva, Israel

Alfonso Leal-Cerro
Endocrinology Unit, Hospital Virgen del Rocio, Seville, Spain

Mauro Maccario
Division of Endocrinology, University of Turin, 10126 Turin, Italy

Karen Kulju McKee
Dept. of Metabolic Disorders, Merck Research Laboratories, Building RY-80-265, P.O. Box 2000, Rahway, NJ 07065, USA

Shlomo Melmed
Academic Affairs, Cedars-Sinai Medical Center, 8700 Beverly Blvd., B-131, Los Angeles, CA 90048, USA

Dragan Micic
Institute of Endocrinology, University Clinical Center, Belgrade, Yugoslavia

Niels Møller
Institute of Experimental Clinical Research, University of Aarhus, and Medical Dep M (Endocrinology & Diabetes), Aarhus Kommunehospital, DK-8000 Aarhus, Denmark

Giampiero Muccioli
Division of Endocrinology, University of Turin, 10126 Turin, Italy

Ravi Nargund
Dept. of Medicinal Chemistry, Merck Research Laboratories, Building RY-80-265, P.O. Box 2000, Rahway, NJ 07065, USA

Ralf M. Nass
Division of Endocrinology and Metabolism, Department of Medicine, University of Virginia, Box 511-66, Charlottesville, VA 22908, USA

Charles Oliver
Service d'Endocrinologie, Maladies Métaboliques et de la Nutrition, Hôpital Nord, Chemin des Bourrely, 13915 Marseille Cedex 20, France, and Laboratoire de Neuroendocrinologie Expérimentale, INSERM U 501, Faculté de Médecine, Boulevard Pierre Dramard, 13916 Marseille Cedex 20, France

Oksana C. Palyha
Dept. of Metabolic Disorders, Merck Research Laboratories, Building RY-80-265, P.O. Box 2000, Rahway, NJ 07065, USA

Arthur A. Patchett
Dept. of Medicinal Chemistry, Merck Research Laboratories, Building RY-80-265, P.O. Box 2000, Rahway, NJ 07065, USA

Angela Peñalva
Department of Medicine, University of Santiago de Compostela, Spain

Manuel Pombo
Department of Pediatrics, University of Santiago de Compostela, Spain

Sheng-Shung Pong
Dept. of Metabolic Disorders, Merck Research Laboratories, Building RY-80-265, P.O. Box 2000, Rahway, NJ 07065, USA

Vera Popovic
Institute of Endocrinology, University Clinical Center, Belgrade, Yugoslavia

Asad Rahim
Department of Endocrinology, Christie Hospital NHS Trust, Wilmslow Road, Manchester M20 4BX, UK

Richard J.M. Ross
Department of Medicine, University of Sheffield, Clinical Sciences Centre, Northern General Hospital, Herries Road, Sheffield S5 7AU, UK

Giuseppe Rossoni
Dept. of Pharmacology, Chemiotherapy & Medical Toxicology, University of Milan, Via Vanvitelli, 32, 20129 Milan, Italy

Nicole Sauze
Laboratoire de Neuroendocrinologie Expérimentale, INSERM U 501, Faculté de Médecine, Boulevard Pierre Dramard, 13916 Marseille Cedex 20, France

Stephen M. Shalet
Department of Endocrinology, Christie Hospital NHS Trust, Wilmslow Road, Manchester M20 4BX, UK

Tamotsu Shibasaki
Department of Physiology, Nippon Medical School, Sendagi 1-1-5, Bunkyo-ku, Tokyo 113-8603, Japan

Ilan Shimon
Institute of Endocrinology, Sheba Medical Center, Tel-Hashomer, Israel

Roy G. Smith
Baylor College of Medicine, One Baylor Plaza, M320, Houston, TX 77030, USA

Axel Steiger
Max Planck Institute of Psychiatry, Department of Psychiatry, Kraepelinstrasse 10, D-80804 Munich, Germany

Hitoshi Sugihara
Department of Medicine, Nippon Medical School, Sendagi 1-1-5, Bunkyo-ku, Tokyo 113-8603 Japan

Johan Svensson
Research Centre for Endocrinology and Metabolism, Gröna Stråket 8, Sahlgrenska University Hospital, S-413 45 Göteborg, Sweden

Carina P. Tan
Dept. of Metabolic Disorders, Merck Research Laboratories, Building RY-80-265, P.O. Box 2000, Rahway, NJ 07065, USA

Michael O. Thorner
Division of Endocrinology and Metabolism, Department of Medicine, University of Virginia, Box 511-66, Charlottesville, VA 22908, USA

Greet Van den Berghe
Department of Intensive Care Medicine, University Hospital Gasthuisberg, University of Leuven, B-3000 Leuven, Belgium

Lex H.T. Van Der Ploeg
Dept. of Metabolic Disorders, Merck Research Laboratories, Building RY-80-265, P.O. Box 2000, Rahway, NJ 07065, USA

Ichiji Wakabayashi,
Department of Medicine, Nippon Medical School, Sendagi 1-1-5, Bunkyo-ku, Tokyo 113-8603, Japan

Richard F. Walker
Director of Pharmaceutical Studies and Research Compliance, University of South Florida, St. Petersburg, FL, USA

Growth Hormone Secretagogues
Edited by E. Ghigo, M. Boghen, F.F. Casanueva and C. Dieguez

Chapter 1

Introduction

STEVEN W.J. LAMBERTS

Department of Medicine, University Hospital Dijkzigt, Rotterdam, The Netherlands

Growth Hormone-Releasing Peptides (GHRP) were discovered by C.Y. Bowers and his group in 1976. Experimenting with the metenkephalin molecule they identified through theoretic low-energy conformational calculations, computer modelling, and structural modification a series of small peptides that are able to stimulate GH secretion (1–3). At that time these compounds were not considered to be endogenous releasing peptides.

In the early eighties the first highly potent GHRP-6 (hexapeptide) was developed (4). This compound increases GH secretion acting both at the hypothalamic and pituitary level (4). Subsequent studies into the nature and activity of GHRP-6 and its follow-up compounds has led to a number of findings which are relevant to clinical practice (5–8).

In retrospect it is interesting to note that GHRPs were constructed well before the isolation and characterization of Growth Hormone Releasing Hormone (GHRH) in 1982 (9). For many years it was thought that the pulsatile secretion of GH by the pituitary somatotrophs was controlled by only two antagonistic hypothalamic peptides: somatostatin which inhibits GH release and GHRH which stimulates GH release. Both peptides had been purified and well characterized, while their specific receptors have been cloned. Both GHRH and somatostatin receptors belong to the family of seven transmembrane receptors coupled to a heterotrimeric GTP-binding protein. The somatostatin receptor is coupled to a Gi protein and its activation inhibits adenylate cyclase. On the other hand, the GHRH receptor is coupled to a Gs protein and its activation stimulates adenylate cyclase activity leading to increased intracellular cyclic AMP levels. The discovery of the GHRPs has led to the hypothesis of a third endocrine pathway controlling GH secretion, and indeed human GHRP receptors eventually have been cloned in the anterior pituitary and hypothalamus (10,11). GHRP pituitary action is mediated via a phosphoinositol-protein kinase-C intracellular pathway, while its hypothalamic action is not yet firmly established (12). The latter might involve the release of endogenous GHRH as well as inhibition of somatostatin release and/or the action of an as yet unknown hypothalamic factor. In Figure 1 a scheme is represented of the regulation of GH secretion in which the three hypothalamic regulatory

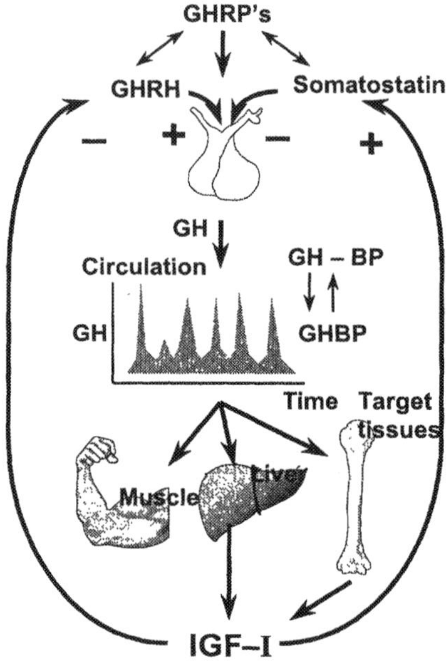

Figure 1. Schematic description of the regulation of the Growth Hormone (GH)-Insulin-like Growth Factor-I (IGF-I)-axis.

systems are depicted. In addition the episodic, pulsatile nature of GH secretion is shown, as well as its binding to a specific GH-binding protein within the circulation (GHBP). Finally in this figure it is indicated that most of the biological effects of GH are mediated via peripherally formed growth factors, the most important being Insulin-like Growth Factor-I (IGF-I).

Over the last twenty years a great number of peptidyl and non-peptidyl GH secretagogues have been developed (6,13). The GHRPs which were initially designed as effective releasers of GH in animals and man had to be administered intravenously or subcutaneously, but subsequently also intranasally and orally active compounds became available. Among these, the non-peptidyl GHRPs (L-692429, L-692585, MK-677) have already been studied extensively in man (6,13). More recently, a number of cyclic peptides, as well as penta-, tetra-, and pseudotripeptides have also been synthesized and tested in animals (6,13).

GH is currently used extensively in the treatment of GH-deficient children, as well as in GH-deficient adults. GH is administered once daily, and therefore does not mimic the normal pulsatile release pattern of GH. Also synthetic IGF-I is available for clinical studies, but a potential disadvantage of its use is the occurrence of hypoglycemia.

In theory both the GHRPs and GHRH would be attractive alternatives for GH and IGF-I in the activation of the GH/IGF-I-axis in patients with absolute or relative (aging, catabolism, burns) GH-deficiency, as long as somatotroph activity is intact (Figure 2). Theoretically an orally active GH secretagogue induces a GH secretory pattern which is close to

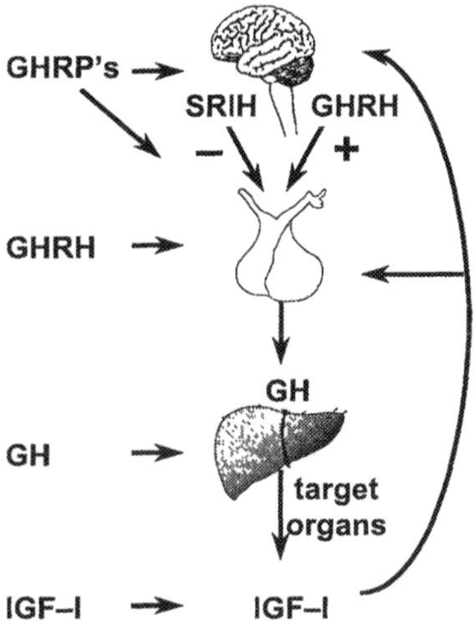

Figure 2. Four potential therapeutic interventions to activate the GH/IGF-I-axis: 1. GHRPs, 2. GHRH, 3. GH, 4. IGF-I.

the physiological GH secretion, inducing IGF-I levels within normal limits (14). Therefore such an oral compound would have major advantages above GH and IGF-I with regard to tolerability, compliance and the incidence of adverse effects (5,6). These aspects will be extensively discussed throughout this volume.

One other aspect of the GHRPs should be mentioned in this Introduction as well. Although GHRPs were initially based on an opioid peptide structure, they are devoid of opioid activity. Still their GH-releasing activity is not in all instances specific.

GH secretagogues also have a stimulatory effect on Prolactin (PRL), Adrenocorticotropin (ACTH) and cortisol secretion both in animals and in man. The mechanism underlying the PRL releasing activity is unclear. A hypothalamus-mediated effect has not been demonstrated, while a direct stimulation of pituitary somatomammotrope cells has been hypothesized (15). Clinical effects of chronic elevated PRL levels during treatment with GH secretagogues might therefore include mastopathy, galactorrhea, and/or a loss of libido. The stimulatory effect of GHRPs on the activity of the hypothalamo-pituitary-adrenal axis in man might be even more cumbersome (16–18). The acute ACTH and cortisol response after the start of GHRP's administration seems to be mediated via the hypothalamus, as it is lost after cutting the pituitary stalk (19). A major unknown factor is whether the stimulatory effect of GHRPs in man will eventually be lost during prolonged administration, but even a minor activation of the hypothalamo-pituitary-adrenal axis would on clinical grounds be unacceptable, as adverse effects in many organ systems have to be expected.

The characterization and availability of the GHRPs is one of the most exciting developments in experimental and clinical neuroendocrinology. Apart from their potential use in diagnosis and therapy of disease, they have given new insight into the physiology and pathophysiology of GH secretion.

REFERENCES

1. Bowers, C.Y., Chang, J., Momany, F., Folkers, K.A. (1977) Effects of the enkephalins and enkephalin analogs on release of pituitary hormones *in vitro*. In: Molecular Endocrinology. I. Macintyre (Ed.). Elsevier/North Holland Biochemical Press, Amsterdam, pp. 287–292.
2. Bowers, C.Y., Momany, F., Reynolds, G.A., Hong, A. (1981) A study on the regulation of growth hormone release from the pituitaries of rats *in vitro*. Endocrinology 108, 1071–1080.
3. Momany, F.A., Bowers, C.Y., Reynolds, G.A., Chang, D., Hong, A., Newlander, K. (1981) Design synthesis and biological activity of peptides which release growth hormone *in vitro*. Endocrinology 108, 31–39.
4. Bowers, C.Y., Momany, F.A., Reynolds, G.A., Hong, A. (1984) On the *in vitro* and *in vivo* activity of a new synthetic hexapeptide that acts on the pituitary to specifically release growth hormone. Endocrinology 114, 1537–1545.
5. Ghigo, E., Arvat, E., Muccioli, G., Camanni, F. (1997) Growth hormone-releasing peptides. Eur. J. Endocrinol. 136, 445–460.
6. Camanni, F., Ghigo, E., Arvat, E. (1998) Growth hormone-releasing peptides and their analogs. Front. Neuroendocrinol. 19, 47–72.
7. Korbonits, M., Grossman, A.B. (1995) Growth hormone-releasing peptide and its analogues. TEM 6, 43–49.
8. Saenger, P. (1996) Editorial: oral growth hormone secretagogues — Better than Alice in Wonderland's growth elixir? J. Clin. Endocrinol. Metab. 81, 2773–2775.
9. Bertherat, J. (1997) Cloning of the growth hormone secretagogues receptor cDNA: new evidence for a third endocrine pathway controlling growth hormone release. Eur. J. Endocrinol. 136, 37–38.
10. Pong, S.S., Chaung, L.Y.P., Dean, D.C., Nargund, R.P., Patchett, A.A., Smith, R.G. (1996) Identification of a new G-protein linked receptor for growth hormone secretagogues. Mol. Endocrinol. 10, 57–61.
11. Howard, A.D., Feighner, S.D., Cully, D.F. et al. (1996) A receptor in pituitary and hypothalamus that functions in growth hormone release. Science 273, 974–977.
12. Bowers, C.Y. (1998) GHRP: Unnatural to natural. Abstr. L9, American Endocrine Society, New Orleans.
13. Smith, R.G., van der Ploeg L.H.T., Howard, A.D. et al. (1997) Peptidomimetic regulation of growth hormone secretion. Endo. Rev. 18, 621–645.
14. Smith, R.G., Cheng, K., Schoen, W.R. et al. (1993) A nonpeptidyl growth hormone secretagogue. Science 260, 1640–1643.
15. Renner, U., Brockmeier, S., Strasburger, G.J. et al. (1994) Growth hormone (GH)-releasing peptide stimulation of GH release from human somatotroph adenoma cells: interaction with GH-releasing hormone, thyrotropin-releasing hormone, and octreotide. J. Clin. Endocrinol. Metab. 78, 1090–1096.
16. Copinschi, G., van Onderbergen, A., L'hermite-Baleriaux, M. et al. (1996) Effects of 7-day treatment with a novel, orally active, growth hormone (GH) secretagogue, MK-677, on 24-h GH profiles, Insulin-like growth factor I, and adrenocortical function in normal young men. J. Clin. Endocrinol. Metab. 81, 2776–2782.
17. Bowers, C.Y., Reynolds, G.A., Durham, D., Barrera, C.M., Pezzoli, S.S., Thorner, M.O. (1990) Growth hormone (GH)-releasing peptide stimulates GH-release in normal men and acts synergistically with GH-releasing hormone. J. Clin. Endocrinol. Metab. 70, 975–982.
18. Ghigo, E., Arvat, E., Gianotti, L. et al. (1994) Growth hormone-releasing activity of hexarelin, a new synthetic hexapeptide, after intravenous, subcutaneous, intranasal and oral administration in man. J. Clin. Endocrinol. Metab. 78, 693–698.
19. Loche, S., Cambiaso, P., Carta, D. et al. (1995) The growth hormone-releasing activity of hexarelin, a new synthetic hexapeptide, in short normal and obese children and in hypopituitary subjects. J. Clin. Endocrinol. Metab. 80, 674–678.

Growth Hormone Secretagogues
Edited by E. Ghigo, M. Boghen, F.F. Casanueva and C. Dieguez

Chapter 2

GHRP: Unnatural Toward the Natural

CYRIL Y. BOWERS

Tulane University Medical School, New Orleans, LA 70112, U.S.A.

The continued interest in the peptidyl and peptidomimetic GHRP-GH secretagogues has resulted from many talented investigators who have demonstrated a practical diagnostic–therapeutic value as well as a theoretical physiological relevance of this new class of GH releasing compounds. Together we are all gradually converting the unnatural into the natural.

Because of the unnatural origin of GHRP and knowing it is not just a GHRH mimic, it seems most reasonable not to have any strong preconceived convictions or inclinations about its action on GH release. This includes that its main action may not be directly on the pituitary and that its hypothalamic action may not always involve the release of endogenous GHRH or inhibition of SRIF release and/or action and could even involve the release of an additional factor(s) from the hypothalamus to mediate in part the pituitary action of GHRP. Knowledge of the normal and abnormal regulation and release of GH appear to be considerably expanded from studies on the GHRPs.

Some of our GHRP findings obtained with various collaborators are listed in chronological order in Table 1. A consistent focus of our studies has concerned the elucidation and understanding of the relationship between GHRP and GHRH on the release of GH. Depending on the experimental study, the GH release induced by GHRP and GHRH can be not only independent and dependent, but also additive and synergistic as well as permissive (1).

A general point about the GHRPs concerns the importance of distinguishing and considering the differences between the pharmacological and the putative physiological actions. They are overlapping but physiologically the hypothalamic paracrine local secretion, distribution and action of the putative GHRP hormone being inside the blood–brain barrier would have special implications as would its presence, amount and timing of secretion into the portal system. Pharmacologically the blood–brain barrier also needs special consideration. This is because the hypothalamic and pituitary actions of

TABLE 1

HISTORY OF GHRPs

1976–80	4 new classes of peptides developed
1980–82	GH release specific *in vitro*/*in vivo* in rats, lambs, calves, monkeys; increased BW in rats
1982–84	Interactions of GHRP, GHRH, SRIF on GH release characterized
1984	Activity reflects putative new hypothalamic (H) hormone different from GHRH
1984–88	Dual H and pituitary (P) action; putative H U-factor (unknown factor) hypothesized
1990–92	Synergistic GH release in humans with GHRH, small rise of cortisol, PRL; oral administration in men and GH-deficient children
1990–94	More active GHRP-1, -2; high affinity binding in H and P
1992–93	Continuous infusion increased pulsatile GH release in men
1993–96	Released GH in acromegalics and P tumors of these patients, increased IP_3
1992–97	Acute and chronic studies in short-statured children with variable GH deficiencies
1991–98	Partial purification of natural GHRP-like hormone from porcine H
1995–97	Evidence obtained for putative U-factor in humans
1994–98	Evidence obtained for possible deficiency of putative GHRP-like hormone in normal older subjects with decreased GH secretion
1996–98	Continuous infusion increased pulsatile GH release in critically ill patients

GHRP are complementary and administration of low dose GHRP peripherally would be outside the blood–brain barrier and may only reach specific hypothalamic anatomical sites.

The first GHRP, TyrDTrpGlyPheMetNH_2 (DTrp^2) was synthesized in 1976 (2) before the isolation of GHRH in 1982 by Guillemin with Ling as well as Vale with Rivier. This pentapeptide, which was derived from Met enkephalin, was only active *in vitro* and was low in potency but it specifically released GH and had no opiate activity.

Four different chemical types of GHRPs with 4 or 5 amino acids were developed between 1976–80 but they were only active *in vitro*. They included DTrp^2 (TyrDTrpGlyPheMetNH_2), DTrp^3 (TyrAlaDTrpPheMetNH_2), DTrp^{2,3} (TyrDTrpDTrpPheNH_2), and DTrp^2,LTrp^4 (TyrDTrpAlaTrpDPheNH_2). These GHRP types became templates for future development of the GHRPs by ourselves as well as other groups. Examples of each type are now highly active *in vitro* and *in vivo*. The presence, number, position and stereochemistry of the Trp residue has been very helpful in appreciating the scope of the structure–activity relationship. There are now 3 general chemical classes of GHRPs which include peptides, partial peptides and nonpeptides.

The sequences of our initial GHRP-6, GHRP-1 and GHRP-2 developed between 1980 and 1989 were HisDTrpAlaTrpDPheLysNH_2, AlaHisDβNalAlaTrpDPheLysNH_2 and DAlaDβNalAlaTrpDPheLysNH_2, respectively (1,3–6). They are partially protected small peptides consisting of 6 or 7 amino acids which are active in various animals and humans.

By 1984, GHRP and GHRH were considered to act on different receptors and GHRP was thought to reflect the activity of a new physiological hormone involved in the regulation of GH (4).

Results of earlier biological studies between 1982 and 1986 revealed a complementary and synergistic action of GHRP and GHRH on GH release when administered together to rats, cows and monkeys (6). Although this synergism supported the independent action of these 2 peptides, the marked inhibition of the *in vivo* GHRP GH response by GHRH antiserum emphasized that the *in vivo* GH response to GHRP greatly depended on endogenous GHRH and that GHRH may even be the mediator of the action of GHRP on GH release (7). Subsequent studies of Pandya et al. in normal young men further supported this conclusion since they demonstrated that a GHRH antagonist markedly inhibited the GH response to GHRP (8).

In order to understand the GHRP–GHRH interrelationship it is necessary to emphasize that GHRP can release GH independently of GHRH (7,9). GHRP acts directly on the pituitary to release GH in the absence of GHRH and GHRP+GHRH additively augments GH release *in vitro* and synergistically *in vivo* indicating the 2 peptides have an independent action on GH release. Furthermore, in contrast to GHRH, GHRP releases a large amount of GH *in vivo* in rats with only a small concomitant rise of pituitary cAMP which demonstrates that endogenous GHRH is not the primary mediator of GH release induced by GHRP (7). When GHRH increases GH release *in vivo*, it concomitantly and markedly raises pituitary cAMP levels.

The *in vitro* demonstration that GHRH but not GHRP acts via the intracellular adenyl cyclase pathway and specific high affinity binding of GHRP sites in crude pituitary and hypothalamic peripheral cell membranes are other findings which underscore the differences between GHRP and GHRH (7). These results forecast the important accomplishment of the cloning of the G protein 7 transmembrane coupled receptor of GHRP (10). Subsequently, direct *in vitro* evidence has been obtained by Adams et al. (11) and Wu et al. (12) that GHRP acts via the phospholipase-C pathway, however, crosstalk does occur between the GHRP and GHRH pathways.

Results in Figure 1 show more direct evidence for the existence of the putative GHRP-like hormone in porcine hypothalami. Recorded are results of a highly purified fraction from porcine hypothalami, which is devoid of GHRH, that releases GH *in vitro* in the rat pituitary dispersed cell culture. When the fraction was added together with 3 different antagonists, 2 GHRP and one GHRH, only the GHRP and not the GHRH antagonist inhibited the GH release induced by this fraction. These results also demonstrate that the unnatural synthetic and the putative natural GHRP GH secretagogue activity parallel each other. Furthermore, in contrast to GHRH, this fraction did not increase pituitary cAMP. From mass spectrometry results of Don Hunt, a small peptide and some of its amino acids have been identified in highly purified but still impure fractions that appear to reflect the activity of a natural GHRP-like hormone. In addition, we have utilized the empirical approach to synthesize the putative GHRP-like hormone de novo. Some of the amino acids found in the highly purified porcine active fractions as well as those in the unnatural synthetic GHRPs were incorporated into these *de novo* peptides. Thus far, small synthetic GHRPs consisting of only L-amino acids have been synthesized with moderate activity that could possibly be related in part to the putative natural GHRP-like hormone. Further isolation studies are currently on-going.

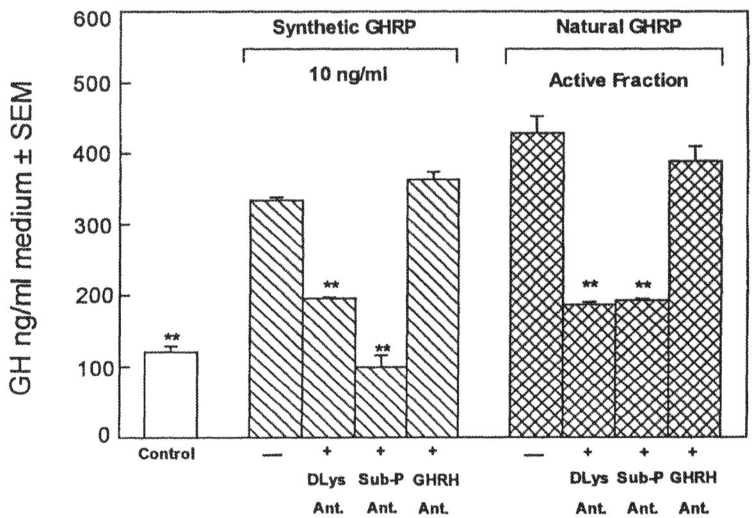

Figure 1. On the isolation of natural GHRP from porcine hypothalami in rat pituitary dispersed cell culture. These are results of a highly purified fraction that are considered to represent the activity of the putative natural endogenous GHRP. GH was released by both the unnatural synthetic GHRP and the putative natural GHRP alone as well as with the GHRH antagonist. Both DLys^3-GHRP-6 and Sub P antagonist ([DArg^{1}DPhe^{5}DTrp^{7,9}Leu^{11}]-substance P) inhibited the GH release induced by the synthetic and natural GHRP. The fraction did not increase cAMP release. Dose of antagonist = 10 μg/ml. ***p* value = <0.01–<0.001.

The difference in the *in vitro–in vivo* results appears to be due to an additional *in vivo* action(s) that GHRP has on the hypothalamus and at times either indirectly or directly on the attenuation of the pituitary action of SRIF. In earlier studies, Thorner et al. (13) described GHRP as a functional SRIF antagonist. Some evidence for the latter was reported in rats (7) and in humans by Massoud et al. (14) especially the effects of the combined GHRP and GHRH.

Although the GHRP hypothalamic action has not been completely elucidated, strong indirect evidence has been obtained in humans to indicate that a new unappreciated action(s) is induced by GHRP. In part from the novel effects on GH release in humans, GHRP has been envisioned to act on the hypothalamus to release the putative hypothalamic U-factor (unknown factor). U-factor plus GHRH and at times GHRP or the putative GHRP-like hormone presumably act on the pituitary to release GH.

Included in this brief publication are select clinical results of normal younger and older men and women considered to reveal some of the unusual actions of GHRP that occur with and without GHRH.

A basic point of general significance is that the endocrine mechanisms involved in the GH releasing action of GHRP are probably dosage dependent. Another general point is that one reason the GH releasing action of GHRP is somewhat confusing and difficult to understand is because GHRP has two anatomical sites of action and the hypothalamic actions are still incompletely understood.

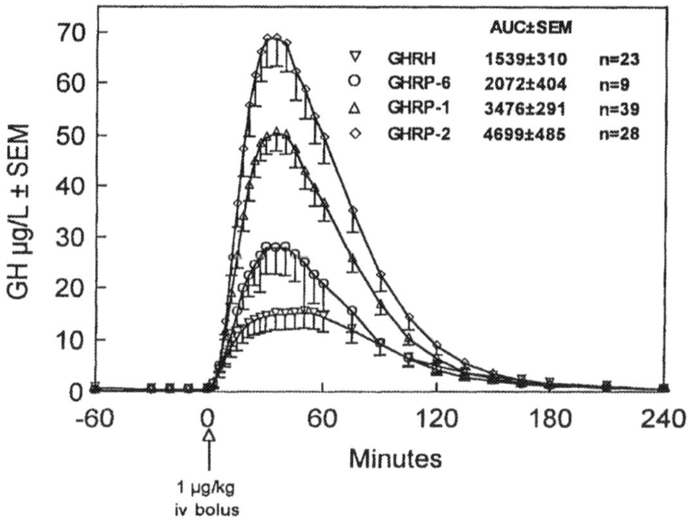

Figure 2. GH responses to GHRP-6, GHRP-1, GHRP-2 and GHRH 1-44NH_2 in normal young men. Values are mean ± SEM. AUC = area under the curve. Used with permission: Springer Verlag, New York.

After encouragement and help from Michael Thorner, we performed the first clinical study with GHRP-6 in 1988 (15). Also, at this same time Ilson et al. (16) of SmithKline Beecham performed a clinical study with GHRP-6. Recorded in Figure 2 is the comparative GH releasing activity of GHRH and 3 synthetic GHRPs, GHRP-6, GHRP-1 and GHRP-2 after i.v. bolus administration of 1 µg/kg to normal young men. The results demonstrate that all the GHRPs released more GH than GHRH. Also, they demonstrate that any endogenous GHRH possibly released by a GHRP hypothalamic action would not, by itself, fully explain the mechanism of how GHRP releases more GH than GHRH.

Results in Figure 3 (men) and Figure 4 (women) again support the independent and complementary action of GHRP and GHRH on GH release, because in normal younger and older men and women, i.v. bolus 1 µg/kg GHRP-2 administered together with 1 µg/kg GHRH, released GH synergistically in all of the subjects. Also, the GH response to GHRP, GHRH and the 2 peptides together in the older subjects, compared to the younger subjects, was decreased.

Several different, important and somewhat surprising results evolved from 2 studies performed by Craig Jaffe and Ariel Barkan (17) as well as Michael Thorner's group (18) in which GHRP-6 was continuously administered i.v. over 36 or 24 hours to normal young men. Since the amplitude but not the frequency of pulsatile GH secretion was increased in both studies, the results in Figure 5 demonstrate that continuous infusion of GHRP augments the physiological pattern of pulsatile GH release. There was essentially no rise of the interpulse GH level indicating lack of an attenuation effect on the release or action of SRIF but there was an increase in the serum IGF-I level of about 40%. These overall results were somewhat surprising because in earlier studies in rats repeated hourly injections of GHRP and continuous infusion of GHRP-6 during perifusion of the rat pituitary *in vitro* markedly desensitized the GH release induced by GHRP.

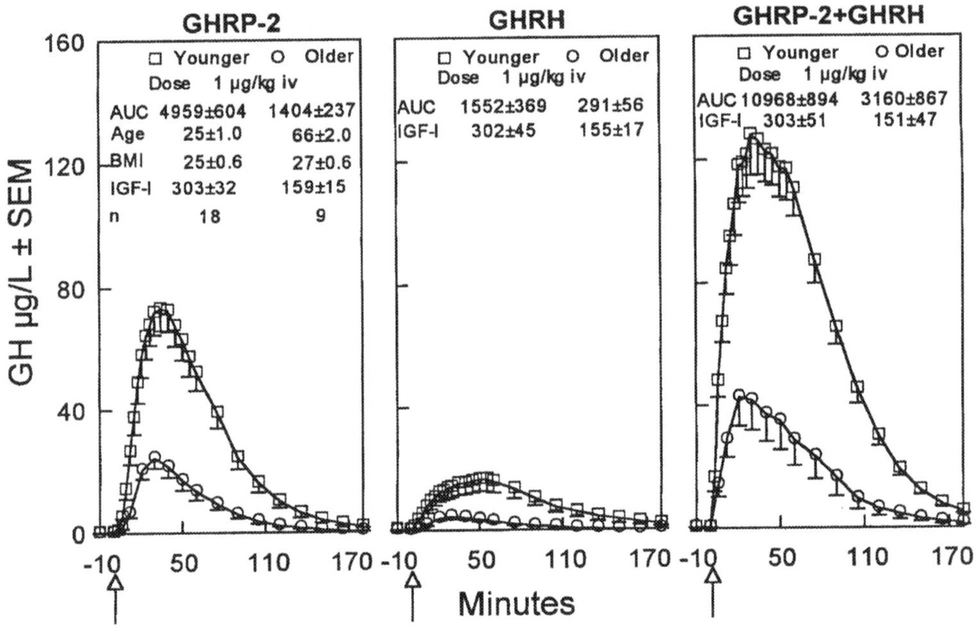

Figure 3. Effect of GHRP-2, GHRH and GHRP-2+GHRH in normal younger and older men. Values are mean ± SEM. Used with permission: Marcel Dekker, New York.

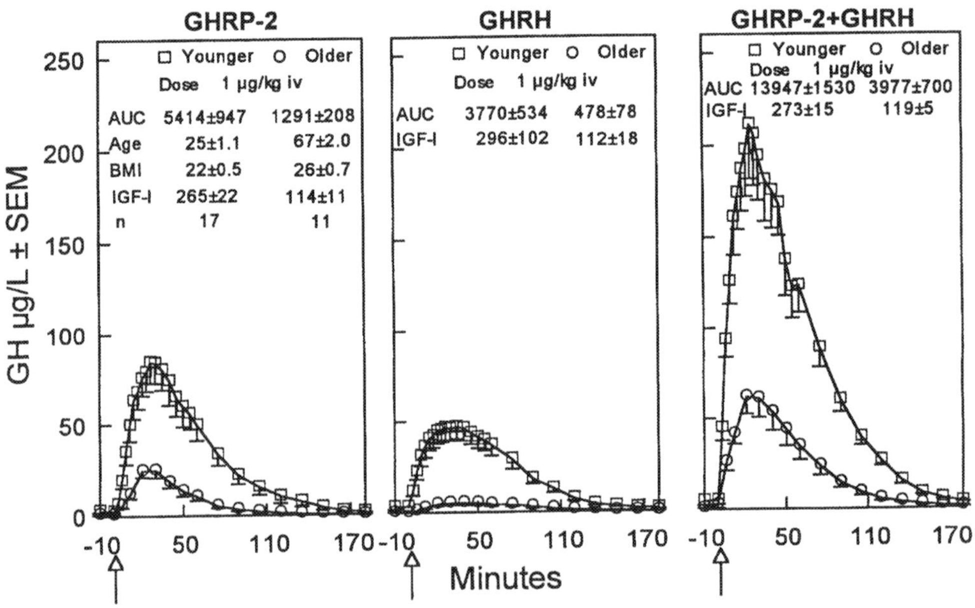

Figure 4. Effect of GHRP-2, GHRH and GHRP-2+GHRH in normal younger and older women. The GH responses to GHRH were higher in younger women than younger men but were about the same in the older subjects. The GH responses to GHRP-2 were about the same in men and women in both the younger and older subjects. Used with permission: Marcel Dekker, New York.

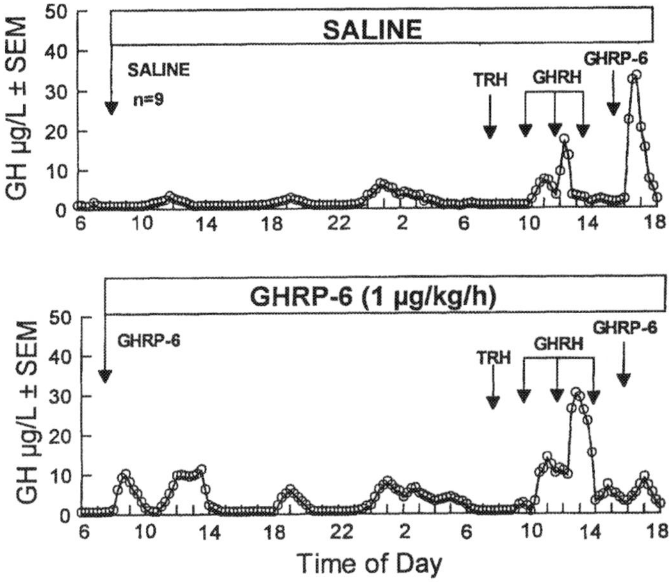

Figure 5. Effect of continuous infusion of saline or GHRP-6 in normal younger men. GHRP-6 was infused at a rate of 1 µg/kg/h for 36 hours. IV bolus TRH (50 µg), GHRH (1 µg/kg×3) and GHRP-6 (1 µg/kg×1) were administered at the end of the infusion period. TRH was injected at 0800 but there was no release of GH. The GHRP-2 GH response was decreased while the GHRH GH response was increased. The amplitude but not the frequency of the GH pulses was increased during the entire GHRP infusion period. Used with permission: The Endocrine Society.

Also, as recorded in Figure 5 in the study by Jaffe and Barkan, the individual acute GH responses to i.v. bolus TRH, GHRH and GHRP were determined at the end of the GHRP infusion period. These results reveal that TRH had no effect on GH release; the GH response to the first two of the three sequential boluses of GHRH were increased while the third was decreased and, as predicted, the GH response to i.v. bolus GHRP was markedly decreased. To explain these results it is postulated that both sensitization and desensitization of the GHRP GH releasing action was induced by continuous GHRP infusion. The i.v. bolus GHRH GH response was sensitized while the i.v. bolus GHRP GH response was desensitized. The sustained sensitization of the GHRP action on the spontaneous pulsatile release of GH during the entire infusion period shows that the desensitization effect of GHRP is only partial. Also, the results reveal that continuous GHRP administration does not inhibit or interfere with the timing of the normal pulsatile secretion of endogenous GHRH from the hypothalamus or appear to enhance the pulsatile secretion of GH by interfering with the hypothalamic release or pituitary action of SRIF. The normal pulsatile GH secretion supports that the negative feedback effect of serum GH and IGF-I on GHRH and SRIF release are still operative. These intact feedback actions help to promote physiological pulsatile secretion of GH and to minimize the super-physiological secretion of GH. The effects of continuous infusion of GHRP *in vivo* are envisioned to enhance the pituitary GH response to exogenous GHRH and endogenous GHRH and, in this study, may be the reason the GH response to GHRH as well as the spontaneous GH pulses are increased. A somewhat confusing issue is that this GHRH sensitization effect is hypothesized to be mediated via a hypothalamic action rather than a direct pituitary action of GHRP presumably via the hypothalamic putative U-factor. Because of the excess GHRH administered i.v. at the end of the GHRP infusion period, it is apparent that the sensitization or augmentation of the GHRH GH response induced by GHRP does not result from any effect it may have on the release of endogenous GHRH. Furthermore, GHRP seems unlikely to augment the GHRH action by inhibiting the release of SRIF from the hypothalamus or the action of SRIF on the pituitary. During the normal physiological secretion of GH, enhanced SRIF tone is the usual reason proposed for a lower interpulse level of GH; however, GHRP at dosages which increase the amplitude of the spontaneous GH pulse has essentially little to no effect on interpulse GH levels. Pertinent is that the GH response to GHRP is increased when SRIF release or action is decreased experimentally by pentobarbital, opiates, SRIF antiserum or pyridostigmine. If GHRP released GH by interfering with the release or action of SRIF, these agents would not be expected to further increase the GH response to GHRP. However, the SRIF inhibitory pituitary action may possibly be attenuated by a high dose of 10 μg/kg GHRP as well as 1 μg/kg GHRP+GHRH.

Of special note is that the study of Clark and Robinson reported in 1989, very possibly forecast the sensitization–desensitization action of continuous GHRP infusion in humans (19). Although the mechanism(s) was not elucidated in this study, continuous infusion of GHRP into conscious rats allowed a GH response each time pulses of GHRH were administered. In these rats, continuous GHRP infusion presumably desensitized the GHRP GH response but sensitized the GHRH action on GH release indicating there is only partial desensitization of the action of GHRP on GH release.

The hypothetical hypothalamic U-factor proposed has been further substantiated by the results of the studies recorded in Figures 6 and 7. In these studies, in contrast to the studies recorded in Figures 3 and 4, unequal dosages of GHRP and GHRH were administered together. A low dose of 0.03 μg/kg GHRP-2 was administered with a maximal dose of 1 μg/kg GHRH to normal young men (Figure 6) and women (Figure 7). Whether GHRP-2 and GHRH are administered together in equal or unequal dosages, GH is released synergistically. When the unequal dosage combination of the 2 peptides was administered in the study recorded in Figures 6 and 7, it is notable that GHRP-2 augmented the release of GH even at a very low dosage of ~2 μg to the subject. Since this is a subthreshold GH releasing dosage, and the combined *in vitro* action of GHRP and GHRH is additive rather than synergistic, the results are considered to support that GHRP acts on the hypothalamus rather than the pituitary to mediate the synergistic release of GH. To explain this novel finding, the putative hypothalamic U-factor has been hypothesized to be released by an action of GHRP on the hypothalamus and to be secreted together with GHRH to release GH synergistically via the combined pituitary action of U-factor and GHRH. The excess exogenous GHRH administered eliminates the possibility that low dose GHRP induces a synergistic GH response because it releases endogenous GHRH from the hypothalamus. Also, for reasons previously stated, it is very unlikely that a GHRP effect on SRIF release or action mediates the GHRP synergistic GH response especially at such a low dosage of GHRP. Because of these low dose GHRP results and the indirect evidence in sheep, guinea

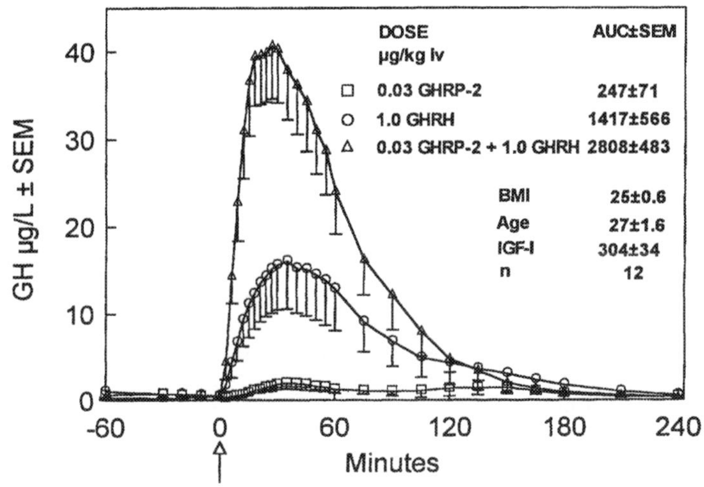

Figure 6. Synergistic release of GH in normal young men to a low dose of 0.03 μg/kg GHRP-2 plus a high dose of 1.0 μg/kg GHRH. Peptides were administered at 0 time. Values are the mean ± SEM. Used with permission: Marcel Dekker, New York.

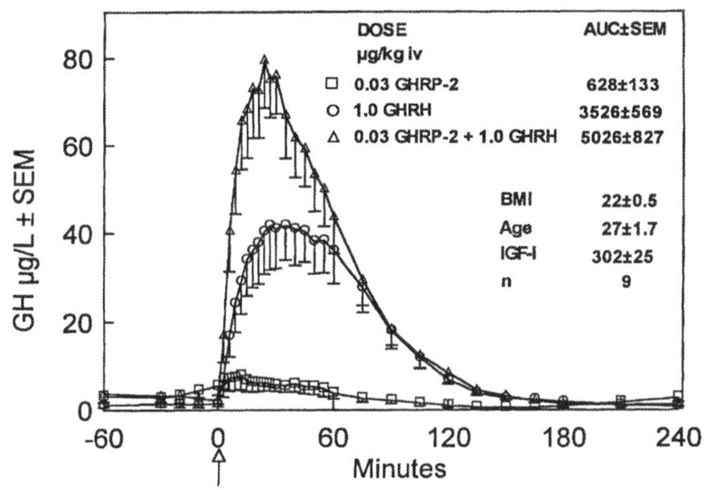

Figure 7. Synergistic release of GH in normal young women to a low dose of 0.03 μg/kg GHRP-2 plus a high dose of 1.0 μg/kg GHRH. Peptides were administered at 0 time. Values are the mean ± SEM. Used with permission: Marcel Dekker, New York.

pigs and rats (20) indicating the limited transport of GHRP into the hypothalamus after peripheral administration due to the blood–brain barrier, we postulated that low dose GHRP acts outside the blood–brain barrier on the median eminence of the hypothalamus to induce this synergistic effect.

The presumed hypothalamic action of GHRP at this low dose (Figures 6 and 7) is considered to possibly reflect a basic physiological action of the putative GHRP-like hormone.

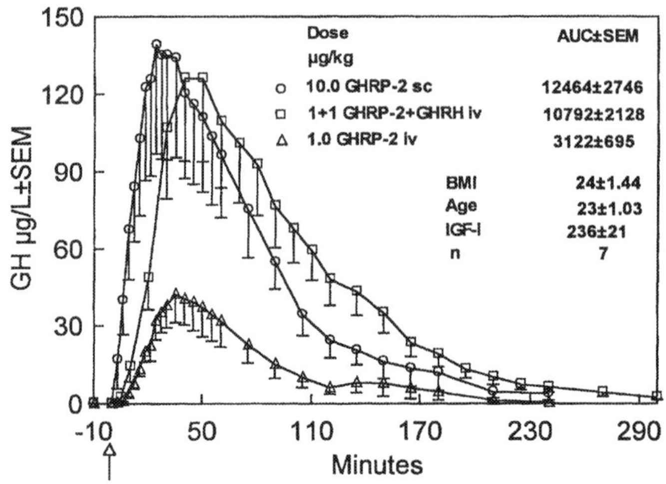

Figure 8. Effect of high dose GHRP-2. Seven normal young men were administered GHRP-2 at 10 μg/kg sc, GHRP-2+GHRH at 1 μg/kg of each peptide by i.v. bolus and GHRP-2 at 1 μg/kg by i.v. bolus. The AUC for the high dose of GHRP-2 and the combination of the peptides was essentially the same.

In the results of the clinical study recorded in Figure 8, the GH response to a high dose of 10 μg/kg GHRP-2 sc was compared to a 1 μg/kg dose of GHRP-2 i.v. which is the standard route and usual dose of GHRP-2 administration clinically. It is apparent that the very high dose of GHRP-2 alone releases an inordinate amount of GH. This GH response is equivalent to the large amount of GH synergistically released by the combined 1 μg/kg dosage of GHRP + GHRH and leads to the hypothesis that the same GH releasing mechanisms may be involved by these two approaches. High dose GHRP is considered to increase the endogenous release of GHRH from the hypothalamus which in turn acts together with U-factor and GHRP on the pituitary to induce a synergistic release of GH. Thus GHRP presumably acts on the hypothalamus to release GHRH but this seems to be dose dependent. In addition, since attenuation of the pituitary inhibitory action of SRIF on GH release may occur, this may be another one of the mechanisms by which the high dose as well as the combined peptides release such a large amount of GH.

Results in the last study reveal possible new insight into the pathophysiology responsible for decreased GH secretion in some normal older subjects. Reasons usually proposed for this decreased GH secretion have been decreased secretion of endogenous GHRH or increased secretion of SRIF. It is now proposed that in some elderly subjects the decreased secretion of GH is due to neither decreased GHRH nor excess SRIF secretion but rather is due to a decreased secretion of the putative hypothalamic GHRP-like hormone.

The results recorded in Table 2 are from the following study. Twenty normal older subjects were randomly selected and the acute GH responses determined to i.v. bolus 1 μg/kg GHRH or GHRP-2, 0.1 μg/kg GHRP-2 as well as 0.1 μg/kg GHRP-2 + 1 μg/kg GHRH. The mean peak GH responses of these 20 subjects have been divided into 3 groups according to the degree of the GH response to the combined low dose GHRP-2 and high dose GHRH, i.e., mild (39 μg/L), moderate (24 μg/L) and severe (2 μg/L). The most

TABLE 2

ON THE PATHOPHYSIOLOGY OF OLDER MEN AND WOMEN WITH DECREASED GH SECRETION

Peptide iv bolus	Dose μg/kg	Mild Peak GH μg/L	Moderate Peak GH μg/L	Severe Peak GH μg/L
GHRH	1.0	14.7±2.7	4.2±0.2	3.3±1.2
GHRP-2	0.1	6.3±2.5	3.5±1.2	1.2±0.4
GHRP-2+GHRH	0.1+1.0	38.6±6.5	24.0±4.2	7.1±0.9
GHRP-2	1.0	41.2±4.7	32.8±5.1	14.3±2.3
			AUC GH μg/L·4 h	
GHRH	1.0	1011±179	352±46	263±67
GHRP-2	0.1	376±95	217±54	103±34
GHRP-2+GHRH	0.1+1.0	2196±402	1402±272	391±27
GHRP-2	1.0	2426±344	1890±327	706±117
IGF-I (μg/L)		151.0±17.0	121.0±11.0	117.0±18.0
Age (years)		63.8±2.3	67.8±1.3	67.3±3.1
BMI		25.3±0.7	26.7±0.2	26.5±1.8
n		5	11	4

Values = mean±SEM, BMI = Body Mass Index..

meaningful findings were obtained in the moderately impaired GH response group. In this group, the GH response to GHRH was markedly decreased possibly due to a decreased pituitary GH content secondary to a decreased secretion of endogenous GHRH or a pituitary insensitivity to the pituitary action of GHRH possibly indicating an excess secretion of SRIF. In regard to these 2 mechanisms, the greater GH response to 1 μg/kg GHRP-2 reveals that a decreased pituitary content of GH is not the reason for the impaired GH response to 1 μg/kg GHRH. Additionally dramatic is that a low dose of GHRP-2 administered together with 1 μg/kg GHRH reverses this impaired GH response to GHRH. As discussed previously, low dose GHRP-2 would be unlikely to be acting to augment the action of high dose GHRH by interfering with SRIF release or action. Furthermore, the relatively high GH response to 1 μg/kg GHRP alone is an indication endogenous GHRH is being secreted in these subjects because as recorded in Figure 9, the GHRH antagonist studies of Pandya et al. (8) demonstrated that in normal young men without endogenous GHRH secretion, the GH releasing action of GHRP is markedly decreased. The excess exogenous GHRH administered and the insensitivity of the pituitary to exogenous GHRH are both reasons to conclude that any release of endogenous GHRH by GHRP-2 would have an inconsequential effect on GH release. Release of endogenous GHRH via the hypothalamic action of GHRP or the direct pituitary action of low dose GHRP simply could not be the mechanism to explain how low dose GHRP reverses the impaired pituitary GH releasing action of GHRH. Rather it is hypothesized to be the result of U-factor released via the hypothalamic action of low dose GHRP.

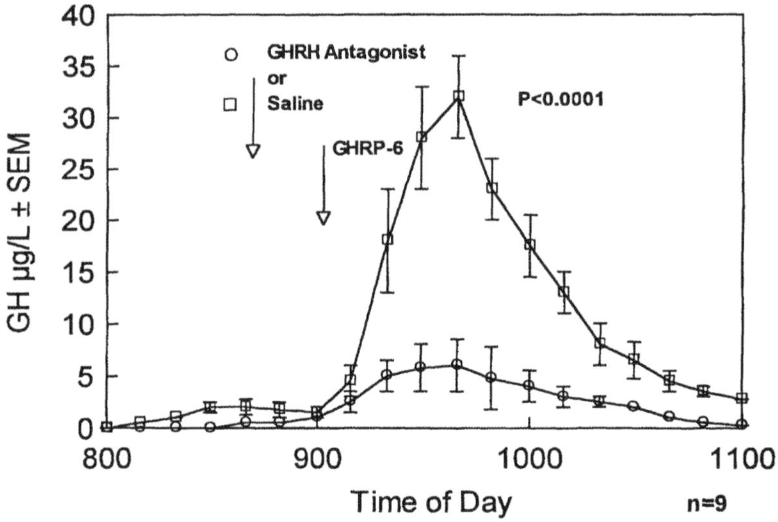

Figure 9. GHRP-6 was administered after saline or 400 µg/kg of GHRH-Antagonist. After the antagonist there was a marked decrease of the GH response to GHRP-6. Used with permission: The Endocrine Society.

Whether a deficiency of the putative GHRP-like hormone plays a role in the mildly and severely impaired GH response groups is more difficult to project because of the smaller number of subjects studied as well as the borderline decreased values of the mildly impaired group and the extreme decreased GH values of the severely impaired group. These studies are currently being expanded to increase the number of subjects in each group in an effort to further prove or disprove the hypothesis that the decreased secretion of GH in some normal older men and women is the result of a deficiency of the putative hypothalamic GHRP-like hormone.

In conclusion, the novel and unique effects of GHRP on GH release observed in humans appear to be immediately relevant to understanding and to distinguishing the actions of GHRP. The special GH releasing actions of GHRP with and without GHRH appear to be valuable for proposing conceptual pharmacological and physiological models of the actions of the unnatural synthetic GHRPs as well as the putative natural GHRP-like hormone on GH secretion. It has not been possible to simulate the unusual GH releasing effects of GHRP observed *in vivo* (Figures 2–8) by *in vitro* studies in which GHRP, GHRH and SRIF have been assessed in many different combinations and dosages and under various experimental conditions. Some of the *in vivo* results support the conclusion that GHRP activates new previously unappreciated mechanisms involved in the release of GH. The GH releasing action of GHRP is varied, dosage dependent as well as overlapping and complementary to the actions of GHRH and/or antagonistic to the actions of SRIF. The broad range of actions of GHRP on GH release appear to evolve from the dual complementary action on the hypothalamus and pituitary. Since evidence increasingly supports that a putative GHRP-like hormone probably does exist, our conceptual model of the physiological role of this hormone is as a modulator of the action of GHRH on the pulsatile secretion of GH. GHRH is viewed as the initiator of the GH pulses and the putative

GHRP-like hormone the modulator by influencing the peak as well as the size and shape of the GH pulses.

ACKNOWLEDGMENT

This work was supported in part by National Institutes of Health grants AM-06164, DK40202 and PHS RR05096-09 (GCRC). Special appreciation is expressed to Dr. Granda-Ayala, to the technicians and the fellows of the Endocrinology and Metabolism Section of the Department of Medicine, and to Robin Alexander for typing the manuscript. Special thanks are expressed to Dr. Fred Wagner at BioNebraska for supplying GHRH 1-44NH_2.

REFERENCES

1. Bowers, C.Y. (1998) GH releasing peptides (GHRPs). In: Handbook of Physiology. J. Kostyo (ed.). Cary, Oxford University Press, pp. 267–297.
2. Bowers, C.Y., Chang J.K., Fong T.T.W. (1977) A synthetic pentapeptide which specifically releases GH, *in vitro*. Proc. 59th Meeting of the Endocrine Society, Chicago, p. 2332.
3. Bowers, C.Y., Momany, F., Reynolds, G.A. (1982) *In vitro* and *in vivo* activity of a small synthetic peptide with potent GH releasing activity. Proc. 64th Meeting of the Endocrine Society, San Francisco, p. 205.
4. Momany, F., Bowers, C.Y., Reynolds, G.A., Hong, A., Newlander, K. (1984) Conformational energy studies and *in vitro* activity data on active GH releasing peptides. Endocrinol. 114, 1531–1536.
5. Bowers, C.Y., Momany, F., Reynolds, G.A., and Hong, A. (1984) On the *in vitro* and *in vivo* activity of a new synthetic hexapeptide that acts on the pituitary to specifically release growth hormone. Endocrinol. 114, 1537–1545.
6. Bowers, C.Y. (1998) GHRP+GHRH synergistic release of GH: scope and implication. In: Growth Hormone Secretagogues. B. Bercu and R. Walker (eds). Marcel Dekker Inc., New York, 1–25.
7. Bowers, C.Y., Veeraragavan, K., Sethumadhavan, K. (1994) Atypical growth hormone releasing peptides. In: Growth Hormone II: basic and clinical aspects. B. Bercu and R. Walker (eds). Springer-Verlag, New York, 203–222.
8. Pandya, N., DeMott-Friberg, R., Bowers, C.Y., Barkan, A.L., Jaffe, C.A. (1998) Growth hormone (GH)-releasing peptide-6 requires endogenous hypothalamic GH-releasing hormone for maximal GH stimulation. J. Clin. Endocrinol. Metab. 83, 1186–1189.
9. Bowers C.Y., Sartor A.O., Reynolds G.A., Badger, T.M. (1991) On the action of the growth hormone releasing hexapeptide GHRP. Endocrinol. 128, 2027–2035.
10. Howard A.D., Feighner S.D., Cully D.F. et al. (1996) A receptor in pituitary and hypothalamus and functions in growth hormone release. Science 273, 974–977.
11. Adams, E.F., Huang, B., Buchfelder, M. et al. (1998) Presence of growth hormone secretagogue-receptor (GHS-R) mRNA in human pituitary tumors and rat GH_3 cells. J. Clin. Endocrinol. Metab. 83, 638–642.
12. Wu, D., Chen, C., Zhang, J., Bowers, C.Y., Clark, I.L. (1996) The effects of GH releasing peptide-6 (GHRP-6) and GHRP-2 on intracellular adenosine 3',5'-monophosphate (cAMP) levels and GH secretion in ovine and rat somatotrophs. J. Endocrinol. 148, 197–205.
13. Thorner, M.O., Vance, M.L., Rogol, A.D., et al. (1990) Growth hormone releasing hormone and growth hormone releasing peptide as potential therapeutic modalities. Acta Paediatr. Scand. 367, 29–32.

14. Massoud, A.F., Hindmarsh, P.C., Brook, C.G.D. (1997) Interaction of the growth hormone releasing peptide hexarelin with somatostatin. Clin. Endocrinol. 47, 537–547.
15. Bowers, C.Y., Reynolds, G.A., Durham, D., Barrera, C.M., Pezzoli, S.S., Thorner, M.O. (1990) Growth hormone releasing peptide stimulates GH release in normal men and acts synergistically with GH-releasing hormone. J. Clin. Endocrinol. Metab. 70, 975–982.
16. Ilson, B.E., Jrokasky, D.K., Curnow, R.T., Stote, R.M. (1989) Effect of a new synthetic hexapeptide to selectively stimulate growth hormone release in healthy human subjects. J. Clin. Endocrinol. Metab. 69, 212–214.
17. Jaffe, C.A., Ho, J., Demott-Friberg, R., Bowers, C.Y., Barkan, A.L. (1993) Effects of a prolonged growth hormone (GH)-releasing peptide infusion on pulsatile GH secretion in normal men. J. Clin. Endocrinol. Metab. 77, 1641–1647.
18. Huhn, W.C., Hartman, M.L., Pezzoli, S.S., Thorner, M.O. (1993) 24-h growth hormone (GH)-releasing peptide (GHRP) infusion enhances pulsatile GH secretion and specifically attenuates the response to a subsequent GHRP bolus. J. Clin. Endocrinol. Metab. 76, 1201–1208.
19. Clark, R.G., Carlsson, L.M.S., Trohnar, J., Robinson, I.C.A.F. (1989) The effects of a growth hormone releasing peptide and growth hormone releasing factor in conscious and anesthetized rats. J. Neuroendocrinol. 1, 249–255.
20. Fairhall, K.M., Mynett, A., Thomas, G.B., Robinson, I.C.A.F. (1996) Central and peripheral effects of peptide and nonpeptide GH secretagogues on GH release *in vivo*. In: Growth Hormone Secretagogues. B. Bercu and R. Walker (eds). Springer-Verlag, New York, 219–236.

Growth Hormone Secretagogues
Edited by E. Ghigo, M. Boghen, F.F. Casanueva and C. Dieguez

Chapter 3

Impervious Peptides as GH Secretagogues

ROMANO DEGHENGHI

Europeptides, 95108 Argenteuil Cedex, France

Growth hormone is a protein, GHRH and the somatomedins family are peptides and are therapeutically available as such. At the time of this writing, none of the more recent Growth Hormone Releasing Peptides and their non-peptidyl mimetics have been approved for treatment, but it is likely that one or more GH secretagogues will eventually become therapeutic agents. Cyril Y. Bowers, the discoverer of the original GHRP series has reviewed their history (1). Other excellent reviews of this new class of GH Secretagogues have been published (2–4).

PEPTIDES VS NON-PEPTIDE MIMETICS

Following the trailblazer, seminal work of Bowers and Momany, ourselves, and groups from Genentech and Novo Nordisk have developed peptidyl analogues of Bowers' GHRP-6.

In the non-peptidyl series, researchers from Merck Research Laboratories are unquestionably in the lead and their spiropiperidine derivative MK-0677 has been the most studied GHS drug candidate. Other groups from Pfizer and Lilly have disclosed in the patent literature their peptidomimetic GH secretagogues.

Medicinal chemists are therefore divided between those who develop non-peptide ligands for peptide receptors and those who continue to favour peptide analogues as potential drugs. The latter have to face the additional problem of how to conveniently deliver their peptide analogues which are poorly absorbed by the oral route.

One of the reasons why peptides are, with few exceptions, not absorbable orally is because of their vulnerability to proteases and peptidases present in the gastro-intestinal tract. In an attempt to minimize this problem, we developed a series of "impervious peptides", so-called because they are poor substrates to peptidases and proteases. Starting from Hexarelin (5), we have downsized the hexapeptide to obtain (see Table 1) a series of smaller peptides of which the pentapeptide derivative EP 51216 and the tripeptide analogue

TABLE 1

COMPARATIVE ACTIVITY OF POTENTIAL PEPTIDE GH SECRETAGOGUES, 300 μg/kg s.c., IN THE 10-DAY RAT MODEL. GH SECRETION WAS ASSESSED 15 MINUTES AFTER ADMINISTRATION OF THE PEPTIDE

Compound	Structure
Inactive (not different from controls)	
L-164,080	Aib-D-Trp-D-HomoPhe-OEt (3)
EP 251	4-Abz-D-Mrp-AlaTrp-D-Phe-Lys-NH_2
EP 252	2-Abz-D-Mrp-AlaTrp-D-Phe-Lys-NH_2
EP 253	3-Abz-D-Mrp-AlaTrp-D-Phe-Lys-NH_2
EP 254	2-N-Methyl-Abz-D-Mrp-Ala-Trp-D-Phe-Lys-NH_2
EP 255	Pip-D-Mrp-Ala-Trp-D-Phe-Lys-NH_2
EP 256	D-Pip-D-Mrp-Ala-Trp-D-Phe-Lys-NH_2
EP 257	4-Amino-Phe-D-Mrp-Ala-Trp-D-Phe-Lys-NH_2
EP 258	D-4-Amino-Phe-D-Mrp-Ala-Trp-D-Phe-Lys-NH_2
EP 7459	His-Ala-D-Trp-Ala-Mrp-D-Phe-Lys-NH_2
EP 7460	His-D-Trp-Ala-Mrp-D-Phe-Ala-Lys-NH_2
EP 9399	His-D-Mrp-Ala-Trp-D-Phe-Lys
EP 40804	His-D-Mrp-Ala-Phe-D-Trp-Lys-NH_2
EP 50887	TXM-D-Mrp-D-βNal-Phe-Lys-NH_2
EP 51322	GAB-D-Mrp-D-βNal-NH_2
EP 51343	Aib-D-Ser(Bzl)-D-Mrp-NH_2
EP 60021	D-Mrp-D-Mrp-NH_2
EP 60022	GAB-D-Mrp-D-Mrp-NH_2
EP 60260	D-Mrp-D-Mrp-Phe-NH_2
EP 70683	His-D-Mrp-Ala-Oia-D-Phe-Lys-NH_2
EP 92439	His-D-Mrp-D-Lys-Trp-D-Phe-Lys-NH_2
EP 92440	His-Ala-D-Trp-D-Lys-Mrp-D-Phe-Lys-NH_2
EP 92441	His-D-Mrp-D-Lys-Mrp-D-Phe-Lys-NH_2
Weakly active (GH range, 30–40 ng/ml)	
EP 51321	GAB-D-Mrp-D-βNal-OEt
EP 60261	D-Mrp-D-Mrp-Mrp-NH_2
EP 60274	GAB-D-Mrp-Mrp-NH_2
EP 60275	D-Mrp-Mrp-NH_2
Active (GH range, 50–90 ng/ml)	
MK-677	Hybrid structure (3)
GHRP-6	His-D-Trp-Ala-Trp-D-Phe-Lys-NH_2 (1)
EP 7458	His-D-Trp-Ala-Mrp-D-Phe-Lys-NH_2
EP 40738	D-Thr-His-D-Mrp-Ala-Trp-D-Phe-Lys-NH_2
EP 40904	Thr-D-Mrp-Ala-Trp-D-Phe-Lys-NH_2

(continued)

TABLE 1 (*continuation*)

Compound	Structure
EP 41616	IMA-D-Mrp-D-Trp-Phe-Lys-NH_2
EP 41617	IMA-D-Mrp-D-βNal-Phe-Lys-NH_2
EP 51390	Aib-D-Mrp-Mrp-NH_2
EP 60761	GAB-D-Mrp-D-Mrp-D-Mrp-Lys-NH_2
Very active (GH range, 100–150 ng/ml)	
Hexarelin	His-D-Mrp-Ala-Trp-D-Phe-Lys-NH_2
GHRP-2	D-Ala-D-βNal-Ala-Trp-D-Phe-Lys-NH_2 (1)
G 7039	INIP-D-βNal-D-βNal-Phe-Lys-NH_2 (3)
G 7509	INIP-D-βNal-D-Trp-Phe-Lys-NH_2 (3)
EP 259	IMA-D-Mrp-Ala-Trp-D-Phe-Lys-NH_2
EP 41614	INIP-D-Mrp-D-Trp-Phe-Lys-NH_2
EP 41615	INIP-D-Mrp-D-βNal-Phe-Lys-NH_2
EP 50477	GAB-D-Mrp-D-Trp-Phe-Lys-NH_2
EP 50886	TXM-D-Mrp-D-Trp-Phe-Lys-NH_2
EP 51215	GAB-D-Mrp-D-Mrp-Phe-Lys-NH_2
EP 92111	His-D-Mrp-Ala-Trp-D-Phe-Lys-OH
EP 92632	Ala-His-D-Mrp-Ala-Trp-D-Phe-Lys-NH_2
Most active (GH range, 160–200 ng/ml)	
EP 40735	His-D-Mrp-Ala-Trp-D-Phe-Lys-Thr-NH_2
EP 40736	His-D-Mrp-Ala-Trp-D-Phe-Lys-D-Thr-NH_2
EP 40737	D-Thr-D-Mrp-Ala-Trp-D-Phe-Lys-NH_2
EP 50885	GAB-D-Mrp-D-βNal-Phe-Lys-NH_2
EP 51216	GAB-D-Mrp-D-Mrp-Mrp-Lys-NH_2
EP 51389	Aib-D-Mrp-D-Mrp-NH_2
EP 71563	Cys-Tyr-GAB-D-Mrp-D-Mrp-Mrp- Lys-NH_2
EP 93183	Tyr-Bpa-Ala-His-D-Mrp-Ala-Trp-D-Phe Lys-NH_2
EP 930497	Tyr-Ala-His-D-Mrp-Ala-Trp-D-Phe-Lys-NH_2
EP 931829	D-Ala-D-Mrp-Ala-Trp-D-Phe-Lys-NH_2

INIP, isonipecotinyl; IMA, imidazolylacetyl; GAB, γ-amino-butyryl; TXM, tranexamyl = 4 (-aminomethyl)-cyclohexanecarbonyl; Mrp, 2-methyl-Trp; Aib, α-aminoisobutyryl, Abz, aminobenzoyl; Pip, pipecolyl, Oia, oxy-indolalanine; Bpa, p-benzoyl-Phe.

EP 51389 have been found to be potent GH secretagogues in the infant rat (6) and in the dog. In the latter species and indeed even in humans, the pentapeptide derivative EP 51216 elicited a GH response when given orally at doses of 0.3 to 0.6 mg/kg.

Oral bioavailability, however, is not only dependent on the "imperviousness" of peptides, or indeed even of non-peptidic molecules. Other important factors are the size of the molecule, its lipid–water partition coefficient and the related propensity of forming hydrogen bonding with the aqueous physiologic environment.

An intriguing possibility is to deliver GH secretagogues by sustained release parenteral devices, such as those successfully employed in the field of LHRH analogues, if the sustained release is compatible with therapeutic efficacy and has an acceptable safety profile.

STRUCTURE–ACTIVITY RELATIONSHIP IN THE HEXARELIN ANALOGUES SERIES

In our 1994 communication (7), we reported our motivation to test, in tryptophan rich peptides, the substitution with the more stable 2-Methyl Trp derivative (Mrp).

Apart from an increased chemical stability, the Mrp substitution was beneficial when a D-Trp was replaced by a D-Mrp, but not when a Trp was substituted with Mrp, at least with the well known GHRP-6 structure (Figure 1):

GHRP-6 : His-D-Trp-Ala-Trp-D-Phe-Lys-NH_2 (active)
Hexarelin : His-D-Mrp-Ala-Trp-D-Phe-Lys-NH_2 (more active)
EP 7458 : His-D-Trp-Ala-Mrp-D-Phe-Lys-NH_2 (less active)

Figure 1.

This observation seemed to indicate the importance of the unencumbered indole N-H of Trp for receptor binding, confirmed by the inactivity of Oxyindolalanine (Oia) derivative of Hexarelin: His-D-Mrp-Ala-Oia-D-Phe-Lys-NH_2 (EP 70683, mixture of two stereoisomers) compared to Hexarelin in the rat (8), in which the indolic N-H is perturbed by the neighbouring oxygen in position 2 (9).

If we take GHRP-6 as the model prototype Figure 1, our investigations have shown that the D-Trp in position 2 can be advantageously substituted with the more stable, more hydrophobic D-2MeTrp (D-Mrp). Bowers had similarly shown that the DTrp could be substituted with a D-Nal (β-Naphthylalanine) in GHRP-2. Some or total loss of activity, as we have seen, occurs when the Trp in position 4 is replaced with the L-2MeTrp or with Oia, the oxidated form of Trp.

Prolongation of the chain on the N terminal side is compatible with retention and even augmentation of activity (cf EP 930497, EP 93183).

It is unlikely that the same hypothalamic, pituitary or peripheral receptors for which GHRP-6 and similar peptides are ligands, show the same specificity for shorter GHS, such as MK 0677 and EP 51389. There is now evidence (10) that this is indeed the case with some of the shorter GHS being unable to fully displace radioligands such as ^{125}I-Tyr-Ala-His-D-Mrp-Ala-Trp-D-Phe-Lys-NH_2.

RESISTANCE TO PROTEASES AND PEPTIDASES

Experimentally the metabolic stability of GHRP-6 (SK&F 110679) or of hexarelin has been confirmed at least in the rat from which more than 50% of these peptides can be recovered unchanged in the bile following their subcutaneous administration. This observation prompted the SK&F group to observe that GHRP-6 "was not designed with metabolic stability in mind [but] it is tempting to speculate that the structural features that are important for receptor binding and pharmacological activity of these peptides may also confer metabolic stability, protecting them from degradation by peptidases" (11). We propose the term *impervious peptides* to describe the metabolic stability characteristic of this series of secretagogues.

The resistance to peptidases and protease of Hexarelin (EP23905), the pentapeptide EP51216 and the tripeptide EP 51389 was measured *in vitro* by incubation at 37°C for one hour in conditions that caused extensive degradation of an LHRH analogue chosen as a reference peptide. The results are summarised in Table 2. This table demonstrates the resistance and high resistance of EP23905 and EP51389 respectively. Not surprisingly, EP51389 is totally resistant because of D amino acids composition. The sensitivity of EP51216 to trypsin and protease is essentially due to the deamidation of the C-terminal amide. Surprisingly, EP 23905 (Hexarelin) is very resistant to these enzymes. Since the primary structure cannot explain this resistance, one can suggest a secondary 'cyclic' structure as having a protective effect.

TABLE 2

	Trypsin	Chymotrypsin	Pepsin	Protease
EP51216	37%	0%	0%	51%
EP51389	0%	0%	0%	0%
EP23905	6%	0%	0%	4.5%

The percentage of degradation is calculated as: 100% of residual peptide.

CONCLUSIONS

The peptide approach to the practical development of GH secretagogues remains a viable one, particularly when such peptides are rendered impervious and are appropriately modified to render them less polar and more absorbable by the oral route. The discovery of peripheral receptors opens new opportunities for medicinal chemists and pharmacologists for the development of organ or tissue specific agents.

ACKNOWLEDGEMENTS

I am deeply indebted to Professors Eugenio Müller, Vittorio Locatelli and co-workers at the University of Milan for most of the animal work done with the novel peptides described in the foregoing. I acknowledge the outstanding contributions from Professor Giampiero

Muccioli, University of Turin and of Professor Huy Ong, University of Montreal, for their important binding studies in human and animal tissues. My colleagues at Europeptides in France, François Boutignon, Hélène Touchet, Sandrine David and Edith Barré have given much of their time and ability to our project. I am particularly indebted to Professors Ezio Ghigo and Franco Camanni and their team at the University of Turin for their innovative, competent and enthusiastic contributions for both basic and clinical aspects of this project.

REFERENCES

1. Bowers, C.Y. (1996) Xenobiotic Growth Hormone Secretagogues: Growth Hormone Releasing Peptides. In: Growth Hormone Secretagogues. B.B. Bercu and R.F. Walker (eds). Springer, New York, pp. 9–25.
2. Ghigo, E., Arvat, E., Muccioli, G., Camanni F. (1997) Growth Hormone-Releasing Peptides. European J. Endocrin. 136, 445–460.
3. Nargund, R.P., Van der Ploeg, L.H.T. (1997). Growth Hormone Secretagogues. Ann. reports in Med. Chem. Vol 32, 221–230.
4. Smith, R.G., Van der Ploeg, L.H.T., Howard, A.D. et al. (1997) Peptidomimetic Regulation of Growth Hormone Secretion. Endocrine Reviews 18, 621–645.
5. Deghenghi, R. (1996) Examorelin. Drugs of the Future 21 (4), 366–368.
6. Deghenghi, R., Cananzi, M.M., Torsello, A. et al. (1994) GH-Releasing Activity of Hexarelin, a new Growth Hormone-Releasing Peptide, in infant and adult rats. Life Sci. 54, 1321–8.
7. Deghenghi, R. (1994) Growth Hormone-Releasing Peptides in Growth-Hormone Secretagogues. Verlag, New York, pp. 85–102.
8. Locatelli, V. (1997) Personal Communication. September 26, 1997.
9. Savige, W.E., Fontana, A. (1980) Oxidation of Tryptophan to Oxindolylalanine by Dimethylsulfoxide-Hydrochloric Acid. Int. J. Peptide Protein Res. 15, 285–297.
10. Muccioli, G., Ghé, C., Ghigo, M.C., et al. (1997) GHRP Receptors in Pituitary, Central Nervous System and Peripheral Human Tissues. Abs. 186, J. Endocrinol. Invest. 20 (suppl. to No. 4), 52.
11. Davis, C.B., Crysler, C.S., Boppana, V.K, et al. (1994) Disposition of Growth Hormone-Releasing Peptide (SK&F 110679) in rat and dog following intravenous or subcutaneous administration. Drug Metab. Dispos. 22, 90–98.

Growth Hormone Secretagogues
Edited by E. Ghigo, M. Boghen, F.F. Casanueva and C. Dieguez

Chapter 4

GHRP Structure–Activity Relationship: An In Vivo Perspective

ROSS CLARK

Research Centre for Developmental Medicine and Biology, School of Medicine and Health Sciences, University of Auckland, Auckland, New Zealand.

INTRODUCTION

The discovery of novel GHRPs and the comparison of their structure and function has been the subject of much recent experimentation (1), debate and discussion in both academic laboratories and within the pharmaceutical industry. However, despite the progress in GHRP chemistry and new GHRP receptor characterisation (2,3) the physiology, pharmacology and biology of the GHRP system is still very poorly understood. This paucity of biological knowledge has in particular affected medicinal chemistry projects aimed at exploiting the activity of the GHRPs for medical benefit. This lack of biological knowledge has impacted the choice of assays for testing for GH-releasing activity and testing for the specificity of pituitary hormone release. There is also very limited published data on GHRP efficacy in animals, beyond data on GH release, and because of this there is also a similar paucity of data on the optimal or most efficacious pattern of GHRP exposure. This lack of efficacy data in animals, and appropriate models for testing molecules and delivery patterns, and therefore a lack of information on optimal routes and patterns of administration, has perhaps contributed to the development of GHRPs in humans being much slower than one would have originally predicted.

Recently, a very promising orally active and potent GHRP, MK-677, failed to show a maintained acceleration of statural growth in children (4). This disappointing result may be another symptom of our poor understanding of the parameters governing GHRP efficacy and therefore our inability to translate knowledge of structure–activity relationships, found using *in vitro* systems, into molecules useful as therapeutic agents. It is possible that GHRP structure–function studies need to be reassessed, especially in terms of the elements needed for a successful drug and clinical development programme. This review therefore describes the factors that affect GHRP efficacy *in vivo* in animals and in humans and identifies the issues that need to be considered in current and future structure–function studies of GHRPs.

HOW TO MEASURE GHRP ACTIVITY?

A key question in the design of an integrated GHRP biology and chemistry programme is which activity of the GHRPs should be followed, and therefore which assay systems should be used? The answer to the question, of which activity to follow, is not obvious. This can be seen from different laboratories using different assays to measure GHRP activity (5–7). One reason for this difference is that at one time it was believed that the GHRPs were similar, or at least equivalent in terms of their activity, to growth hormone releasing factor (GRF) in that their major site of action, or even their sole site of action, was at the level of the pituitary gland. This activity on the pituitary gland was thought to be due to a direct effect of GHRP on somatotrophs that was independent of the hypothalamic releasing factors GRF and somatostatin. This was based on GHRP having direct GH-releasing effects *in vitro* on cultured somatotrophs. In contrast early *in vivo* data suggested that the GH-releasing activity of GHRP injections was dependent on GRF and somatostatin activity (8). This experimental data, that the majority of the GH-releasing activity of GHRP *in vivo* depended on it having direct hypothalamic activity on the GRF and somatostatin systems, was met with some doubt (9). The effects of GHRP *in vivo* are complex and to account for this it has been suggested that an additional hypothalamic factor (U factor) is released by GHRP (9) *in vivo*. But in the early 1990s doubt existed regarding the relative importance, and relevance, of the pituitary and hypothalamic activities of GHRPs. This, of course, affected the choice of the appropriate assays, *in vitro* or *in vivo*, with which to measure GHRP activity. These debates, in turn, clearly affected the choice of assays to use in a chemistry programme for GHRP drug design. Some chemistry programmes appear to have chosen to use only *in vitro* assays based on the direct GH-releasing activity of GHRPs on pituitary cells in culture. Other programmes routinely used *in vivo* assays, or a combination of both methods.

ACTIVITY MEASURES: *IN VITRO* OR *IN VIVO*, OR BOTH?

The earliest studies of GHRP structure–function showed the importance of the choice of models for measuring GHRP activity. For example, the initial studies by Bowers (5), of synthetic GHRP peptides, showed that structure–activity relationships *in vitro* did not necessarily translate into *in vivo* activity. The largest subsequent studies of GHRP structure–activity seem to have used routinely only an *in vitro* pituitary assay of GH release (6,10,11). It remains unclear whether the large number of compounds described in this work (6,10,11) would show similar activity *in vitro* and *in vivo*. The largest published comparison of such activity is shown in Figure 1. This data shows a very poor relationship between *in vitro* and *in vivo* potency (7). The *in vivo* data from such comparisons, despite being based on intravenous injections, will be confounded by different classes of compounds having different pharmacokinetics or biodistribution rather than having different receptor binding. Despite such a caveat, some of the scatter shown in Figure 1 (7) probably reflects a structurally specific and mechanism-based difference in GHRP activity. It seems clear that the use of different but complementary assays of biological activity is of value in dissecting the structure–activity relationships of GHRPs. A biological basis for these differences can

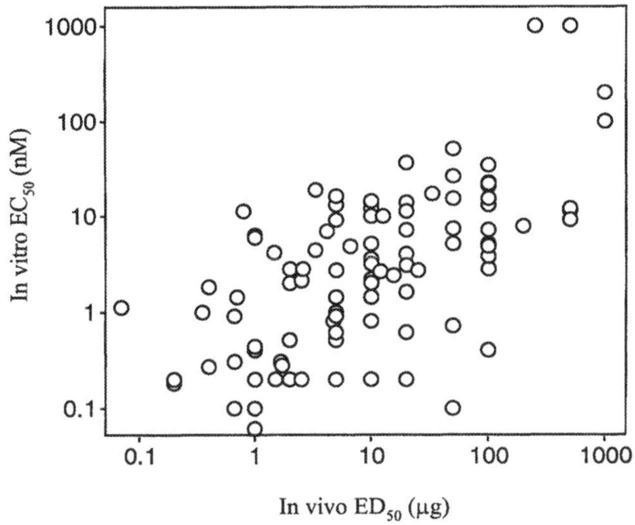

Figure 1. The GH release *in vitro* (EC_{50} in nM) plotted against the GH secretion *in vivo* (ED_{50} in µg injected/rat) for 107 structurally diverse GH secretagogues. It is clear that activity *in vitro* is a poor predictor of activity *in vivo*. From (7).

now be established, as GHRP receptors are now being discovered (2,3), but it will remain necessary to confirm the activity of GHRPs *in vivo*.

SPECIFICITY OF GHRPs FOR HORMONE RELEASE

Initial *in vitro* studies suggested an absolute specificity of GHRP for GH secretion, compared to its effects on the release of other pituitary hormones. In particular the studies by Bowers (5), and later by the Merck group (12), showed that the GHRPs do not release adrenocorticotrophic hormone (ACTH) *in vitro*. However, *in vivo*, it is clear that the GHRPs do affect the release of other hormones, especially ACTH and prolactin. However in terms of *in vivo* activity it is clear that the GHRPs do release ACTH (13) and prolactin (14) in the rat, and ACTH and prolactin release in humans (15). Very few studies have addressed specifically the effect of GHRP on prolactin and ACTH secretion, despite the effects *in vivo* of GHRPs on ACTH release being comparable in magnitude to the effect of corticotrophin releasing factor in rats (13) and in humans (16). There is as yet no clear evidence that a GHRP has been produced that has full GH-releasing activity yet lacks ACTH-releasing activity, or conversely, one which only possesses ACTH-releasing activity. The possibilities inherent in a small molecule with specific ACTH releasing activity should not be ignored.

We tested the potential problems caused by GHRP stimulating the hypothalamic–pituitary–adrenal (HPA) axis in an animal model of obesity and Type II diabetes, the Zucker Diabetic Fatty (ZDF) rat (17). The obesity and diabetes of the ZDF rat is known to be sensitive to adrenal hormones and therefore in these animals we thought that the diabetogenic effects of GHRPs should be revealed. In ZDF rats we found that a GHRP

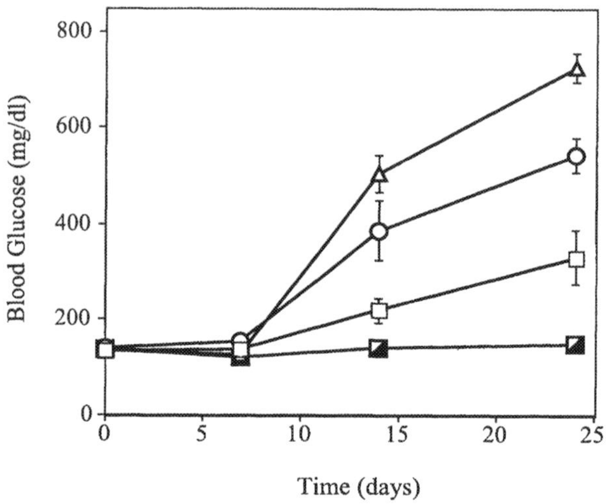

Figure 2. Non-fasting blood glucose levels in young obese Type II diabetic ZDF rats. The progression of diabetes is shown for obese control rats (open squares) and for non-obese rats of the same strain (half-squares). For 24 days, obese rats were given twice daily injections of vehicle (open squares), another group of rats was given a GHRP (G-7039, circles) at 100 μg/day (s.c., bid, at 50 μg/injection), which had a very clear diabetogenic effect. A third group of rats (triangles) was given recombinant human GH (500 μg/day, s.c, bid, 250 μg/injection). The GHRP clearly accelerated the progression of diabetes in these animals. From (17).

analogue and hGH both stimulated body weight gain. We expected that hGH would worsen the diabetic state, which it did, as shown in Figure 2. The dramatic diabetogenic effect of GHRP was a surprise. Most of the responses to GHRP in this experiment could be explained by it causing GH release, but the increases in blood lipids and body fat, which were not seen with hGH, probably reflect an activation of the HPA axis by the GHRP. In rodents it is well known (18) that the diabetogenic effects of GH are amplified in the presence of glucocorticoids and that all known GHRPs after acute administration raise glucocorticoid levels. The activation of the HPA axis in combination with a stimulation of GH probably explains the dramatic diabetogenic effects of GHRP in ZDF rats shown in Figure 2.

This study (17) suggests that GHRP analogs with glucocorticoid-releasing activity should be given with caution to obese or diabetes-prone individuals. In fact elderly normal human subjects given an orally active GHRP for 4 weeks have shown increased fasting blood glucose levels (19). Impairments in glucose control without a reduction in body fat have also been reported in obese subjects after 8 weeks treatment with this orally active GHRP (20), and are similar to that seen in ZDF rats, and are probably caused by the GHRP causing both GH and adrenal steroid release. As seen in the glucocorticoid sensitive ZDF rats, even a small but consistent rise in adrenal stimulation, accompanying GH stimulation, seems to provide an undesirable extra diabetogenic drive in susceptible individuals. The hope (1, 6) that GHRP administration by being more "physiological" in causing a pulsatile release of GH, would have less side effects than treatment with hGH, may not in fact be the case.

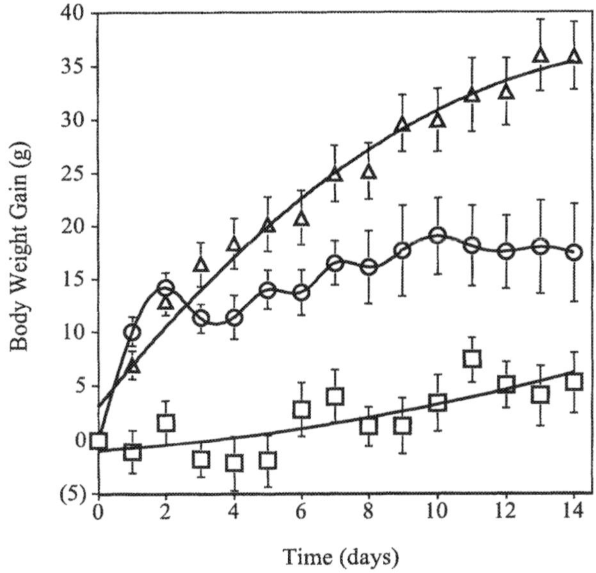

Figure 3. Body weight gain in normal adult female rats (150 days of age) treated with either saline (squares) or 100 μg/d of a GHRP (G-7039) given as either a continuous s.c. minipump infusion (circles) or as twice daily s.c. injections (triangles). It is clear that the injections of GHRP cause a maintained stimulation of weight gain while the response to the continuous exposure to the GHRP shows tachyphlaxis. From (7).

Therefore structure–activity studies of GHRPs clearly need to address many issues in addition to the ability of GHRP to directly release GH from pituitary cells *in vitro*.

GHRP DOWN-REGULATION AND SYNERGY

GHRP down-regulation following prolonged GHRP administration *in vivo* in the rat was described by Bowers in his original description of GHRP efficacy (5). In the short-term, the administration of GHRP *in vivo* causes an immediate down-regulation of responsiveness to subsequent exposures to GHRP. For example, an infusion of GHRP results in a large initial release of GH, then after several hours a long-term down-regulation of GH secretion (8). We have published the only comparison of different patterns of GHRP exposure on long-term efficacy. The body weight gains from this experiment are shown in Figure 3. The experiment showed a dramatic waning of anabolism after infusions of GHRP, while anabolism was maintained with GHRP injections (7). Such observations should influence GHRP structure–activity considerations for GHRP candidate drugs, GHRP formulations and GHRP delivery. However it is clear that continuous exposure to GHRPs has not been seen as of key importance in GHRP drug design programmes. For example MK-677 was chosen (11) for clinical development because of its "superior oral potency and duration of action". Given the persistence and thus long-lasting activity of this molecule it was always possible that this molecule may not have long lasting efficacy in humans. This possibility was recently confirmed when MK-677 failed to show a maintained acceleration of statural growth in children (4).

A synergistic interaction between administered GRF and GHRP occurs *in vivo*, but apart from one study (12), such synergism has not been seen *in vitro* (21). It is likely that the dramatic effects of GHRPs on GH secretion *in vivo*, which are much more impressive than those *in vitro* (22), may be a reflection of this synergism and of an inhibition of somatostatin secretion (8). However the dramatic synergistic effects of GHRP and GRF on GH secretion have yet to be translated into dramatic synergistic effects on efficacy endpoints (growth, anabolism, reduced fat depot size) *in vivo*. Why this has not been the case is difficult to understand or discern.

ARE THERE SURROGATE MARKERS OF GHRP EFFICACY?

A rise in IGF-1 concentrations in blood, caused by a rise in GH secretion, has been assumed to be the "surrogate" marker of GHRP activity that would inevitably lead to efficacy in terms of GHRP mimicking the effects of administering GH on anabolism, growth or lipolysis. It has been a salutary lesson in endocrinology that increases in IGF-1, the surrogate marker of increased GH activity, have not resulted in long-term effects of GHRP administration that are comparable to the effects of administering GH or IGF-1 (4,17). Thus, IGF-1 levels do not seem to be a good surrogate marker for GHRP efficacy on anabolism, growth or lipolysis. The reason for this is unclear. It could be that GHRP administration changes the pattern of GH exposure, rather than the amount of GH exposure, causing a rise in IGF-1 levels. A more continuous GH exposure causes larger rises in blood IGF-1 levels than intermittent GH exposure (23), suggesting that a rise in IGF-1 levels following GHRP exposure is not predictive of GH-like efficacy. The pediatric literature now suggests that the growth response to treatment with injections of IGF-1 in children does not match the growth response that can be caused by GH treatment (24). This confirms the animal data which has shown that IGF-1 administration does not show the full anabolic and growth promoting efficacy of treatment with GH (25). It is even unclear as to whether or not the blood IGF-1 response to GH treatment is predictive of the statural growth response in GH-deficient children (26). This series of experimental observations clearly illustrates the tenuous nature of the use of blood IGF-1 concentrations as a surrogate marker for GHRP efficacy. Therefore, the fact that the administration of a particular GHRP causes a rise in IGF-1 concentrations in animals cannot be taken as evidence pre-clinically that a molecule will show long-term efficacy in animals or humans. In addition, a rise in IGF-1 concentrations in humans following the administration of GHRP should not be taken as predictive of long-term efficacy.

TESTING GHRPs FOR ANABOLIC AND GROWTH PROMOTING EFFICACY

The above discussion underscores what seems to be obvious, but has been perhaps ignored in the transition of the GHRPs from pre-clinical to clinical studies. This is that the actual endpoints that are intended to be achieved by GHRPs in humans should be shown in animals rather than relying on surrogate endpoints. For example, for GHRP use in pediatrics for statural growth, animal data showing a robust increase in epiphyseal cartilage growth or longitudinal bone growth would seem to be necessary. Given that these are the "classic" endpoints for GH action and activity, it is surprising that there is no literature showing dramatic "GH-like" effects of GHRP *in vivo* on bone growth in animals. The recent human data, that GHRPs are relatively poor stimulators of bone growth (4), may be the reason why robust growth promoting animal data is lacking. In comparison, many more experiments have shown that GRF induces classic GH-like growth responses in animals (27). Clear structure–activity data of GHRP efficacy on long-term efficacy endpoints such as bone growth in animals, is needed. Such data should be predictive of superior GHRP efficacy in humans.

STRUCTURE–ACTIVITY OF GHRP: THE FUTURE

Solutions to the issues identified above need to be found if future structure–activity analyses of GHRP are to identify clinically useful GHRPs. It is clear that the dramatic GH release that can be induced by GHRPs should be amenable to being harnessed for long-term clinical benefit. However, current GHRP molecules and/or modes of administration have not been able to translate effects on GH secretion into long-term efficacy on clinically desirable endpoints. It is also apparent that the lack of dramatic long-term GHRP efficacy in animals suggests that there are basic flaws in widely held concepts of GHRP activity and use. Key issues that need to be addressed are the importance of GHRP-induced ACTH release and long-term down-regulation, both of which may limit long-term GHRP efficacy. The relationship between the down-regulation, the mode of GHRP administration, and therefore of GHRP receptor exposure, are key related issues. Although the acute GH-releasing activity of the GHRPs is tantalizing, much more research is needed before a GHRP will become a pharmaceutical with acceptable GH-like efficacy in humans.

ACKNOWLEDGEMENTS

The help and advice of my many friends and colleagues who I have worked with on GHRP is gratefully acknowledged. This review would not have been possible without the combined skills in chemistry of Todd Somers, Bob McDowell and John Burnier and the biological skills of Mike Cronin, Deborah Mortensen and Iain Robinson.

REFERENCES

1. Smith, R.G., Pong, S.-S., Hickey, G., et al. (1996) Modulation of pulsatile GH release through a novel receptor in hypothalamus and pituitary gland. Rec. Prog. Horm. Res. 51, 261–268.
2. Howard, A.D., Feighner, S.D., Cully, D.F., et al. (1996) A receptor in pituitary and hypothalamus that functions in growth hormone release. Science 273, 974–977.
3. Pong, S.S., Chaung, L.Y.P., Dean, D.C., Nargund, R.P., Patchett, A.A., Smith, R.G. (1996) Identification of a new g-protein-linked receptor for growth-hormone secretagogues. Mol. Endocrinol. 10, 57–61.
4. Yu, H., Cassorla, F., Tiulpakov, A., et al. (1998) A double blind placebo-controlled efficacy trial of an oral growth hormone (GH) secretagogue (MK-0677) in GH deficient (GHD) children.

Program and Abstracts 80th Annual Meeting US Endocrine Society, New Orleans, Louisiana, OR24-6, 84.
5. Bowers, C.Y., Momany, F., Reynolds, G.A., Hong, A. (1984) On the *in vitro* and *in vivo* activity of a new synthetic hexapeptide that acts on the pituitary to specifically release growth hormone. Endocrinology 114, 1537–1545.
6. Smith, R.G., Cheng, K., Schoen, W.R., et al. (1993) A nonpeptidyl growth hormone secretagogue. Science 260, 1640–1643.
7. McDowell, R.S., Elias, K.A., Stanley, M.S., et al. (1995) Growth-hormone secretagogues — characterization, efficacy, and minimal bioactive conformation. Proc. Natl. Acad. Sci. 92, 11165–11169.
8. Clark, R.G., Carlsson, L., Trojnar, J., Robinson I. (1989) The effects of a growth hormone-releasing peptide and growth hormone-releasing factor in conscious and anaesthetized rats. J. Neuroendocrinol. 1, 249–255.
9. Bowers, C.Y., Sartor, A.O., Reynolds, G.A., Badger, T.M. (1991) On the actions of the growth hormone-releasing hexapeptide, GHRP. Endocrinology 128, 2027–2035.
10. Schoen, W.R., Ok, D.R.J.D., et al. (1994) Structure–activity relationships in the amino acid side-chain of L-692,429. Bioorg. Med. Chem. Lett. 4, 1117–1122.
11. Patchett, A.A., Nargund, R.P., Tata, J.R., et al. (1995) Design and biological activities of l-163,191 (mk-0677) — a potent, orally-active growth-hormone secretagogue. Proc. Natl. Acad. Sci. 92, 7001–7005.
12. Cheng, K., Chan, W.W.S., Barreto, A., Convey, E.M., Smith, R.G. (1989) The synergistic effects of His-DTrp-Ala-Trp-DPhe-LysNH$_2$ on growth hormone (GH)-releasing factor-stimulated release and intracellular adenosine 3′5′-monophosphate accumulation in rat primary pituitary cell culture. Endocrinology 124, 2791–2798.
13. Thomas, G.B., Fairhall, K.M., Robinson, I.C.A.F. (1997) Activation of the hypothalamo-pituitary-adrenal axis by the growth hormone (GH) secretagogue, GH-releasing peptide-6, in rats. Endocrinology 138, 1585–1591.
14. Thomas, G.B., Carmignac, D.F., Bennett, P.A., Robinson, I.C.A.F. (1998) Stimulation of prolactin release by GHRP-6 in dwarf rats is estrogen dependent. 80th Annual Meeting of The Endocrine Society, New Orleans, LA, P2–225.
15. Bowers, C.Y., Reynolds, G.A., Durham, D., Barrera, C.M., Pezzoli, S.S., Thorner, M.O. (1990) Growth hormone (GH)-releasing peptide stimulates GH release in normal men and acts synergistically with GH-releasing hormone. J. Clin. Endocrinol. and Metab. 70, 975–982.
16. Arvat, E., di Vito, L., Maccagno, B., et al. (1997) Effects of GHRP-2 and hexarelin, two synthetic GH-releasing peptides, on GH, prolactin, ACTH and cortisol levels in man. Comparison with the effects of GHRH, TRH and hCRH. Peptides 16, 885–891.
17. Clark, R.G., Thomas, G.B., Mortenson, D.L., et al. (1997) Growth hormone secretagogues stimulate the hypothalamic–pituitary axis and are diabetogenic in the Zucker Diabetic Fatty rat. Endocrinology 138, 4316–4323.
18. Kostyo, J.L. (1987) Diabetogenic effects of growth hormone. In: Growth hormone: Basic and clinical aspects. O. Issaksson et al. (eds). Elsevier, Amsterdam, 217–225.
19. Chapman, I.M., Bach, M.A., Van Cauter, E., et al. (1996) Stimulation of the growth hormone (GH)-insulin-like growth factor I axis by daily oral administration of a GH secretogogue (MK-677) in healthy elderly subjects. J. Clin. Endocrinol. Metab. 81, 4249–4257.
20. Svensson, J., Lonn, L., Jansson, J.O., et al. (1998) Two-month treatment of obese subjects with the oral growth hormone (GH) secretagogue MK-677 increases GH secretion, fat-free mass, and energy expenditure. J. Clin. Endocrinol. Metab. 83, 362–369.
21. Bowers, C. (1993) GH releasing peptides — structure and kinetics. J. Paed. Endocrinol. 6, 21–31.
22. Goth, M.I., Lyons, C.E., Canny, B.J., Thorner, M.O. (1992) Pituitary adenylate-cyclase activating polypeptide, growth-hormone (GH)-releasing peptide and GH-releasing hormone stimulate GH release through distinct pituitary receptors. Endocrinology 130, 939–944.
23. Clark, R., Mortensen, D., Carlsson, L., Carmignac, D., Robinson, I. (1995) Growth responses to patterned GH delivery. Endocrine 3, 717–723.

24. Rosenfeld, R. (1998) IGF-1 treatment of growth hormone insensitivity. In: Molecular mechanisms to regulate the activities of insulin-like growth factors. K. Takano et al. (eds). Elsevier, Amsterdam, 359–364.
25. Skottner, A., Clark, R., Fryklund, L., Robinson, I. (1989) Growth responses in a mutant dwarf rat to human growth hormone and recombinant human insulin-like growth factor I. Endocrinology 124, 2519–2526.
26. Lee, P.D., Durham, S.K., Martinez, V., Vasconez, O., Powell, D.R., Guevara-Aguirre, J. (1997) Kinetics of insulin-like growth factor (IGF) and IGF-binding protein responses to a single dose of growth hormone. J. Clin. Endocrinol. Metab. 82, 2266–2274.
27. Clark, R., Robinson, I. (1987) The control and significance of the secretory pattern of growth hormone in the rat. In: GH, Basic and Clinical. K. Binder et al. (eds). Elsevier, Amsterdam, vol ICS 748.

Growth Hormone Secretagogues
Edited by E. Ghigo, M. Boghen, F.F. Casanueva and C. Dieguez

Chapter 5

Molecular Analysis of the Growth Hormone Secretagogue Receptor

ANDREW D. HOWARD, SHENG-SHUNG PONG, KAREN KULJU MCKEE, OKSANA C. PALYHA, DONNA L. HRENIUK, CARINA P. TAN, RAVI NARGUND, ARTHUR A. PATCHETT, LEX H.T. VAN DER PLOEG, ROY G. SMITH and SCOTT D. FEIGHNER

Merck Research Laboratories, Rahway, NJ 07065, U.S.A.

SUMMARY

The molecular cloning of a receptor for growth hormone secretagogues (GHSs) from humans and other species provides evidence that a third neuroendocrine pathway exists, in addition to growth hormone releasing hormone and somatostatin, that aids in the control of pulsatile growth hormone (GH) release from the pituitary gland, presumably regulated by an as yet unidentified hormone.

MOLECULAR CLONING

Expression cloning (Figure 1) was adopted to isolate a cDNA encoding the growth hormone secretagogue receptor (GHS-R; (1)). The isolation of cDNA and genomic clones encoding the GHS-R has been the subject of recent reviews (2,3). In brief, earlier investigations aimed at elucidation of the signal transduction pathway activated by GHSs demonstrated that GHSs activate phospholipase C resulting in a rise in inositol triphosphate and intracellular Ca^2. In contrast, growth hormone releasing hormone (GHRH) signal transduction occurs through the activation of adenylate cyclase and subsequent elevation of intracellular cAMP (4,5). Initially, *Xenopus* oocytes injected with swine pituitary poly $(A)^+$ mRNA as a source of GHS-R mRNA occasionally gave a modest activation of Ca^{2+} activated Cl^- currents in response to MK-0677. To improve reproducibility, assay throughput, and reliability of the response, swine poly $(A)^+$ mRNA was supplemented with various G_α subunit mRNAs. In addition, rather than to measure the activation of phospholipase C by electrophysiological methods, mRNA encoding the Ca^{2+} sensitive luminescent protein aequorin

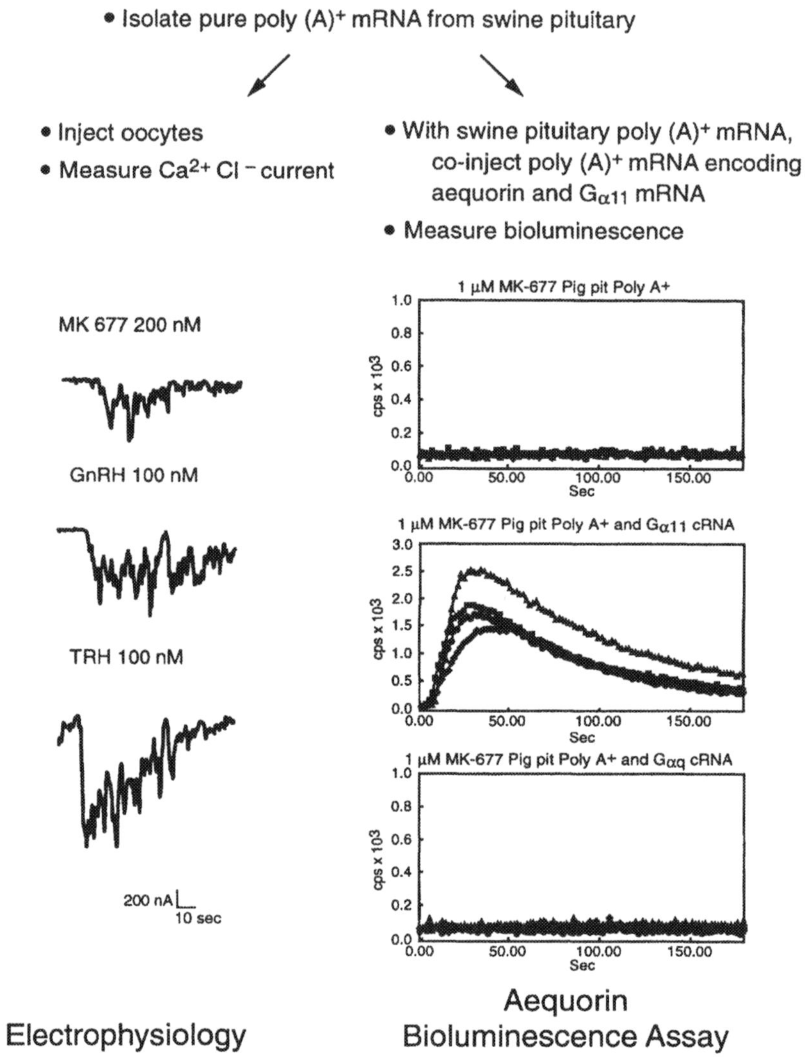

Figure 1. Expression cloning of the GHS-R. *Xenopus* oocytes were injected with swine pituitary poly (A)+ mRNA, G_{a11} cRNA (with or without) and aequorin mRNA for measurement of bioluminescence, or only swine pituitary poly (A)+ mRNA for detection of Ca^{2+} activated Cl^- currents. Following a 36 hour incubation the *Xenopus* oocytes were challenged with 1 µM MK-0677.

was co-injected (6). Using a luminometer, responses were measured generally 2-days post-injection following challenge with MK-0677. As outlined in Figure 1, MK-0677 responsiveness was dependent on the co-injection of the $G_{\alpha q}$ family member G_{11}. Other G_{α} subunit tested failed to rescue MK-0677-induced bioluminescence. This observation provided a key biochemical breakthrough in our expression cloning protocol by enhancing assay sensitivity and reproducibility. Once this assay was in place, we injected complex pools of ~10,000 cRNAs from unfractionated swine pituitary cDNA libraries and identified a single cDNA which encoded the GHS-R (Figure 1).

STRUCTURE

To date, only a single type of GHS-R has been identified at the molecular level (Figure 2), though additional G-protein coupled receptors (GPC-Rs) that may confer GHS sensitivity have been postulated to exist (see below). The GHS-R identified by expression cloning is a classical GPC-R containing seven putative alpha-helical membrane spanning segments (7-TM) and three intracellular and three extracellular loops (7). In addition, a highly conserved motif responsible for G protein interaction (D/ERY) found in the second intracellular loop immediately following TM-3 is present in the GHS-R. As noted in other GPC-Rs, consensus sequences for N-linked glycosylation, phosphorylation are present, and cysteines located in the first two extracellular loops capable of disulfide bonding (Cys-115 and Cys-197) are also found.

Molecular analysis of GHS-Rs from swine, human, rat, mouse and dog revealed that the GHS-R is strongly conserved in evolution. Comparison of the amino acid sequences among these species (Figures 2 and 6) indicates remarkable overall sequence identity, with only few amino acid substitutions in the TM domains and loop regions. The most divergent GHS-R protein was found in the dog, in which a 17 amino acid segment in the N-terminal extracellular domain are lacking when compared to the other species tested. As both multiple genomic and cDNA clones were isolated, it is unlikely that this is due to a cloning artifact. Despite numerous attempts, other dog GHS-R cDNAs or genomic clones could not be identified suggesting that this clone represents an authentic dog GHS-R. This conclusion is further supported by the pharmacological characterization of the dog GHS-R (see next sections). The full length human GHS-R gene encodes a 366 amino acid protein, except in rodents where the protein measures 364 residues with a loss of one amino acid in the N-terminal extracellular domain and one amino acid in the second intracellular loop. The rat and mouse forms differ by only two amino acids from each other.

From both human and swine pituitary, two types of cDNAs were isolated: type 1a, encoding a functional protein containing 7-TM domains, and type 1b, encoding a protein containing TM-1 through 5 with no measurable functional activity in cell based assays. These two forms most likely arise from transcription of a single gene by alternative mRNA processing (8). This assertion was confirmed by determination of the nucleotide sequence for the proposed human exon–intron boundaries and the complete intron of the human gene, as diagrammed in Figure 3. As seen in many GPC-Rs, introns are usually placed between TM domains and often following TM-5. Type 1a cDNA encodes the complete 7-TM GHS-R and results from a splicing event which removes the intron. In type 1b cDNA, the intron is not removed and an alternative polyadenylation signal is presumably utilized in the intron. As a result, the human and swine type 1b cDNA contain a short, 24-amino acid open reading frame fused to leucine-263 which is conserved in human and swine. Southern blot analysis (Figure 3) to search for GHS-R related genes again indicates that the GHS-R is highly conserved since a simple hybridization pattern is observed when high stringency post-hybridizational washing conditions are utilized. Our analysis included numerous mammalian and non-mammalian species, including Drosophila, Ceanorhabditis elegans and teleost fish. The existence of the GHS-R can apparently be extended to Precambrian

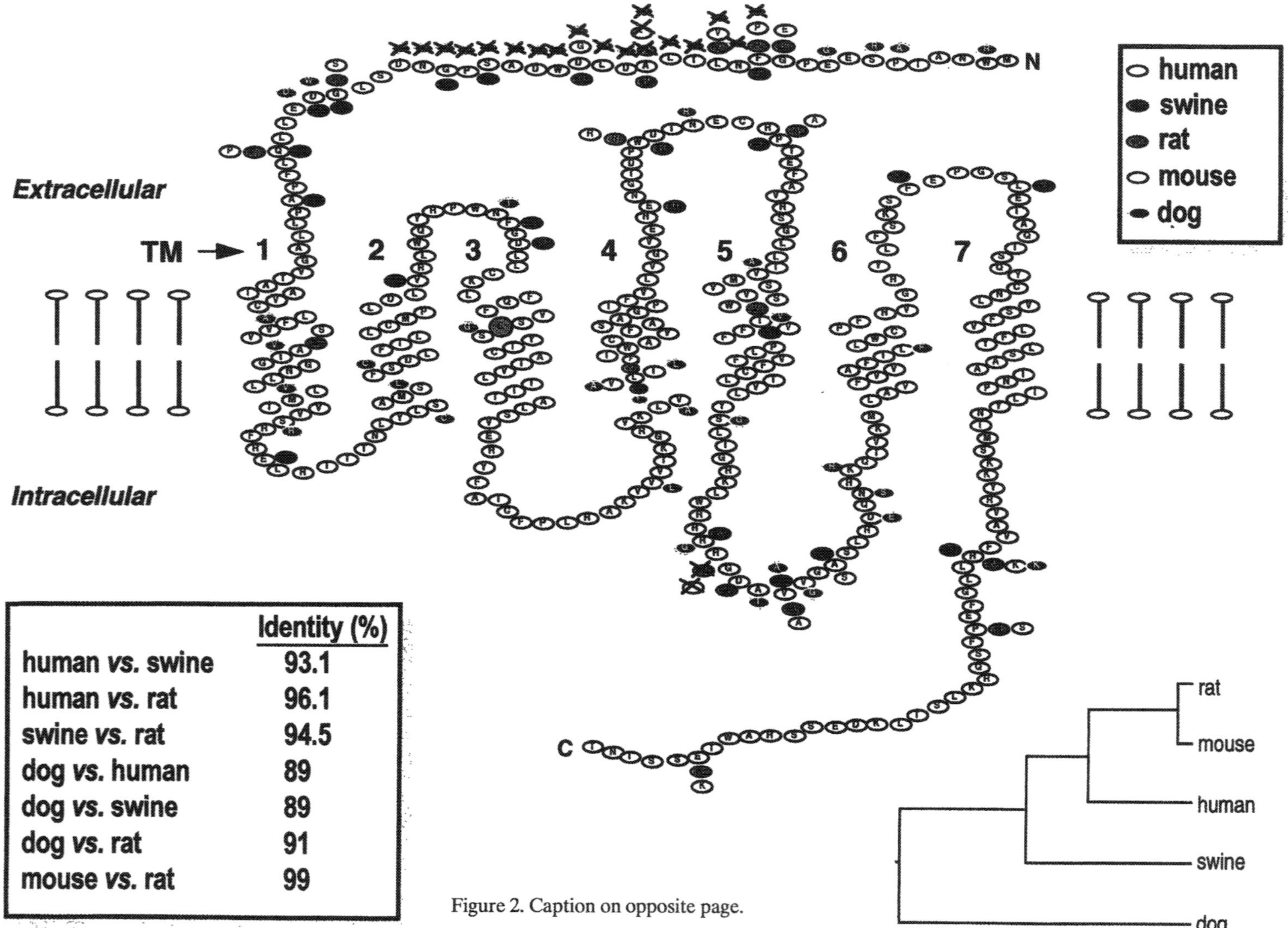

	Identity (%)
human *vs.* swine	93.1
human *vs.* rat	96.1
swine *vs.* rat	94.5
dog *vs.* human	89
dog *vs.* swine	89
dog *vs.* rat	91
mouse *vs.* rat	99

Figure 2. Caption on opposite page.

Figure 2 (*Opposite*). Deduced amino acid sequences of GHS-Rs from swine, human, rat, mouse and dog. Topological representation of the GHS-R with transmembrane domains numbered from 1 to 7 and the N-terminal extracellular domain and C-terminal intracellular domain. Individual amino acid residues are shown (single letter amino acid code). The human sequence is given in yellow and used as the reference where different residues are given the color red (swine), blue (rat), white (mouse) and green (dog). Amino acids which are not present in the rat, mouse and dog orthologs are shown as an **X**. The inset at the bottom left reveals the overall amino acid identities of the GHS-Rs and results from a Pileup alignment (Wisconsin Package Version 9.1, Genetics Computer Group (GCG), Madison, WI.; gap extension 4, gap creation 12) are represented as a dendrogram (bottom right). The essential residue for ligand binding E124 in TM-3 is shown (purple).

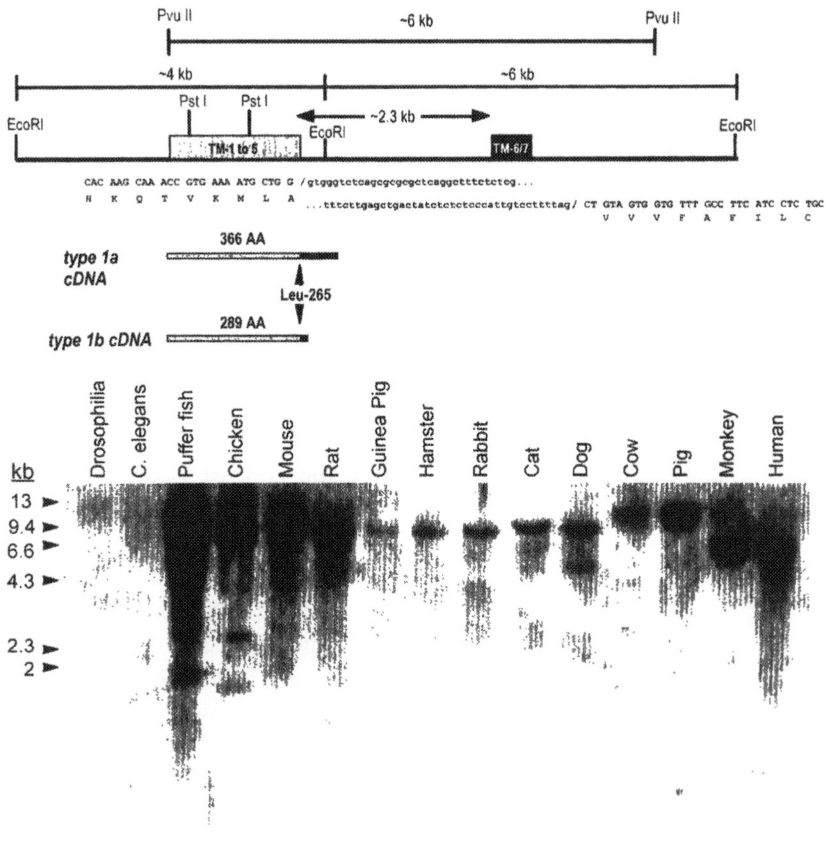

Figure 3 (*above*). The GHS-R gene and Southern blot analysis in different species. Shown in the top panel is the physical map and nucleotide sequence of the human GHS-R gene with its exon–intron boundaries. The open box represents the coding sequence of TM-1 to 5, the shaded box the coding sequence of TM-6 and 7. The single intron of the GHS-R is outlined by the thin line separating the coding exons. Sizes of the restriction enzyme fragments (in kb) are indicated above the physical map. The nucleotide sequence at the exon–intron (upper and lower case, respectively) boundaries is shown just below the physical map. The structure of the type 1a and type 1b cDNAs, which diverge at amino acid 265 is shown below the physical map. For the Southern blot (lower panel) BamHI restriction enzyme digested DNA was fractionated and hybridized with a ^{32}P-labeled human GHS-R probe (complete coding sequence). The blot was washed at high stringency following hybridization.

times (~400,000 million years) as amino acid sequences strongly related to the human GHS-R have been identified in teleost fish (unpublished results). This observation points to the importance of the GHS ligand system in normal physiological function.

Evidence for a second GHS-R subtype has been suggested by studies with a photoactive, radiolabeled derivative of hexarelin, a potent and specific growth hormone releasing protein (GHRP) (9). Covalent crosslinking of this analog to membranes labeled a 57 kDa protein in anterior pituitary and an 85 kDa protein in heart membranes. The protein may be distinct from the GHS-R1a since labeling of the protein was not potently displaced by MK-0677. Molecular cloning is required to unequivocally demonstrate these proteins as potential GHRP receptors.

TISSUE DISTRIBUTION

To date, the GHS-R has been localized at the mRNA level by a combination of *in situ* hybridization, RNase protection and RT-PCR techniques (10–15). These strategies were required due to the extreme low abundance of GHS-R transcripts which prevented the use of Northern blotting analysis. The accumulated expression data (summarized in Table 1) suggests that the GHS-R gene is actively transcribed in the anterior pituitary gland and the brain, two known sites of GHS action. It will be of considerable interest to map the tissue distribution of GHS-R expression using antibody probes when these become available.

Several reports have appeared detailing the expression pattern of the GHS-R by *in situ* hybridization. Collectively these results show that the GHS-R is expressed in the brain, present in both hypothalamic and non-hypothalamic regions. Initial results were obtained in the rhesus monkey using non-overlapping radiolabeled oligonucleotide hybridization probes, which showed specific labeling of the arcuate-ventromedial hypothalamus and infindibular hypothalamus (1). The cloning of the rat GHS-R (8) facilitated detailed mapping studies in the rat brain, again using radiolabeled oligonucleotides (10) or full-length riboprobes (11). Both studies show that the GHS-R is prominently expressed in several hypothalamic nuclei, including the arcuate nucleus, the ventromedial hypothalamic, and supraoptic nucleus. Both the arcuate and ventromedial hypothalamic nuclei are thought to play a key role in the regulation of GH secretion (2,16,17). Significant hybridization signals were also noted in the dentate gyrus and CA2 layers of the hippocampus, a brain region which has not been strongly implicated in growth control but is often associated with memory and learning. Attempts to finely determine the precise localization of the GHS-R within the hypothalamus in comparison to other neuronal markers has revealed that the GHS-R is localized to a region of the arcuate nucleus and ventromedial nucleus distinct from GHRH and NPY containing domains (11). Additionally, in the same study, GHS-R expression was markedly regulated by GH. In the GH-deficient *dw/dw* rat, GHS-R expression was increased while in *dw/dw* rats treated with GH, GHS-R expression decreased. These observations are in accord with the notion that GH regulates its own release through negative feedback loops in the hypothalamus involving GH, and in addition, GHS-Rs. A recent study also attempted to identify the cell type in the

TABLE 1

SUMMARY OF THE DISTRIBUTION OF GHS-R EXPRESSION IN DIFFERENT TISSUES

Data is compiled from Refs. 10–15 which utilized the techniques of *in situ* hybridization, RNase protection and RT-PCR to generate an outline of GHS-R expression patterns. Abbreviations utilized are: FE, functional expression cloning; ISH, *in situ* hybridization; RPA, RNase protection assay; RT-PCR, reverse transcription– polymerase chain reaction; Arc, arcuate nucleus; VMH, ventro-medial hypothalamus, INF, infundibular hypothalamus; DG, dentate gyrus; SO, supraoptic nucleus; EW, Edinger-Westphal nucleus; TM, tuberomammaliary nucleus; LA, lateroanterior nucleus; dorsal raphe nucleus; MnR, median raphe nucleus; Sch, suprachiasmatic nucleus.

Species	Tissue	Technique	Result
Swine	pituitary	FE	+
Human	pituitary	FE	+
Rat	pituitary	FE	+
Rat	hypothalamus	FE	+
Swine	brain	ISH	+Arc, VMH, INF
Rat	brain	ISH	DG, VMH, ARC AHA, AVPO, SO, Sch, LA, TM, EW, DR, MnR
Human	pituitary subthalamic n. hippocampus pancreas	RPA	+++ + + +?
Human	amygdala caudate n. thalamus substantia nigra mammary gland corpus callosum cerebellum whole brain stomach liver heart fetal whole brain testis thymus adrenal gland uterus spinal cord bone marrow thyroid lung	RPA	–
Rat	anterior pituitary	ISH	+ –
Human	somatotropinoma	RT-PCR	+
Human	prolactinoma	RT-PCR	–
Rat	GH cells	RT-PCR	+
Human	adenomas (GH, prolactin and ACTH secreting)	RT-PCR	+
Human	bronchial tumours	RT-PCR	+

(continued)

TABLE 1 (*continuation*)

Species	Tissue	Technique	Result
Rat	anterior pituitary hypothalamus brain stem cerebral cortex kidney–pelvis	RT-PCR	+
Rat	cerebellum posterior pituitary thyroid heart lung stomach duodenum small intestine colon liver pancreas spleen adrenal gland kidney–cortex kidney–medulla ureter testis epididymis uterus ovary placenta skeletal muscle adipose tissue	RT-PCR	–

arcuate and ventromedial hypothalamus which expresses the GHS-R (12). Using dual chromogenic and *in situ* hybridization, the GHS-R was found to co-localize to both arcuate and ventromedial hypothalamic cells which express GHRH. This result provides evidence that GHRH release into hypophyseal portal blood may be regulated by GHSs.

RNase protection and RT-PCR analysis of GHS-R expression in normal tissues are generally consistent with expression confined to brain and pituitary gland. Three recent reports (13–15) show by RT-PCR that transcripts for the GHS-R can be found in a variety of human tumors of pituitary origin including, somatotrophinomas, the rat pituitary cell line GH3 and corticotrope adenoma (Cushing's disease). This result confirms the presence of GHS-R in corticotrope adenomas and provides an explanation why GHRPs stimulate the release of ACTH in patients with Cushing's syndrome. The presence of the GHS-R transcripts in non-pituitary tumors (bronchial carcinoid) and in the renal pelvis is not firmly established at this time. The expression of the GHS-R in the brain suggests broader functions, beyond the control of GH release, for the GHS-R and its natural ligand in normal physiology.

PHARMACOLOGY

The availability of cloned GHS-Rs from several species allowed for a detailed comparison of their pharmacological properties in comparison with their naturally-occurring counterparts. Full-length cDNAs encoding swine, human, rat, mouse or dog GHS-R were placed into mammalian expression vectors and transiently or permanently expressed in monkey kidney cells (COS-7) or human embryonic kidney fibroblasts (HEK-293). Both cell types do not express detectable endogenous GHS-Rs. Binding and functional activation assays were utilized to characterize the cell lines and to compare the intrinsic potency of various GHS ligands. For the cloned GHS-Rs, two assays were utilized to measure receptor expression: [^{35}S]-MK-0677 binding to cell membranes and aequorin bioluminescence in whole cells for a determination of the efficacy of functional coupling. Characterization of the native GHS-R isolated from primary pituitary and hypothalamic tissue relied on [^{35}S]-MK-0677 binding. In addition, GH release was measured in rat primary pituitary culture. All cloned GHS-Rs can be expressed at high levels in fibroblast recipient cell lines using both the radioligand and aequorin functional assays as a measure of GHS-R expression. Figure 4 shows an example of an [^{35}S]-MK-0677 saturation isotherm for the cloned dog GHS-R, and a dose-response curve for functional activation of the cloned mouse GHS-R. Binding is saturable, of high affinity, identifying two classes of binding sites when expressed in heterologous cells. The identification of two classes of binding sites is most likely due to an excess of uncoupled receptors over available G proteins, leading to the presence of high and low affinity state GHS-Rs, as often observed when cloned receptors are overexpressed (18). This assertion is supported by the observation that only ~50% of the binding can be blocked by GTPγS (Figure 4). Table 2 compares the binding properties of the cloned GHS-Rs to the values of these same parameters obtained for GHS-R identified in tissue. In general, there is good agreement between intrinsic potency for binding and activation of the GHS-R in functional assays when compared across species for the cloned and native GHS-Rs.

Site-directed mutagenesis of the human GHS-R has provided insights into key residues that are essential for GHS-R function (19). While overall amino acid sequence identity in the family of GPC-Rs is low, the seven transmembrane helices contain recognizable motifs which help to identify conserved structural elements that can be used in nucleating 2D GPC-R sequence alignments. Diversity in the transmembrane region (especially the presence of a charge in the hydrophobic helices) can give clues to the location of amino acid residues that are potentially relevant for interaction with ligands. The location of essential amino acid residues of the GHS-R was initially based on a functional 2D sequence alignment of highly conserved motifs in related GPC-Rs and homology modeling based on the helical footprint of bacteriorhodopsin. In the 2D sequence alignment a comparison was made with the angiotensin-II, β_2-adrenergic, neurotensin, somatostatin-2 receptor and human, swine and rat GHS-Rs. Prior to mutagenesis, these data were combined to form a preliminary 3D GHS-R model docked with representative members of three classes of GHSs: the peptide, GHRP-6 and the non-peptides biphenyl benzolactam, L-692,585 and the spiroindolane, MK-0677.

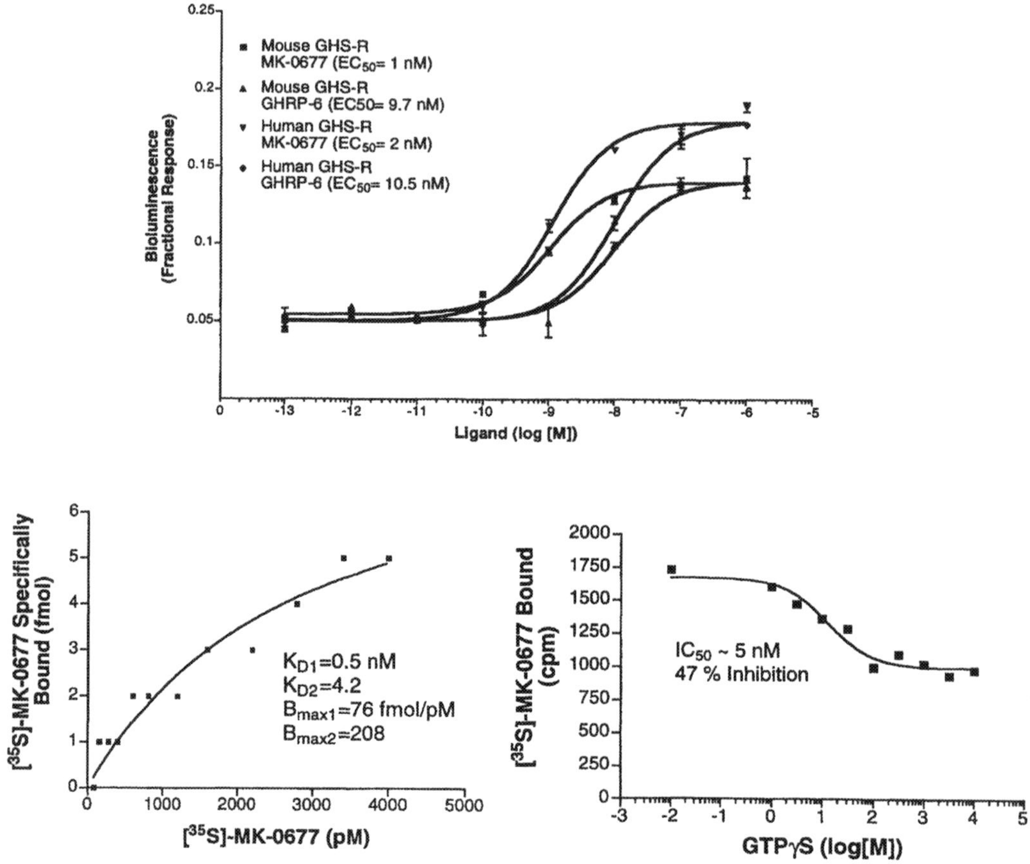

Figure 4. Pharmacological characterization of the cloned mouse GHS-R (functional activation) and cloned dog GHS-R (radioligand binding). *(top panel)* The HEK-293 cell line stably expressing aequorin (293-AEQ17) was transfected with a mouse GHS-R expression plasmid and expression was assessed 2 days later. Two days following transfection, cells were "charged" with the chromophore coelentrazine, scraped and centrifuged in plastic tubes for luminometer measurements. Data for each point on the dose–response curve represent the average of triplicate measurements for each sample ($\sim 3 \times 10^5$ cells/tube). *(bottom panels)* For the dog GHS-R, binding of [^{35}S]-MK-0677 to crude cell membranes was conducted (29). Shown are the saturation isotherm of specific binding (>90% as defined the difference in binding in the absence and presence of 100 nM unlabeled MK-0677), and a dose–response curve to assess the effect of GTPγS on binding.

An important feature for GHS agonist bioactivity is the presence of a basic amino group (20–22). Based on conservation between human, swine and rat GHS-Rs and the 3D docking model, E124 in TM-3 (noted in Figure 2) was proposed to serve as the counter ion to the basic N^+ found in GHS agonists such as MK-0677, L-692,585, and the amino terminus of GHRP-6. An analogous residue is found in TM-3 of the β_2 adrenergic receptor (23). When E124 is changed to Q, consistent with their basic side-chain, MK-0677, GHRP-6 and L-692,585 are no longer able to signal in the E to Q mutant. To confirm the role of this residue in receptor activation, L-168,740, an analog of MK-0677, was synthesized in which the primary amine side chain was modified to the corresponding alcohol which could then

TABLE 2

SUMMARY OF PHARMACOLOGICAL CHARACTERIZATION OF CLONED AND NATIVE GHS-RS FROM VARIOUS SPECIES

[^{35}S]-MK-0677 binding on native and recombinant receptor were utilized to generate K_d, B_{max} and IC_{50} values (28). EC_{50} values in the aequorin bioluminescence assay are shown as a measure of receptor activation.

Receptor	Competitor IC_{50} (nM)					
	MK-0677	L-692,585	L-692,429	GHRP-6	GHRP-1	GHRP-2
Cloned Human	0.15	50	900	6	ND*	ND*
Cloned Rat	0.1	1.5	200	ND*	0.2	ND*
Cloned Porcine	0.2	3.0	150	4	0.1	0.26
Cloned Canine	2.0	ND*	300	18	10	ND*
Porcine Pituitary	0.2	2.0	90	9	0.8	0.4
Canine Pituitary	0.2	3.0	70	ND*	ND*	ND*

Receptor	K_d (nM)	B_{max} (fmol/pM)
Cloned Human	0.036	86
	0.32	183
Cloned Rat	0.08	100
	0.31	201
Cloned Dog	0.5	76
	4.2	208
Cloned Swine	0.4	800
Swine Pituitary	0.16	6.3
Rat Pituitary	0.18	2.3

serve as the counterion to E124, though hydrogen bonding interactions. As shown in Figure 5, L-168,740 did not activate the wild type receptor at concentrations as high as 10 μM. However, functional rescue of the E124Q mutant was achieved by L-168,740, with an EC_{50} of 1.3 μM. Shown for comparison from the same experiment are dose–response curves for the activation of wild type and the E124Q mutant receptor by MK-0677, which gives a >200-fold decrease in potency against the E124Q. We conclude that the basic amine on the GHS agonist forms a salt bridge to the negative charge on the GHS-R. In addition, when the conservative change, E124>D was made, no effect on functional activation was observed, suggesting that a productive GHS-R confirmation was maintained.

GHS-R RELATED RECEPTORS

A comprehensive search was initiated for GHS-R related GPC-Rs. Overall, this search was undertaken to further elucidate the mechanism of action of MK-0677 as it proceeds through clinical evaluation in humans. Our rationale for the notion that GHS-R family members

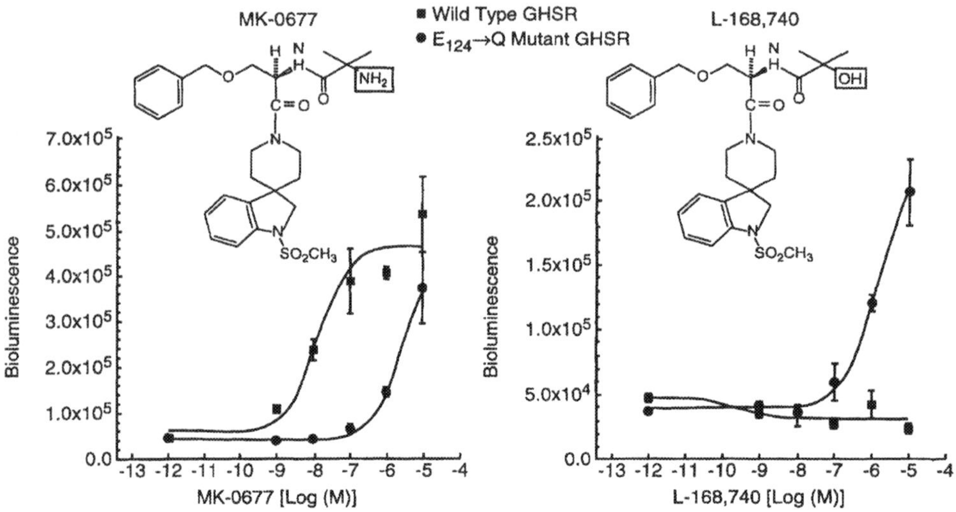

Figure 5. Activation of the E124 human GHS-R mutant. E124 of the human GHS-R was mutated to Q and its functional properties were assessed in the aequorin assay. Shown are dose–response curves for the wild-type (E124) and mutant (Q124) GHS-R in response to challenge by MK-0677 and an alcohol analog L-168,740. Chemical structures are also shown.

may exist was several-fold. First, the GHS-R protein sequence was sufficiently divergent from any previously cloned GPC-R, with the closest relative (neurotensin-R) being only 32% identical at the protein level, that one could expect other members of this sub-family to have gone unnoticed. Secondly, the GHS-R was found in regions of the brain not previously associated with GH release suggesting that GHSs may have additional pharmacological roles. In turn diverse pharmacology is often associated with action of a ligand at distinct receptor subtypes. In the strictest sense, the GHS-R can be classified as an orphan GPC-R (oGPC-R) since its natural ligand has not been identified, therefore MK-0677 being a synthetic molecule may be selecting for only one subtype of a putative receptor family that could be revealed once the natural ligand is known.

Our strategy for identifying GHS-R family members involved three overlapping approaches: (1) Hybridization screening of genomic DNA and cDNA libraries; (2) PCR of genomic and cDNA templates including the use of Gene Trapper technology; (3) daily searching of public nucleic acid and protein sequence databases, particularly the dbEST subset database of Genbank. Using the above approaches, we isolated both new and previously cloned GPC-Rs as well as orthologs of known GPC-Rs.

In the first phase of our search we focused on: (i) isolating human GHS-R family members from genomic DNA by PCR and reduced stringency genomic library screening, and (ii) searching for related receptors that are expressed (cDNA templates) in brain, pituitary, stomach and thyroid by cDNA library screening, and using Gene Trapper technology (Life Technologies) in human fetal brain. The Gene Trapper technique is a positive cDNA selection system that involves the hybridization of a biotinylated oligonucleotide designed against a target sequence. Sequences that are complementary to the designed oligonucleotide in the target are then recovered after binding to streptavidin beads. In the second phase, we concentrated on the isolation of GHS-R related sequences from lower vertebrates. Interestingly, GHS-R related sequences were indeed detected in the puffer fish *Spheroides nephelus*, one of the oldest known vertebrates. Several investigators have shown the utility of identifying homologous genes of mammalian origin in puffer fish, most notably from *Fugu rubripes* (24). Puffer fish are estimated to contain ~90% of mammalian genes but have a genome size ~8-fold smaller than mammals because intron size and repeat sequences are reduced in size and abundance. In studies in progress, three full-length GPC-R genes with good sequence identity (58%, 47% and 34% at the protein level) to the human GHS-R have been isolated and are currently being characterized (unpublished results).

Hybridization Screening

The search for GHS-R family members first involved identifying a human genomic DNA segment with strong sequence conservation to the human GHS-R. Determination of the nucleotide sequence from this fragment revealed that it partially encodes GHS-R family member 1 (FM-1, GPR38), and served to provide broad access to both new members of the previously unknown GHS-R family as well as other related GPC-Rs. This fragment shares ~71% nucleic acid and 67% amino acid sequence identity to the human GHS-R (25). This hybridization probe was used to successfully isolate clone FM-2 (GPR39), the type 2 galanin receptor (26), and the GHS-R and type 2 neurotensin-R from rat hypothalamus (unpublished results).

Database Mining

Genbank databases were monitored daily using the Tblastn program (27) with amino acid sequence from the human GHS-R TM domains 6–7 (residues 265–366). Two significant "hits" were detected. A mouse EST derived from a T-cell library was identified with a significant homology score (63% DNA, 36% amino acid sequence identity) to the 3′ end of the gene for the human GHS-R. Full length cDNA were then obtained for both the mouse and human forms (28). The human and murine FM-3 exhibit strong protein sequence identity (73%). In addition, a cosmid clone (K10B4) from the worm *C. elegans* contains an open reading frame encoding a full-length GPC-R with strongest protein sequence identity to the human GHS-R (~29%). The open reading frame is contained on five exons.

Protein Sequence Analysis

An alignment (Pileup) was generated using the deduced amino acid sequences for GHS-R and related sequences and is shown as a dendrogram in Figure 6. The intron in GPR38 is positioned between TM 5 and 6 and its location is conserved when compared to the intron in the GHS-R. GPR38 shares 52% identity and 73% similarity with the human GHS-R (73% to 89% similarity in the transmembrane helical domains). GPR39's overall similarity (based on the predicted amino acid sequence of the cDNA) to the type 1a GHS-R is 52%. All of the GHS-R family members are closely related to the neurotensin receptors (53–58%

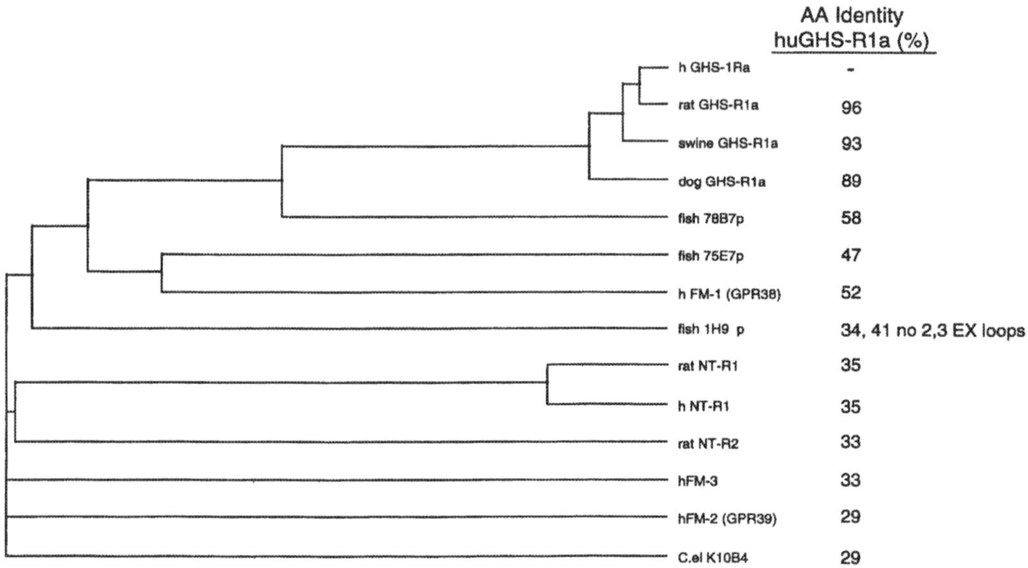

Figure 6. Alignment of GHS-R to related protein sequences. The PileUp program was utilized to compare the GHS-Rs to related sequence including the receptors for neurotensin NT-R1 and NT-R2 and is shown graphically as a dendrogram. For GPC-Rs in which naturally occurring ligands have been identified, the NT-Rs have the closest sequence identity to the GHS-R. The GAP program was used to calculate percent amino acid sequence identity.

similarity) with TM2 showing the highest level of conservation (71–91% similarity). Database searching indicates that FM-3 is most closely related to the GHS-R and NT-R type 1 with 33 and 29% protein sequence identity. As shown in Figure 6, FM-3 groups as a separate branch distinct from both GHS-R and NT-R sequences and the orphan GPC-Rs GPR38 and GPR39.

Expression Profile

The apparent rarity of GPR38 transcripts precluded their detection by standard Northern blotting, while GPR39 mRNA could readily be observed by this method. The more sensitive technique of RNase protection was therefore used to assess the expression pattern of GPR38. RNAse protection with GPR38 revealed expression in human stomach, thyroid and bone marrow. GPR38 transcripts could not be detected in whole brain, hippocampus, pituitary gland, subthalamic nucleus, caudate nucleus, cerebellum, thalamus, placenta, testes, uterus, lung, kidney, or spleen. Other tissues which did not contain detectable GPR38 transcripts included prostate, pancreas, skeletal muscle, thymus, small intestine, adrenal and salivary gland. Unlike the GHS-R where expression in neuroendocrine tissues is consistent with a role in growth hormone function, the significance of GPR38 expression in the thyroid, stomach and bone marrow is yet to be elucidated.

Whereas the expression profile of GPR38 was restricted, GPR39 transcripts were detected in many tissues and brain regions. Most, if not all brain regions tested gave a single hybridizing species of ~1.8–2 kb. However several peripheral tissues, such as stomach and small intestine contain a transcript, in addition to the ~1.8–2 kb band, at ~3 kb which may have arisen by alternative mRNA processing. Interestingly, this ~3 kb species is the only hybridizing band noted in other tissues such as pancreas, thyroid and colon.

FM-3 was cloned from a mouse T-cell library and its mRNA was readily detected by Northern blot analysis. The predominate transcript size was ~5 kb, which was detected in all tissues examined. A band of ~2 kb was also noted in most of the tissues (with high abundance in testis), suggesting alternative mRNA processing of the FM-3 primary transcript. Southern blot analysis (Fig. 3B) of EcoR1-digested genomic DNA using the mouse form as a radiolabeled probe gave a simple hybridization pattern in all species tested, indicative of a single, highly conserved gene encoding FM-3.

Ligand Identification

Attempts to identify activating ligands for any orphan GPCR first involve expression of the full-length contiguous open reading frame for the protein in a suitable heterologous cell type. This can usually be accomplished by placing the ORF in a suitable mammalian expression vector which drives high-level protein expression. It is pivotal to validate that the orphan GPC-R is being expressed appropriately, that is, that a protein of predicted molecular weight is observed and expression at cell membrane is achieved. In the absence of anti-peptide antibodies against the protein, epitope tagging (usually without any negative effect at the N-terminus) the protein can be used to document expression. Then, based on amino acid sequence as a rough guide for its closest relative, functional and ligand binding assays can be utilized to identify a cognate ligand for the orphan GPCR. Experiments performed to identify the natural ligands for GPR38, 39 and FM-3 have as yet been unsuccessful. Transfected HEK-293 cells expressing each clone separately (cell membrane expression confirmed using epitope-tagged protein) failed to bind radiolabeled MK-0677 or neurotensin. In addition, several peptides including endothelin, VIP, gastrin, growth hormone-releasing hormone, somatostatin, TRH, calcitonin, and galanin did not activate FM-3 as measured by the aequorin bioluminescence assay which senses IP_3-induced Ca^{2+} mobilization as a result of phospholipase C activation (data not shown). The identification of an activating ligand for these novel GPC-Rs remains to be discovered.

ACKNOWLEDGMENTS

We thank Jennifer Anderson, Michael Chou, Doris Cully, Carmen Diaz, Mike Dashkevicz, Alex Elbrecht, Michael Hamelin, Paul Liberator, Philip Paress, Charles Rosenblum, Xiao-Ming Guan, Bret Defranco, Keith Judd, Eve Szekely, Don Thompson, Joe Arena, Lee Chaung, Ken Liu, Anna Pomes, Sheng-Shung Pong, Jim Schaeffer, Dennis Dean, Dave Melillo, Jim Tata, John Woods, Patrick Griffin, Traci Clark, Julie DeMartino, Sunil Gupta, Robert Havens, Mike Rigby, Dalip Sirinathsinghji, Kristine Prendergast, Dennis Underwood, Michael S. Phillips, and Molecular Cell Science, Inc. for their participation and invaluable input into the GHS-R program.

REFERENCES

1. Howard, A.D., Feighner S.D., Cully D.F., Arena J.P., Liberator P.L., Rosenblum C.I., Hamelin M., Hreniuk D.L., Palyha O.C., Anderson J., Paress P.S., Diaz C., Chou M., Liu K.K., McKee K.-K., Pong S.-S., Chaung L.-Y., Elbrecht A., Dashkevicz M., Heavens R., Rigby M., Sirinathsinghji D., Dean D.C., Melillo D.G., Patchett A.A., Nargund R., Griffin P.R., DeMartino J.A., Gupta S.K., Schaeffer J.A., Smith R.G. and Van der Ploeg L.H.T. (1996) A receptor in pituitary and hypothalamus that functions in growth hormone release. Science 273, 974–977.
2. Smith, R.G., Van Der Ploeg L.H.T., Howard A.D., Feighner S.D., Cheng K., Hickey G.J., Wyvratt M.J., Fisher M.H., Nargund R.P. and Patchett A.A. (1997) Peptidomimetic regulation of growth hormone secretion. Endocrine Reviews 18, 621–645.
3. Van Der Ploeg, L.H.T., Howard A.D., Smith R.G. and Feighner S.D. (1998) Molecular cloning and characterization of human swine and rat growth hormone secretagogue receptors. In: Growth hormone secretagogues in clinical practice. B. Bercu and R.F. Walker (eds). Marcel Dekker, Inc., New York, pp. 57–76.
4. Thorner, M.O., Chapman, I.M., Gaylinn, B.D., Pezzoli, S.S. and Hartman, M.L. (1997) Growth hormone-releasing hormone and growth hormone-releasing peptide as therapeutic agents to enhance growth hormone secretion in disease and aging. Recent Prog. Horm. Res. 52, 215–244.
5. Smith, R.G., Pong, S.-S., Hickey, G., Jacks, T., Cheng, K., Leonard, R., Cohen, C.J., Arena, J.A., Chang, C.H., Drisko, J., Wyvratt, M., Fisher, M., Nargund, R. and Patchett, A.A. (1996) Modulation of pulsatile GH release through a novel receptor in hypothalamus and pituitary gland. Rec. Prog. Horm. Res. 51, 261–286.
6. Button, D. and Brownstein, M. (1993) Aequorin-expressing mammalian cell lines used to report Ca^{2+} mobilization. Cell Calcium 14, 663–671.
7. Probst, W.C., Snyder, L.A., Schuster, D.I., Brosius, J. and Sealfon, S.C. (1992) Sequence alignment of the G-protein coupled receptor superfamily. DNA Cell Biology 11, 1–20.
8. McKee, K.-K., Palyha, O.C., Feighner, S.D., Hreniuk, D.L., Tan, C., Phillips, M., Smith, R.G., Van der Ploeg, L.H.T. and Howard, A.D. (1997) Molecular analysis of rat pituitary and hypothalamic growth hormone secretagogue receptors. Molecular Endocrinology 11, 415–423.
9. Ong, H., McNicoll, N., Escher, E., Collu, R., Deghenghi, R., Locatelli, V., Ghigo, E., Muccioli, G., Boghen, M. and Nilsson, M. (1998) Identification of a pituitary growth hormone-releasing peptide (GHRP) receptor subtype by photoaffinity labeling. Endocrinology 139, 432–435.
10. Guan, X.-M., Yu, H., Palyha, O.C., Kulju McKee, K., Feighner, S.D., Sirinathsinghji, D.J.S., Smith, R.G., Van der Ploeg, L.H.T. and Howard, A.D. (1997) Distribution of mRNA encoding the growth hormone secretagogue receptor in brain and peripheral tissues. Mol. Brain Res. 84, 23–29.
11. Bennett, P.A., Thomas, G.B., Howard, A.D., Feighner, S.D., Van der Ploeg, L.H.T., Smith, R.G. and Robinson, I.C.A.F. (1997) Hypothalmic growth hormone secretagogue-receptor (GHS-R) expression is regulated by growth hormone in the rat. Endocrinology 138, 4552–4557.
12. Tannenbaum, G.S., Lapointe, M., Beaudet, A. and Howard, A.D. (1998) Expression of growth hormone secretagogue-receptors by growth hormone-releasing hormone neurons in the mediobasal hypothalamus. Endocrinology (in press).
13. de Keyzer, Y., Lenne, F. and Bertagna, X. (1997) Widespread transcription of the growth hormone-releasing peptide receptor gene in neuroendocrine human tumors. European J. of Endocrinol. 137, 715–718.
14. Adams, E.F., Huang, B., Buchfelder, M., Howard, A.D., Smith, R.G., Feighner, S.D., Van Der Ploeg, L.H.T., Bowers, C.Y. and Fahlbusch, R. (1998) Presence of growth hormone secretagogue receptor messenger RNA in human pituitary tumors and rat GH3 cells. J. Clinical Endocrinol. Metabolism 83, 638–642.
15. Yokote, R., Sato, M., Matsubara, S., Ohye, H., Niimi, M., Murao, K. and Takahara, J. (1998) Molecular cloning and gene expression of growth hormone-releasing peptide receptor in rat tissues. Peptides 19, 15–20.

16. Robinson, I. (1997) Hypothalamic targets for growth hormone secretagogues. Acta Paediatr. Suppl. 423, 88–91.
17. Tannenbaum, G.S. (1991) Neuroendocrine control of growth hormone secretion. Acta Paediatr. Scand. Suppl. 372, 5–16.
18. Kenakin, T. (1996) The classification of seven transmembrane receptors in recombinant expression systems. Pharmacol. Rev. 48, 413–451.
19. Feighner, S.D., Howard, A.D., Prendergast, K., Palyha, O.C., Hreniuk, D.L., Nargund, R., Underwood, D., Tata, J.R., Dean, D.C., Tan, C.T., McKee, K.K., Woods, J.W., Patchett, A.A., Smith, R.G. and Van Der Ploeg, L.H.T. (1998) Structural requirements for the activation of the human growth hormone secretagogue receptor by peptide and nonpeptide secretagogues. Molecular Endocrinology 12, 137–145.
20. Smith, R.G., Cheng, K., Schoen, W.R., Pong, S.-S., Hickey, G., Jacks, T., Butler, B., Chan, W. W., Chaung, L.-Y.P., Judith, F., Taylor, J., Wyvratt, M.J. and Fisher, M.-H. (1993) A nonpeptidyl growth hormone secretagogue. Science 260, 1640–1643.
21. Patchett, A.A., Nargund, R.A., Tata, J. R., Chen, M.-M., Barakat, K.J., Johnston, D.B.R., Cheng, K., Chan, W.W.-S., Butler, B., Hickey, G., Jacks, T., Scheim, K., Pong, S.-S., Chaung, L.-Y.P., Chen, H.Y., Frazer, E., Leung, K.H., Chiu, S.-H.L. and Smith R.G. (1995) Design and biological activities of L-163, 191 (MK-0677): a potent and orally active growth hormone secretagogue. Proc. Natl. Acad. Sci. USA 92, 7001–7005.
22. Momany, F.A., Bowers, C.Y., Reynolds, G.A., Hong, A. and Newlander, K. (1984) Conformational energy studies and *in vitro* and *in vivo* activity data on growth hormone-releasing peptides. Endocrinology 114, 1531–1536.
23. Strader, C.D., Sigal, I.S., Candelore, M.R., Rands, E., Hill, W.S. and Dixon, R.A.F. (1988) Conserved aspartate residues 79 and 113 of the B-adrenergic receptor. Journal of Biological Chemistry 263, 10267–10271.
24. Macrae, A.D. and Brenner, S. (1995) Analysis of the dopamine receptor family in the compact genome of the puffer fish *Fugu rubripes*. Genomics 25, 436–446.
25. McKee, K.K., Tan, C.P., Palyha, O.C., Liu, J., Feighner, S.D., Hreniuk, D.L., Smith, R.G., Howard, A.D. and Van Der Ploeg, L.H.T. (1997) Cloning and characterization of two human G protein-coupled receptor genes (GPR38 and GPR39) related to the growth hormone secretagogue and neurotensin receptors. Genomics 46, 426–434.
26. Howard, A.D., Tan, C., Shiao, L.-L., Palyha, O.C., McKee, K.-K., Weinberg, D.H., Feighner, S.D., Cascieri, M.A., Smith, R.G., Van Der Ploeg, L.H.T. and Sullivan, K.A. (1997) Molecular cloning and characterization of a new receptor for galanin. FEBS Lett. 405, 285–290.
27. Altschul, S.F., Gish, W., Miller, W., Myers, E.W. and Lipman, D.J. (1990) Basic local alignment search tool. J. Mol. Biol. 215, 403–410.
28. Tan, C.P., McKee, K.K., Liu, J., Palyha, O.C., Feighner, S.D., Hreniuk, D.L., Smith, R.G. and Howard, A.D. (1998) Cloning and characterization of a human and murine T-cell orphan G protein-coupled receptor similar to the growth hormone secretagogue and neurotensin receptors. Genomics (in press).
29. Pong, S.-S., Chaung, L-Y.P., Dean, D.C., Nargund, R.P., Patchett, A.A. and Smith, R.G. (1996) Identification of a new G protein-linked receptor for growth hormone secretagogues. Molecular Endocrinol. 10, 57–61.

Growth Hormone Secretagogues
Edited by E. Ghigo, M. Boghen, F.F. Casanueva and C. Dieguez

Chapter 6

Intracellular GHRP Signalling

CHEN CHEN and IAIN J. CLARKE

Prince Henry's Institute of Medical Research, Clayton, Victoria 3168, Australia

INTRODUCTION

The original growth hormone-releasing peptide (GHRP), which was synthesised by Bowers et al. (1) and named GHRP-6, stimulates GH secretion in a relatively specific manner although it is only a weak GH secretagogue. Many more potent GHRPs have now been developed including hexarelin, GHRP-1 and GHRP-2 (2). The most recent development in this field has been the identification of a GHRP receptor in several animal species including man (3–6). In spite of the efficacy and relative specificity of GHRPs for the stimulation of GH release, and the existence of a specific receptor for GHRPs, no endogenous ligand has yet been identified for these synthetic peptides. There is good evidence, however, that such a ligand exists since activity can be found in the hypophysial portal blood of sheep using a bioassay that utilises cells transfected with the cloned receptor (7). This activity is closely correlated with the secretion of GH (*ibid.*).

Since it is only a matter of time before the endogenous GHRP(s) is/are identified and purified, it is timely to revise our understanding of the way that GH synthesis and secretion is regulated. Our present understanding of these processes relies on the integrated actions of the stimulatory factor, growth hormone releasing factor (GRF) and the inhibitory factor, somatostatin (SRIF) (8–10). The current model, which largely depends upon data from the male rat, is one which dictates that GH release is solely determined by the sum of the combined effects of GRF and SRIF (11). This is supported by experimental data of reciprocal alterations in the hypophysial portal blood levels of GRF and SRIF in the rat (12), suggesting that these patterns relate to the pulsatile secretion of GH in an explicit manner. On the other hand, studies in sheep (13–15) have not found such a close relationship between GRF, SRIF and GH in the sheep, calling into question the general applicability of the male rat model. The possible existence of an endogenous GHRP would further complicate the issue.

The signal transduction systems mediating the responses to GHRPs in pituitary somatotrophs have been studied in some detail. This chapter discusses these whilst clearly recognising that there are likely to be actions of GHRPs at other levels (hypothalamus) (16) and in other pituitary cell types (e.g. corticotrophs) (17).

G-PROTEIN COUPLED RECEPTORS

The GHRP receptors that were recently identified in the human, swine, and rat pituitary have a high degree of similarity (3,4,18). These receptors clearly belong to a family which is different to that of the GRF receptor and can be blocked only by specific antagonists (19). A series of non-peptidergic compounds (such as MK 0677) have a very high affinity for this receptor (20).

Contrary to the data obtained in human, swine and rat somatotrophs, we have found that GRF receptor antagonist ([Ac-Tyr^1, D-Arg^2] GHRH$_{1\text{-}29}$) prevents GHRP-2 stimulated GH release in sheep pituitary cells *in vitro* (21). This suggests some relationship between the GHRP-2 binding sites and the GRF receptor, at least in this species. It has been reported that GHRP-6 interacts with a novel low-affinity GHRH binding site in the rat anterior pituitary (22) and Sethumadhavan et al. (23) identified two classes of GHRP-binding sites using [^{125}I]Tyr-Ala-GHRP as a ligand. It seems possible, therefore, that an unidentified receptor exists that has a low affinity for GRF and a high affinity for GHRP-2. Another possibility that we considered was that GHRP might bind to a site on the GRF receptor different to that employed by GRF. This was tested using a GC cell line with over-expression of the GRF receptor (24). In these cells, cAMP levels were increased by GRF but not GHRP-2 or GHRP-6, and the effect of GRF was blocked by a GRF receptor antagonist. This strongly suggests that the GHRPs do not act via GRF receptors. It remains possible, however, that there is more than one type of GHRP receptor, one (GHRP-R1) that is blocked by the GRF antagonist ([Ac-Tyr^1, D-Arg^2] GHRH$_{1\text{-}29}$) and one (GHRP-R2) that is not. One could speculate that GHRP-R1 has a similar binding affinity for GHRP-6, GHRP-1 and GHRP-2 whereas GHRP-R2 would have a higher affinity for GHRP-2 than for GHRP-6 or GHRP-1. GHRP-R2 may also have a low affinity for GRF-binding which would explain the effects of blockade with [Ac-Tyr^1, D-Arg^2] GHRH$_{1\text{-}29}$. Recently, a 57kD GHRP receptor was found in human, bovine and porcine pituitary glands using Hexarelin as binding ligand (25) and this appears to differ from the 41kD receptor identified by Howard et al. (3). The binding affinities of the two receptors are quite different with the 57 kD type having a high affinity to Hexarelin and low affinity to MK 0677 (25). This provides realistic evidence that there is more that one type of GHRP receptor.

The GHRP receptors couple to G proteins based on evidence of activation of adenylyl cyclase by GHRP-2 (26), activation of PKC (27,28), release of intracellular Ca^{2+} ($[Ca^{2+}]i$) (29,30) and increased phosphatidylinositol (PI) turnover following treatment with GHRP-6, non-peptidergic secretagogues and GHRP-2 (31,32). Because both adenylyl cyclase and phospholipase C are activated by GHRP, Gs and Gq are most likely to be involved in these responses.

INVOLVEMENT OF PHOSPHOLIPASE C AND INOSITOL TRIPHOSPHATE ($INSP_3$)

The GHRP receptor identified by Howard et al. (3) is coupled to Gq which is known to mediate the activation of phospholipase C (PLC) (19). In accordance with this, co-expression of the GHRP receptor and Gq11 protein increased the $[Ca^{2+}]i$ response to GHRP in Xenopus oocytes (3). It has also been reported that GHRPs and non-peptidergic GH secretagogues increase PI turnover via an activation of PLC in human acromegalic tumour cells (31,32). Whether PLC is activated in ovine or rat somatotrophs by any GHRP is still an open question. Investigation of this issue would be useful given the difference in response of sheep cells to GHRP-2 and other versions of GHRP.

EFFECT OF GHRP ON INTRACELLULAR CALCIUM ($[Ca^{2+}]i$) AND CELL MEMBRANE CALCIUM CHANNELS

GH secretion is directly related to the $[Ca^{2+}]i$ (33) and influx of Ca^{2+} through voltage-dependent Ca^{2+} channels is increased by GRF and reduced by SRIF (33–38). There is no clear evidence, however, that either GRF or SRIF mobilises Ca^{2+} from intracellular Ca^{2+} storage sites. On the other hand, GHRP-6 causes the release of intracellular Ca^{2+} as well as Ca^{2+} influx through the cell membrane (29,30). In isolated rat somatotrophs, GHRP-6 evoked dual-phase increases in $[Ca^{2+}]i$; an initial transient increase due to intracellular Ca^{2+} release and a second longer lasting phase due to the influx of Ca^{2+} (29,30). In ovine pituitary cells, GHRP caused a subtle and transient increase in $[Ca^{2+}]i$ even when extra-cellular Ca^{2+} was chelated to zero (39). This probably involves the generation of inositol trisphosphate (31,32) but this has not yet been demonstrated. In spite of the mobilisation of Ca^{2+} from intracellular stores, the major contribution to the elevation of $[Ca^{2+}]i$ is caused by influx of extracellular Ca^{2+} (39). It appears that this is an integral factor in the release of GH in response to GHRP because blockade of membrane Ca^{2+} channels abolishes the secretory response (21,40,41).

In somatotrophs, the major membrane Ca^{2+} channels are of the voltage-gated T- and L-types (36,38). Studies of rat and sheep cells have respectively shown that GHRP-6 and GHRP-2 depolarize the cell membrane leading to the opening of these channels (29,39). Since this depolarization can only be recorded with the nystatin-perforated patch clamp configuration (which does not disturb intracellular systems), this implies that an intact second messenger signalling systems are required for ion channel function. Figure 1 shows the effect of GHRP-2 on transmembrane Ca^{2+} current and $[Ca^{2+}]i$ in ovine somatotrophs. GHRP-2 increases voltage-gated T- and L- type Ca^{2+} current (Figure 1A). The measurement of $[Ca^{2+}]i$ was performed on somatotroph-enriched cell populations and it is clear that GHRP-2, but not GnRH or TRH, increases $[Ca^{2+}]i$ levels in these cells (Figure 1B). All of these effects of GHRP on the electrophysiological properties of the somatotroph cell membrane resemble those of GRF but are opposite to the effects of SRIF (42).

The ion channels that are involved in depolarization of somatotroph cell membrane are not defined but it appears that Na^{+} channels do not play a major role in the response to

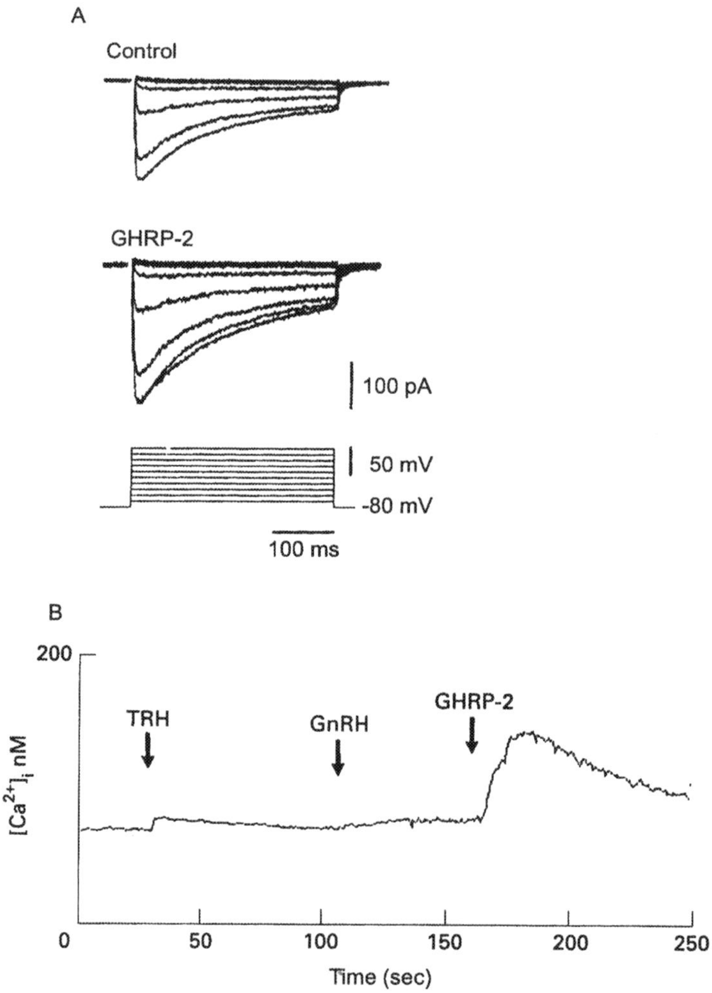

Figure 1. The effect of GHRP-2 on transmembrane Ca^{2+} current and $[Ca^{2+}]i$ in ovine somatotrophs. (A) From a holding potential of –80 mV, depolarising pulses were applied as indicated at the bottom of the figure with a time interval of 3 seconds between pulses. Upper current traces were recorded in control bath solution and the lower current traces were recorded after the addition of GHRP-2 (10 nM) to the bath solution. (B) $[Ca^{2+}]i$ in somatotroph-enriched cell suspensions was measured using Fura-2 method with a spectrofluorimeter. Addition of TRH (10 nM) or GnRH (10 nM) to the extracellular solution induced a very slight increase. GHRP-2 (10 nM) caused a significant increase in $[Ca^{2+}]i$. (Adapted with permission from Ref. 39).

GHRP (29,39). It is thought that voltage-gated Ca^{2+} channel activation is partly responsible for the depolarization caused by GHRP-2 (39), but K^+ channels may also be involved (39,43); these have not been characterised in detail.

In summary, the available data suggest that GHRP first causes the release of intracellular Ca^{2+} and then causes Ca^{2+} influx by an increase in membrane Ca^{2+} permeability of the cell. The latter is due to membrane depolarization most likely via action of second messengers on Ca^{2+} channel protein(s). Possible second messenger pathways involved in the action of GHRP are discussed below.

THE PROTEIN KINASE C PATHWAY

It has been suggested that the action of GHRP-6 and the non-peptidergic analogue L692,429 to stimulate GH release from rat pituitary cells is mediated by protein kinase C (PKC) (27,44). The synergistic action of GHRP-6 and GRF on cAMP accumulation and GH secretion in rat pituitary cells may also be mediated by PKC (27). It should be noted however, that the specificity of the inhibitor (phloretin), used in the latter study, has not been widely tested and the effect of this agent on other kinase systems is not defined. In particular, over the same dose range (10–200 μM), phloretin increased the opening probability of Ca^{2+}-activated K^{+} channels (45) which can hyperpolarize the cell membrane and prevent the stimulation of GH secretion by GHRP-6. Down-regulation of PKC with long-term treatment by phorbol, 12-myristate, 13-acetate (PMA, 1 μM) partially blocked the effect of GHRP-6 on GH secretion (27), suggesting some involvement of PKC in the response. It was shown by Akman et al. (40) however, that GHRP-1 still causes GH release following maximal stimulation of cells with PMA. In ovine pituitary cells GHRP-6 does not cause PKC translocation (28). Furthermore, down-regulation of PKC with PMA does not block GH release in response to GHRP-6 in sheep cells whereas PMA-stimulated GH release is totally abolished by the same treatment (28).

In contrast, GHRP-2 stimulates PKC translocation from cytosol to membrane in ovine somatotrophs in primary culture (28). As shown in Figure 2, the stimulation of PKC translocation by GHRP-2 is dose-dependent, with a maximal response reached at 10^{-7} M. PKC inhibitors (Calphostin C, Chelerythrine, Staurosporine) and down-regulation of PKC by phorbol,12,13-dibutyrate (PDBu) causes only partial attenuation of this response. It seems likely therefore, that PKC is at least partially involved in the action of GHRP-2 (but not GHRP-6) in sheep cells. It is interesting to note that GRF also causes PKC translocation in ovine somatotrophs (28). It is possible to account for the response of ovine somatotrophs to GRF and GHRP-2 by activation of the cAMP/PKA pathway, but the concomitant activation of PKC may play a role to enhance the action of GHRP-2 on cAMP/PKA pathway.

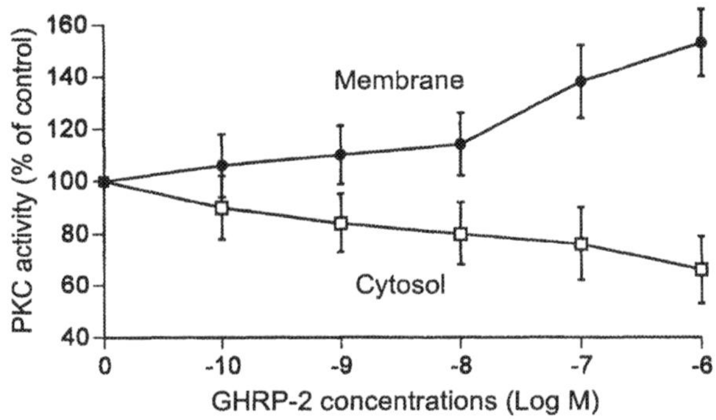

Figure 2. The effect of GHRP-2 on PKC translocation in ovine somatotrophs. Dose–response relationship for the effects of GHRP-2 on the translocation of PKC from the cytosol (□) to the cell membranes (●). Cells were treated with GHRP-2 (10^{-10} to 10^{-6} M) for 4 min. Values are means ± S.E.M.; n=4. (Adapted with permission from Ref. 28.)

THE cAMP AND PROTEIN KINASE A (PKA) PATHWAY

It is well established that GRF activates the cAMP/PKA pathway in somatotrophs and that this is fundamental to the release of GH (46,47). Part of the effect of SRIF is through inhibition of cAMP formation (48). In contrast, GHRP-6 and GHRP-1 have no direct effect on intracellular cAMP levels in rat and ovine somatotrophs (26,40,49). Nevertheless, GHRP-6 may synergise with GRF to increase intracellular cAMP levels (49). Similar results are also found for the non-peptidergic analogue L-692,429 (44), but synergy could not be demonstrated with GHRP-1 and GRF (40). This discrepancy suggests that the cAMP/PKA pathway is not the primary signalling route for GHRP-6 and GHRP-1.

GHRP-2 increases intracellular cAMP levels in ovine (but not rat) somatotrophs (26). There is a dose-dependent increase in cAMP levels in response to GHRP-2 stimulation with corresponding effects on GH secretion (Figure 3). GHRP-2 stimulated GH secretion was blocked in these cells by an inhibitor of adenylyl cyclase and a cAMP binding antagonist (26). Thus, in sheep cells at least, GHRP-2 activates adenylyl cyclase leading to an increase in cAMP levels and activation of cAMP-dependent protein kinase A (PKA). Thus, PKA could phosphorylate transmembrane Ca^{2+} channels to modify their properties in the manner observed by electrophysiological means (see above). A significant species difference appears to exist between sheep and rat somatotrophs in terms of GH release in response to GHRP-2 (21). Neither GHRP-6 nor GHRP-1 increases adenylyl cyclase activity in ovine pituitary cells (26). We propose therefore, that in ovine somatotrophs, the pathway

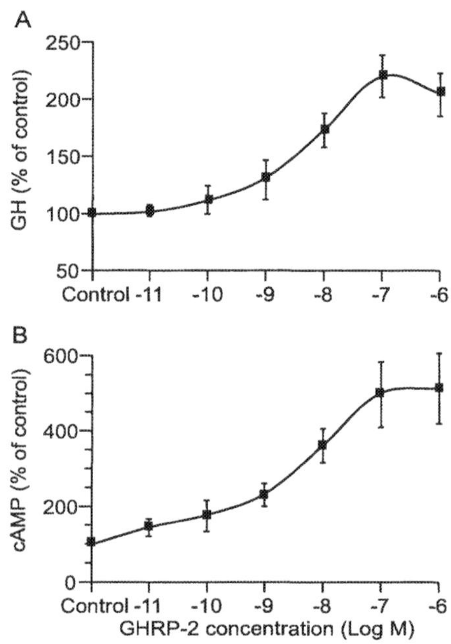

Figure 3. The effect of GHRP-2 on cAMP levels and GH secretion in ovine somatotrophs. Dose–response curves of GHRP-2 stimulation for 30 min on (A) GH release and (B) cAMP accumulation in ovine somatotrophs. Each point represents the mean (± S.E.M.) of 5 experiments. (Adapted with permission from Ref. 26).

employed by GHRP-2, resulting in an increase in cAMP levels, is different to that employed by either GHRP-6 or GHRP-1 in both rat and ovine pituitary cells (21,40,44,49).

The mechanism by which GHRP-2 activates adenylyl cyclase is not clear. Although some subtypes of adenylyl cyclase can be activated by Ca^{2+} (50), these do not appear to mediate the response to GHRP-2 since blockade of Ca^{2+} influx does not affect the cAMP response (26). Although it is clear that GRF elevates cAMP levels in ovine somatotrophs, it may act through a cyclase which is different to that used by GHRP-2, since GHRP-2 and GRF have additive effects on both cAMP accumulation and GH secretion when both secretagogues are applied at maximal doses (21,26).

In summary, GHRP-2 stimulates cAMP accumulation in ovine somatotrophs via activation of adenylyl cyclase, but this response is not seen in rat somatotrophs. This appears to be the pathway responsible for the stimulation of GH secretion by GHRP-2 in the sheep. GHRP-6 and GHRP-1 elevate cAMP levels in ovine, rat and human somatotrophs and amplify the cAMP response to GRF in rat pituitary cells. This suggests activation of different cyclases by the different secretagogues.

CROSSTALK BETWEEN DIFFERENT SIGNALLING SYSTEMS

As discussed above, both the cAMP/PKA and the PLC/PKC systems appear to be involved in the GHRP-2 stimulation of GH release in ovine somatotrophs. GHRP-2 stimulates adenylyl cyclase activity, resulting in cAMP production, and also activates PKA which, in turn, leads to an increase in intracellular Ca^{2+} and GH secretion (26,39). Activation of the PKC pathway by GHRP-2 (28) may positively potentiate the cAMP-PKA pathway by stimulating adenylyl cyclase activity (51) and this potentiation of adenylyl cyclase activity may further increase the accumulation of $[Ca^{2+}]i$. A selective inhibitor for PKA (H89) had no effect on PMA- or GHRP-induced PKC translocation, but inhibited GH secretion in response to either PMA or GHRP-2, suggesting that the effect on GH release is dependent upon the cAMP-PKA pathway. It is therefore suggested that the stimulation of PKC translocation by GHRP-2 is not the major signalling system employed. PKC inhibitors only partially reduce GH secretion in response to GHRP-2, which also suggests that the PKC pathway is not mandatory for the action of GHRP-2 to induce GH secretion in ovine somatotrophs. Thus, we propose that the activation of PKC *potentiates* the action of GHRP-2. This is summarised in diagrammatic form in Figure 4.

REFERENCES

1. Bowers, C.Y., Monany, F.A., Reynolds, G.A., Chang, D., Hong, A. and Chang, K. (1980) Structure–activity relationships of a synthetic pentapeptide that specifically releases growth hormone *in vitro*. Endocrinology 106, 663–667.
2. Momany, F.A., Bowers, C.Y., Reynolds, G.A., Hong, A. and Newlander, K. (1984) Conformational energy studies and *in vitro* and *in vivo* activity data on growth hormone-releasing peptides. Endocrinology 114, 1531–1536.

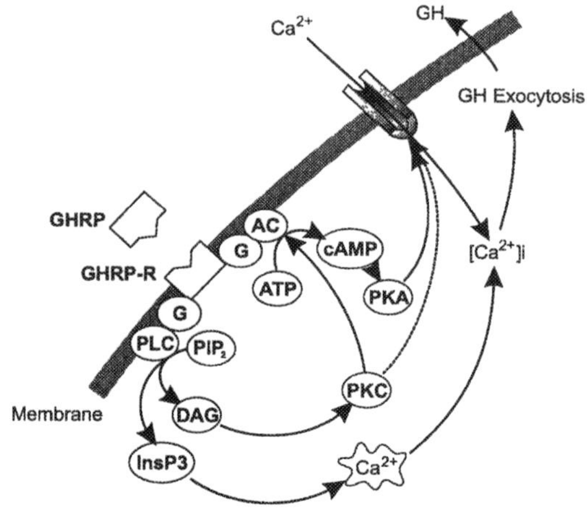

Figure 4. The proposed signalling pathways for GHRP in somatotrophs. The binding of GHRP to a putative receptor activates the phospholipase C (PLC) and adenylyl cyclase pathways via G-proteins, leading to an increase in inositol (1,4,5)-triphosphate ($InsP_3$), and the activity of protein kinase C (PKC) and protein kinase A (PKA). $InsP_3$ then releases Ca^{2+} from $InsP_3$-sensitive Ca^{2+} pool and protein kinases phosphorylate ion channels to increase Ca^{2+} influx. All of these events would lead to an increase in intracellular Ca^{2+} concentration ($[Ca^{2+}]i$) and GH secretion. G = GTP-binding proteins, PIP_2 = phosphatidylinositol (4,5)-bisphosphate; DAG = diacylglycerol.

3. Howard, A.D., Feighner, S.D., Cully, D.F., Arena, J.P., Liberator, P.A., Rosenblum, C.I., Hamelin, M., Hreniuk, D.L., Palyha, O.C., Anderson, J., Sparess, P.S., Diaz, C., Chou, M., Liu, K.K., McKee, K.K., Pong, S.S., Chaung, L.Y., Elbrecht, A., Dashkevicz, M., Heavens, R., Tigby, M., Sirinathsinghji, D.J.S., Dean, D.C., Melillo, D.G., Patchett, A.A., Nargund, R., Griffin, P.R., DeMartino, J.A., Gupta, S.K., Schaeffer, J.M., Smith, R.G. and Van der Ploeg, L.H.T. (1996) A receptor in pituitary and hypothalamus that functions in growth hormone release. Science 273, 974–977.
4. McKee, K.K., Palyha, O.C., Feighner, S.D., Hreniuk, D.L., Tan, C.P., Phillips, M.S., Smith, R.G., Vander Ploeg, L.H.T. and Howard, A.D. (1997) Molecular analysis of rat pituitary and hypothalamic growth hormone secretagogue receptors. Mol. Endocrinol. 11, 415–423.
5. Adams, E.F., Huang, B., Buchfelder, M., Howard, A., Smith, R.G., Feighner, D., Van der Ploeg, L.H.T., Bowers, D.Y. and Fahlbusch, R. (1998) Presence of growth hormone secretagogues receptor messenger ribonucleic acid in human pituitary tumors and rat GH3 cells. J. Clin. Endocrinol. Metab. 83, 638–642.
6. Shimon, I., Yan, X.M. and Melmed, S. (1998) Human fetal pituitary expresses functional growth hormone-releasing peptide receptors. J. Clin. Endocrinol. Metab. 83, 174–178.
7. Leong, D.A., Pomes, A., Veldhuis, J.D. and Clarke, I.J. (1998) A novel hypothalamic hormone measured in hypophyseal portal plasma drives rapid bursts of GH secretion. Abstract for 1998 US Endocrine Society Conference.
8. Brazeau, P., Vale, W., Burgus, R., Ling, N., Butcher, M., Rivier, J. and Guillemin, R. (1973) Hypothalamic peptide that inhibits the secretion of immunoreactive pituitary growth hormone. Science 197, 77–79.
9. Rivier, J., Spiess, J., Thorner, M. and Vale, W. (1982) Characterization of a growth hormone-releasing factor from a human pancreatic islet tumour. Nature 300, 276–278.
10. Guillemin, R., Brazeau, P., Bohlen, P., Esch, F., Ling, N. and Wehrenberg, W.B. (1982) Growth hormone-releasing factor from a human pancreatic tumour that caused acromegaly. Science 218, 585–587.
11. Tannenbaum, G.S. and Ling, N. (1984) The interrelationship between growth hormone (GH)-releasing factor and somatostatin in the generation of the ultradian rhythm of GH secretion. Endocrinology 115, 1952–1957.

12. Plotsky, P.M. and Vale, W. (1985) Patterns of growth hormone-releasing factor and somatostatin secretion into the hypophysical-portal circulation of the rat. Science 230, 461–463.
13. Frohman, L.A., Downs, T.R., Clarke, I.J. and Thomas, G.B. (1990) Measurement of growth hormone-releasing hormone and somatostatin in hypothalamic-portal plasma of unanesthetized sheep: Spontaneous secretion and response to insulin-induced hypoglycemia. J. Clin. Invest. 86, 17–24.
14. Thomas, G.B., Cummins, J.T., Francis, H., Sudbury, A.W., McCloud, P.I. and Clarke, I.J. (1991) Effect of restricted feeding on the relationship between hypophysial portal concentrations of growth hormone (GH)-releasing factor and somatostatin, and jugular concentrations of GH in ovariectomized ewes. Endocrinology 128, 1151–1158.
15. Mangan, E., Cataldi, M., Guillaume, V., Mazzocchi, L., Dutour, A., Razafindraibe, H., Sauze, N., Renard, M. and Oliver, C. (1994) Role of growth hormone (GH)-releasing hormone and somatostatin in the mediation of clonidine-induced GH release in sheep. Endocrinology 134, 562–567.
16. Guillaume, V., Magnan, E., Cataldi, M., Dutour, A., Sauze, N., Renard, M., Razafindraibe, H., Conte-Devolx, B., Deghenghi, R., Lenaerts, V. and Oliver, C. (1994) Growth hormone (GH)-releasing hormone secretion is stimulated by a new GH-releasing hexapeptide in sheep. Endocrinology 135, 1073–1076.
17. Smith, R.G., Cheng, K., Schorn, W.R., Pong, S.S., Hickey, G., Jacks, T., Butler, B., Chan, W.W.S., Chaung, L.Y.P., Judith, F., Taylor, J., Wyvratt, M.J. and Fisher, M.H. (1993) A nonpeptidyl growth hormone secretagogue. Science 260, 1640–1643.
18. Yokote, R., Sato, M., Matsubara, S., Ohye, H., Niimi, M., Murao, K. and Takahara, J. (1998) Molecular cloning and gene expression of growth hormone-releasing peptide receptor in rat tissues. Peptides 19, 15–20.
19. Pong, S.S., Chaung, L.Y.P., Dean, D.C., Nargund, R.P., Patchett, A.A. and Smith, R.G. (1996) Identification of a new G-protein-linked receptor for growth hormone secretagogues. Mol. Endocrinol. 10, 57–61.
20. Patchett, A.A., Nargund, R.P., Tata, J.R., Chen, M.H., Barakat, K.J., Johnston, D.B.R., Cheng, K., Chen, W.W.S., Butler, B.S., Hickey, G.J., Tacks, T.M., Scleim, K., Pong, S.S., Chaung, L.Y.P., Chen, H.Y., Frazier, E., Leung, K.H., Chui, S.H.L. and Smith, R.G. (1995) The design and biological activities of L-163,191 (MK-0677): a potent orally active growth hormone secretagogue. Proc. Natl. Acad. Sci. USA 92, 7001–7005.
21. Wu, D., Chen, C., Katoh, K., Zhang, J. and Clarke, I.J. (1994b) The effects of GH-releasing peptide-2 (GHRP-2 or KP 102) on GH secretion from primary cultured ovine pituitary cells can be abolished by a specific GH-releasing factor (GRF) receptor antagonist. J. Endocrinol. 140, R9–R13.
22. Lau, Y.S., Camoratto, A.M., White, L.M. and Moriarty, C.M. (1991) Effect of lead on TRH and GRF binding in rat anterior pituitary membranes. Toxicology 68, 169–179.
23. Sethumadhaven, K., Veeraragavan, K. and Bowers, C.Y. (1991) Demonstration and characterization of the specific binding of growth hormone-releasing peptide to rat anterior pituitary and hypothalamic membranes. Biochem. Biophys. Res. Commun. 178, 31–37.
24. Chen, C., Farworth, P., Petersenn, S., Musgrave, I., Canny, B.J. and Clarke, I.J. (1998) Growth hormone-releasing peptide-2 (GHRP-2) does not act via the human growth hormone-releasing factor receptor in GC cells. Endocrine. 9, 59–65.
25. Ong, H., Mcnicoll, N., Escher, E., Collu, R., Deghenghi, R., Locatelli, V., Ghigo, E., Muccioli, G., Boghen, M. and Nilsson, M. (1998) Identification of a pituitary growth hormone-releasing peptide (GHRP) receptor subtype by photoaffinity labeling. Endocrinology 139, 432–435.
26. Wu, D., Chen, C., Zhang, J., Bowers, C.Y. and Clarke, I.J. (1996) The effects of growth hormone-releasing peptide-6 (GHRP-6) and GHRP-2 on intracellular adenosine 3′,5′-monophosphate (cAMP) levels and GH secretion in ovine and rat somatotrophs. J. Endocrinol. 148, 197–205.
27. Cheng, K., Chan, W.W.S., Barreto, A., Butler, B. and Smith, R.G. (1991) Evidence for a role of protein kinase-C in His-D-Trp-Ala-Trp-D-phe-Lys-NH_2-induced growth hormone release from

rat primary pituitary cells. Endocrinology 129, 3337–3342.
28. Wu, D., Clarke, I.J. and Chen, C. (1997) The role of protein kinase C in GH secretion induced by GH-releasing factor and GH-releasing peptides in cultured ovine somatotrophs. J. Endocrinol. 154, 219–230.
29. Herrington, J. and Hille, B. (1994) Growth hormone-releasing hexapeptide elevates intracellular calcium in rat somatotropes by two mechanisms. Endocrinology 135, 1100–1108.
30. Bresson-Bepoldin, L. and Dufy-Barbe, L. (1994) GHRP-6 induces a biphasic calcium response in rat pituitary somatotrophs. Cell Cal. 15, 247–258.
31. Lei, T., Buchfelder, M., Fahlbusch, R. and Adams, E.F. (1995) Growth hormone releasing peptide (GHRP-6) stimulates phosphatidylinositol (PI) turnover in human pituitary somatotroph cells. J. Mol. Endocrinol. 14, 135–138.
32. Adams, E.F., Petersen, B., Lei, T., Buchfelder, M. and Fahlbusch, R. (1995) The growth hormone secretagogue, L-692,429, induces phosphatidylinositol hydrolysis and hormone secretion by human pituitary tumors. Biochem. Biophy. Res. Commun. 208, 555–561.
33. Lussier, B.T., French, M.B., Moor, B.C. and Kraicer, J. (1991) Free intracellular Ca^{2+} concentration ($[Ca^{2+}]i$) and GH release from purified rat somatotrophs. III. Mechanism of action of GH-releasing factor and somatostatin. Endocrinology 128, 592–603.
34. Chen, C., Israel, J.M. and Vincent, J.D. (1989a) Electrophysiological responses to somatostatin of rat hypophysial cells in somatotroph-enriched primary cultures. J. Physiol. 408, 493–510.
35. Chen, C., Israel, J.M. and Vincent, J.D. (1989b) Electrophysiological responses of rat pituitary cells in somatotroph-enriched primary culture to human growth hormone-releasing factor. Neuroendocrinology 50, 679–587.
36. Chen, C., Zhang, J., Vincent, J.D. and Israel, J.M. (1990) Two types of voltage-dependent calcium currents in rat somatotrophs are reduced by somatostatin. J. Physiol. 425, 29–42.
37. Chen, C., Zhang, J., McNeill, P., Pullar, M., Cummins, J. and Clarke, I.J. (1992) Growth hormone releasing factor modulates calcium currents in human growth hormone secreting adenoma cells. Brain Res. 604, 345–348.
38. Chen, C. and Clarke, I.J. (1995a) Modulation of Ca^{2+} influx in the ovine somatotroph by growth hormone-releasing factor. Am. J. Physiol. 268, E204–E212.
39. Chen, C. and Clarke, I.J. (1995b) Effect of growth hormone-releasing peptide-2 (GHRP-2) on membrane Ca^{2+} permeability in cultured ovine somatotrophs. J. Neuroendocrinol. 7, 179–186.
40. Akman, M.S., Girard, M., Q'Brien, L.F., HO, A.K. and Chik, C.L. (1993) Mechanism of action of a second generation growth hormone-releasing peptide (Ala-His-D- Nal-Ala-Trp-D-Phe-Lys-NH_2) in rat anterior pituitary cells. Endocrinology 132, 1286–1291.
41. Wu, D., Chen, C., Zhang, J., Katoh, K. and Clarke, I.J. (1994a) Effects *in vitro* of new growth hormone releasing peptide (GHRP-1) on growth hormone secretion from ovine pituitary cells in primary culture. J. Neuroendocrinol. 6, 185–190.
42. Chen, C., Vincent, J.D. and Clarke, I.J. (1994) Ion channels and signal transduction pathways in the regulation of growth hormone secretion. *Trend. Endocrinol. Metab.* 5, 227–233.
43. Pong, S.S., Chaung, L.Y. and Smith, R.G. (1991) GHRP-6 (His-D-Trp-Ala-Trp-D-Phe-Lys-NH_2) stimulates growth hormone secretion by depolarization in rat pituitary cell cultures [abst 230]. In 73th Annual Meeting of the Endocrine Society, Washington, DC, The Endocrine Society.
44. Cheng, K., Chan, W.W.S., Butler, B. and Smith, R.G. (1993) A novel non-peptidyl growth hormone secregogue. Horm. Res. 40, 109–115.
45. Koh, D.S., Reid, G. and Vogel, W. (1994) Effect of the flavoid phloretin on Ca^{2+}-activated K^+ channels in myelinated nerve fibres of xenopus laevis. Neurosci. Lett. 165, 167–170.
46. Frohman, L.A., Downs, T.R. and Chomczynski, P. (1992) Regulation of growth hormone secretion. Front. Neuroendocrin. 13, 344–405.
47. Harwood, J.P., Grew, C. and Aguilera, G. (1984) Action of growth hormone-releasing factor and somatostatin on adenylate cyclase and growth hormone release in rat anterior pituitary. Mol. Cell. Endocrinol. 37, 277-84.

48. Schonbrunn, A. (1990) Somatostatin action in pituitary cells involves two independent transduction mechanisms. Metabolism 39 (Suppl. 2), 96–100.
49. Cheng, K., Chan, .., Barreta, A., Convey, D.M. and Smith, R.G. (1989) The synergistic effects of His-D-Trp-Ala-Trp-D-Phe-Lys-NH_2 on growth hormone (GH)-releasing factor-stimulated GH release and intracellular adenosine 3′,5′-monophosphate accumulation in rat primary pituitary cell culture. Endocrinology 124, 2791–2798.
50. Cooper, E.M.F., Mons, N. and Karpen, J.W. (1995) Adenylyl cyclases and the interaction between calcium and cAMP signalling. Nature 374, 421–424.
51. Cronin, M.J. and Canonico, P.L. (1985) Tumor promoters enhance basal and growth hormone releasing factor stimulated cyclic AMP levels in anterior pituitary cells. Biochem. Biophy. Res. Commun. 129, 404–410.

Growth Hormone Secretagogues
Edited by E. Ghigo, M. Boghen, F.F. Casanueva and C. Dieguez

Chapter 7

The Effects of GH-Secretagogues on Human Pituitary Cells in Culture and on Rat Hypothalamic Tissue

MÁRTA KORBONITS[1], ERIC F. ADAMS[2] and ASHLEY B. GROSSMAN[1]

[1]*Department of Endocrinology, St. Bartholomew's Hospital, West Smithfield, London EC1A 7BE, U.K.*
[2]*Department of Neurosurgery, Erlangen, Germany (Present address: Pharmaceutical Sciences Institute, Aston University, Birmingham, U.K.)*

INTRODUCTION

Growth hormone secretagogues (GHSs) release growth hormone (GH) via both the hypothalamus and the pituitary gland, and also stimulate ACTH and prolactin release. The presence of a seven transmembrane, G protein-coupled specific receptor has been described (1) both in the hypothalamus and the pituitary of several species, including the human. Recently, the presence of the GH-secretagogue receptor (GHS-R) mRNA has also been described in human adenomatous and fetal pituitary tissue (2–5) as well as in ectopic endocrine tumours (Figure 1) (4), and it has been shown to be functionally active in a GHRH- and ACTH-secreting carcinoid (6). Nevertheless, the effects of GHSs and their non-peptide analogues on *in vitro* cultured human pituitary somatotrophs and on rat hypothalamic incubations were investigated well before the specific GHS-R had been identified.

HUMAN PITUITARY CELLS

Growth hormone secretion and second messenger systems

It has been shown that GHSs stimulate *in vitro* growth hormone (GH) release from somatotroph cells (Figure 2) (7–9). As shown with rat pituitary cells (10,11), GHRP-6 induces protein kinase C (PKC)-dependent GH secretion and a biphasic increase in intracellular Ca^{2+} by tumorous human somatotrophs in culture (7,8). The Ca^{2+} response involves a rapid but transient increase followed by an attenuated response and then a longer lasting PKC-dependent plateau phase. Additionally, direct involvement of the adenyl cyclase-coupled GHRH-receptor in the actions of GHRP-2 and GHRP-6 was excluded (9). Because of these

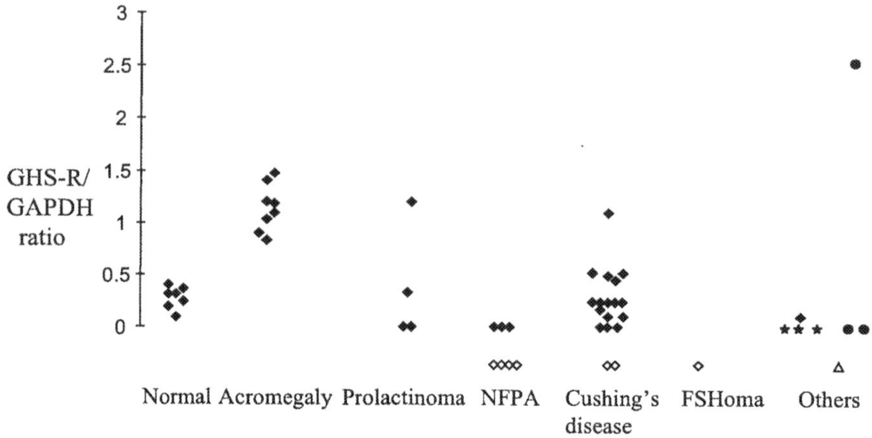

Figure 1. Relative expression of the GHS-R gene at 28 cycles in a duplex PCR using GHS-R and GAPDH (housekeeping-gene) primers, in tissue from 7 normal pituitaries, 8 somatotroph tumours, 4 prolactinomas, 7 non-functioning pituitary adenomas, 18 corticotroph adenomas and an FSHoma, and in non-pituitary tumours including 3 ectopic ACTH-secreting tumours (●), three insulinomas (★) a gastrinoma (♦) and a non-secreting thymic carcinoid tumour (◆). Open symbols represent samples with no detectable expression under any conditions (single PCR with 34 cycles using GHS-R primers only), while filled symbols at "zero" GHS-R/GAPDH ratio represent samples with a very low level of expression.

findings, it was suspected that GHRP-6 could induce hydrolysis of membrane phosphatidyl inositol (PI), since this transduction system yields diacylglycerol (DAG) and inositol trisphosphate (IP_3); this in turn leads to activation of PKC, mobilisation of intracellular Ca^{2+} stores and opening of ion channels (8,12). This assertion was confirmed by an *in vitro* cell culture study in which GHRP-6 and its methylated derivative, hexarelin, were shown to powerfully stimulate PI hydrolysis in human pituitary somatotrophinomas removed from patients with acromegaly (Figure 2) (13). The effects were dose-dependent, maximal stimulation being observed with 100 nmol/l after two hours of exposure, and GH secretion increased in parallel (Figure 2). The non-peptide GHRP analogue, L-692,429, exerted identical effects on PI hydrolysis and GH secretion (14). It thus appears that the primary mechanism of action of GHSs on human pituitary cells is via activation of the PI-PKC/Ca^{2+} transduction system. The stimulatory effects of GHSs on GH secretion can be reduced or abolished by phloretin, an inhibitor of PKC, and also by W7, which inhibits the intracellular Ca^{2+}-binding messenger, calmodulin (Figure 3) (9). In contrast, inhibition of protein kinase A with Rp-adenosine-3′,5′-cyclic monophosphothioate (Rp-cAMPS) and blockade of the GHRH receptor with specific antagonists both failed to significantly alter the effects of GHRP-2 and GHRP-6 and hexarelin, whereas they were able to inhibit the stimulatory effects GHRH (Figure 4) (9). The recent identification and characterisation of the specific receptor to which GHSs bind further supports the hypothesis that GHSs act via the PI-PKC/Ca^{2+} transduction system, since it was shown to be coupled to G_{11}, a G protein capable of inducing phospholipase C activity and hence PI hydrolysis (1).

The GHS receptor is probably over-expressed in somatotroph adenomas (Figure 1) (4). The stimulatory effects of GHSs on both PI hydrolysis and GH secretion by tumorous

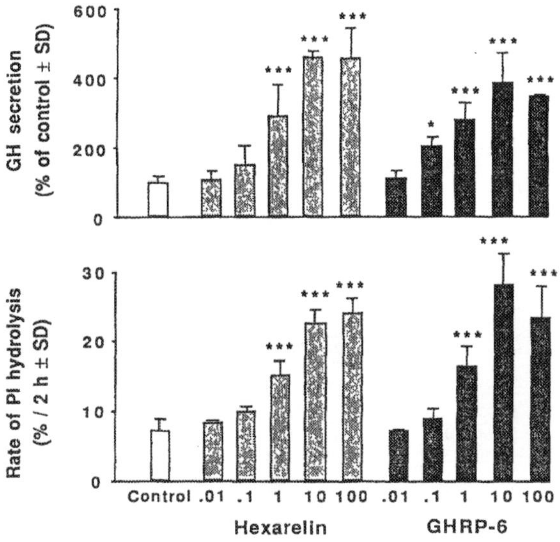

Figure 2. Stimulatory effect of hexarelin and GHRP-6 on GH secretion (upper panel) and rate of PI hydrolysis (lower panel) by cell cultures of a human pituitary somatotrophinoma. *$P < 0.05$, ***$P < 0.001$ vs control.

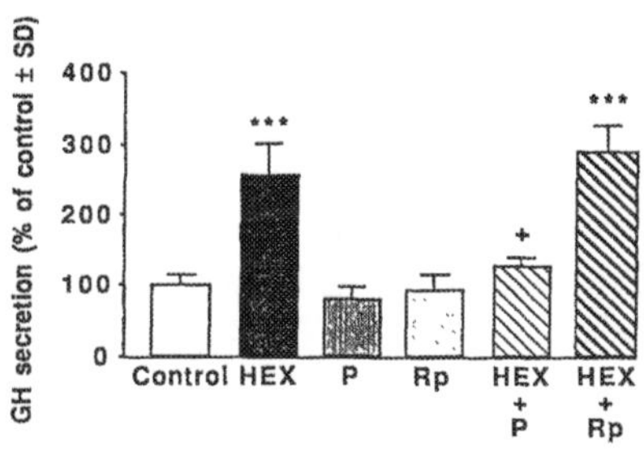

Figure 3. Effects of phloretin-induced (P, 10 μmol/l) and Rp-cAMPS (Rp, 20 μmol/l) on hexarelin-induced (Hex, 1 nmol/l) GH secretion by cell cultures of a human pituitary somatotrophinoma. ***$P < 0.001$ vs control; $^{+}P < 0.001$ vs hexarelin alone.

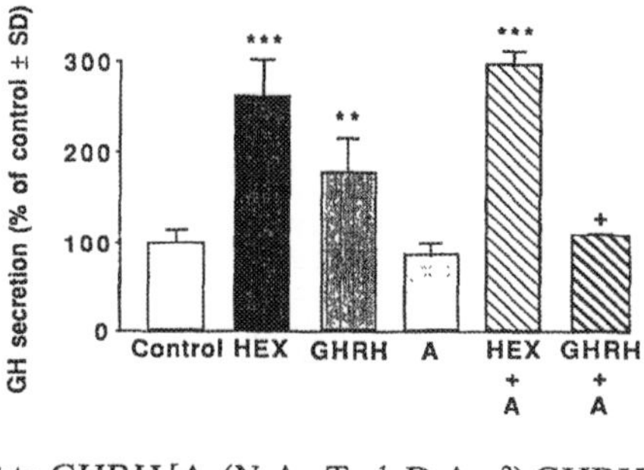

Figure 4. Effect of an antagonist to GHRH [A, (N-Ac-Tyr[1], D-Arg[2]) GHRH-(1–29)-NH_2, 60 nmol/l] on GH secretion by cultured human pituitary somatotrophinoma cells stimulated by hexarelin (Hex, 1nmol/l) and GHRH (2 nmol/l); **$P < 0.01$, ***$P < 0.001$ vs control, $^{+}P < 0.001$ vs GHRH alone.

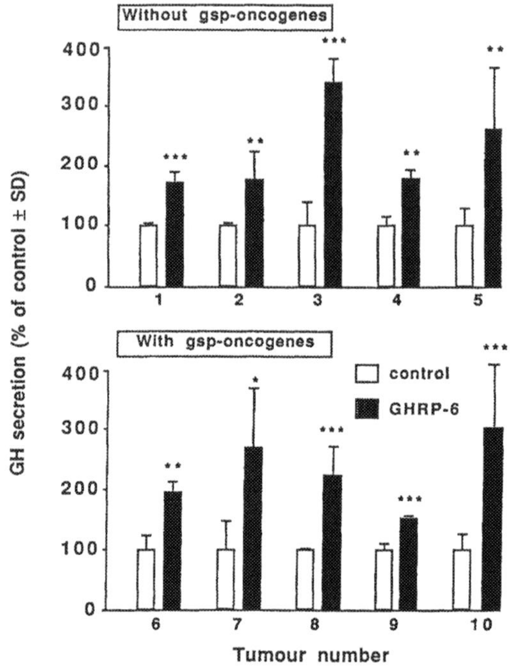

Figure 5. Effect of GHRP-6 (100 nmol/l) on GH secretion by cultured human somatotrophinomas with and without *gsp* oncogenes. $^{**}P < 0.01$, $^{***}P < 0.001$ vs control.

human pituitary somatotrophs is far more consistent than those exerted by GHRH (8,15,16). In our most comprehensive series, only 4 (9.3%) of 43 GH-secreting tumours failed to respond to GHRP-6 in culture (E.F. Adams, unpublished observation), while about 35% showed a failure to respond to GHRH (15). This finding is similar to that observed when using activators of PKC (17). Additionally, expression of *gsp* oncogenes, which leads to constitutive adenyl cyclase activity and elevated cAMP production, does not appear to influence the effects of GHSs on PI hydrolysis and GH secretion, emphasising their independent mechanism of action (Figure 5) (13,16). Nevertheless, GHSs are still able

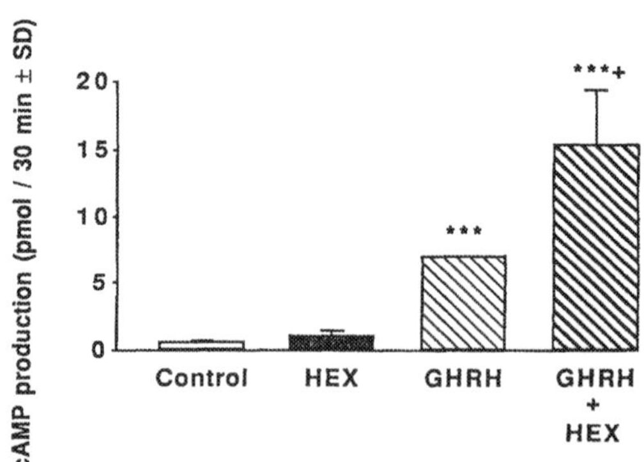

Figure 6. Synergistic effect of hexarelin (Hex, 10 nmol/l) on GHRH-induced (2 nmol/l) cAMP production by cultured human pituitary somatotrophinoma cells. $^{***}P < 0.001$ vs control.

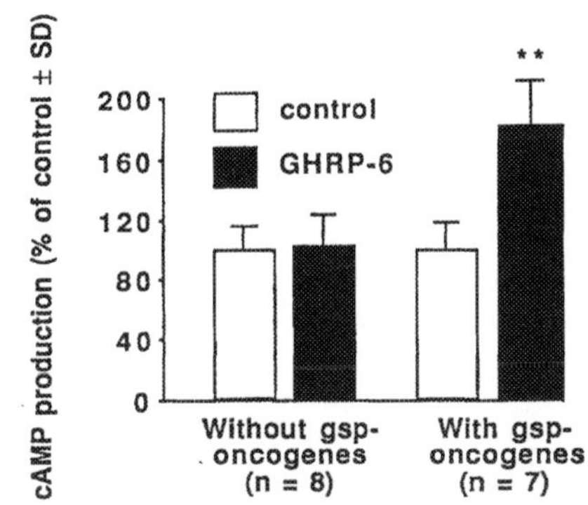

Figure 7. Effect of GHRP-6 (100 nmol/l) on cAMP secretion by cultured human somatotrophinomas with and without *gsp* oncogenes. $^{**}P < 0.01$ vs control.

to influence cAMP production under some circumstances. Both GHRP-2 and GHRP-6 are able to potentiate the stimulatory effects of GHRH on cAMP production by tumorous human somatotrophs in culture (Figure 6) (9), similar to findings with normal rat pituitary cells (18). Moreover, GHRP-2 and GHRP-6 can slightly but significantly stimulate basal cAMP production in those tumours expressing *gsp* oncogenes (Figure 7) (16). Such results suggest that there may be intracellular cross-talk between the PI and adenyl cyclase transduction pathways, as shown to occur in other cell types (19). This might be achieved by PKC-induced phosphorylation of the catalytic subunit of adenyl cyclase and the inhibitory protein, G_i (9,19).

Prolactin secretion

The effects of synthetic GH-secretagogues on human prolactin secretion have also been studied. The non-peptide GHRP analogue, L-692,429, powerfully stimulated prolactin secretion by a mixed somatotrophic-lactotrophic human pituitary tumour in culture, confirming the *in vivo* effects observed in acromegalic patients (14,20). Additionally, some, but not all, pure prolactin-secreting tumours expressed the GHS-R mRNA and also responded to GHSs, both in terms of prolactin secretion and PI hydrolysis (Figure 8) (3). In contrast, in patients with prolactinoma no further rise was observed in circulating prolactin levels after GHS stimulation (20,21). Such findings indicate that tumorous human lactotrophs may also possess the GHS-R (Figure 1), but further studies are required to fully elucidate the role of GHSs in normal prolactin secretion. Nevertheless, it is worth noting that prolactin secretion is also influenced by both the PKA and PKC transduction pathways (22) and thus parallels can be drawn with the control of GH secretion.

Corticotroph adenomas

As corticotroph adenomas are usually very small compared to other hormone-secreting adenomas, it is extremely difficult to study these tumours *in vitro*. The presence of GHS-R mRNA has been shown in some, but not all, corticotroph adenomas (Figure 1) (2,4). Similarly, a response in terms of ACTH release and/or intracellular calcium concentration

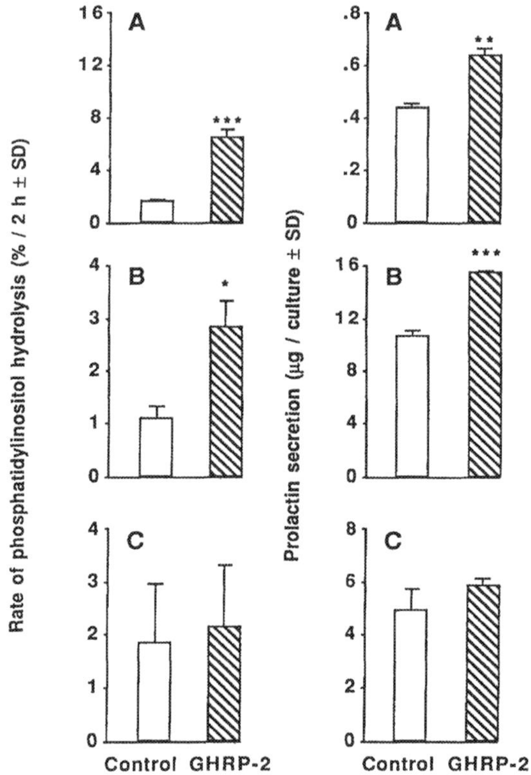

Figure 8. Effect of GHRP-2 (100 nmol/l) on PI hydrolysis (left panels) and prolactin secretion (right panels) by cell cultures of three human pituitary prolactinomas (tumours A, B and C). $*P < 0.05$, $**P < 0.01$, $***P < 0.001$ vs control.

has been found in one but not in another preliminary report (23,24). The effect observed on ACTH release was actually greater than that seen after CRH stimulation on cells derived from the same adenoma (23). Although normal rat pituitary cells do not respond to GHS stimulation with ACTH release (5,18), this strong positive effect on human corticotroph adenoma cells is in accordance with *in vivo* data of ACTH responses in patients with Cushing's disease (25), and with data describing the presence and indeed over-expression of GHS-R in corticotroph adenomas (4).

Non-functioning pituitary adenomas

Some non-functioning pituitary adenomas express GHS-R, while others do not (Figure 1) (3,4). In a preliminary study elevation of intracellular calcium levels has been observed in response to GHRP-6 by non-functioning adenomas in culture (24).

In summary, the use of *in vitro* cell culture has led to a considerable increase in the understanding of the intracellular mechanisms of the action of GHSs on human pituitary cells. Specifically, the studies indicate that the PI-PKC/Ca^{2+} transduction system plays a pivotal role in GH secretory control, perhaps involving a natural counterpart to GHSs. Additionally, the studies have raised the possibility that elevated GH secretion in acromegaly may be due to defects in this second messenger system, at least in some cases (9,15,16), while in lactotroph and corticotroph adenomas the expression of GHS-R may influence the hormone response to specific stimuli.

THE EFFECT OF GHSs ON HYPOTHALAMIC HORMONE RELEASE

In spite of the large number of experimental and clinical studies with different GHS analogues, the exact mode of their GH-releasing action has yet to be fully established. A number of studies have suggested a possible direct activation of GHRH-secreting neurons in the arcuate nucleus: increased electrical activity and *c-fos* expression has been shown after the administration of GHSs (26); GHRH antisera attenuates the GH-releasing effect of GHSs (27); and increased hypophysial portal blood GHRH has been shown following GHS administration in sheep (28). Recently, the presence of GHS-R mRNA has been shown to co-localize to GHRH cells (29). Others have been unable to confirm the elevated GHRH levels in hypophysial portal blood samples, but have reported increased frequency of GHRH release after GHS administration (30). In the rat, but not in the guinea pig, very high doses of intracerebroventricular GHRP-6 produced a paradoxical *decrease* in circulating GH levels (31,32). GH secretagogues not only stimulate GH release but also stimulate prolactin and activate the hypothalamo-pituitary– adrenal (HPA) axis in both animal and human studies (33–38), an effect which is not inhibited by somatostatin (SS) (39). Since GHSs do not stimulate ACTH release directly from pituitary cell cultures (18,40), it is probable that GHSs affect the HPA via either one of the two major ACTH stimulators in the hypothalamus, corticotrophin-releasing hormone (CRH) and/or arginine vasopressin (AVP).

In our *in vitro* hypothalamic incubation system (41), none of the GHRP analogues in the dose range of 10^{-6}–10^{-4} M had any effect on basal GHRH release at 20, 40, 60 or 80 min incubation. However, at 10^{-3} M a strong *inhibitory* effect was shown on both basal (Figure 9) and potassium-chloride-stimulated GHRH release using GHRP-6, hexarelin and L-692,585. Somatostatin release was not inhibited by GHSs; on the contrary, a small, unexpected increase in somatostatin secretion was observed (Figure 10). Interestingly, an

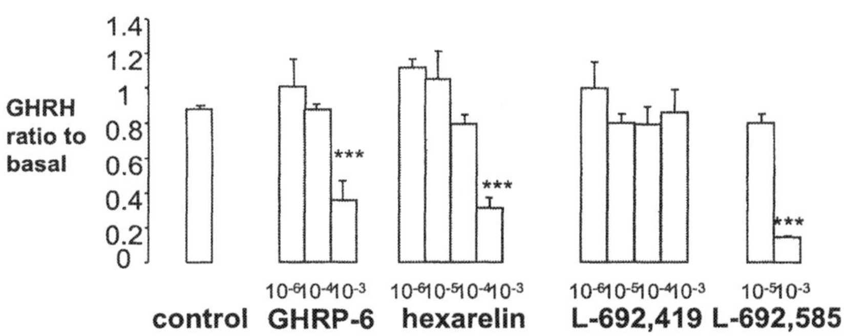

Figure 9. Dose–response curve of GHSs on GHRH release at 80 minutes incubation in rat hypothalamic cultures.

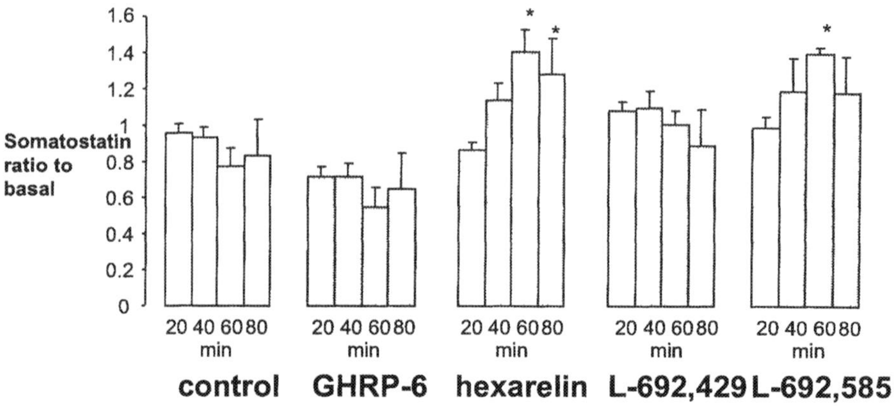

Figure 10. Dose–response curve of GHSs on somatostatin release at 80 minutes incubation in rat hypothalamic cultures.

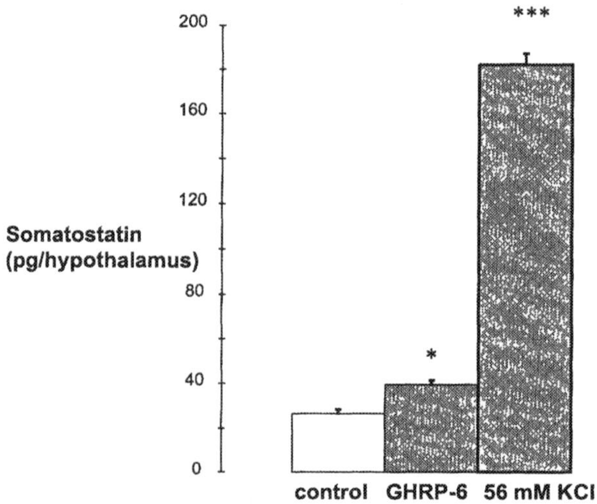

Figure 11. The effect of GHRP-6 (1 μmol/l) and 56 mM potassium-chloride on somatostatin release in rat fetal hypothalamic cell culture (G. Gillies, unpublished observations). $^{*}P < 0.05$, $^{***}P < 0.001$ vs control.

increase in somatostatin release was also observed by two other groups: in a rat fetal hypothalamic incubation system stimulation with 10^{-6} M GHRP-6 significantly increased the release of somatostatin (Figure 11) (G. Gillies unpublished observations), confirming our results in the adult rat (42). In a similar system to ours, significant somatostatin release was observed after stimulation with high concentrations of GHRP-6 (Figure 12) (43).

The release of AVP was strongly stimulated by GHSs (Figure 13); however, we could not detect any change in CRH levels (data not shown). The effects of active GHS analogues were not parallelled by using an inactive non-peptide analogue, L-692,428, ruling out a non-specific effect of high drug concentrations.

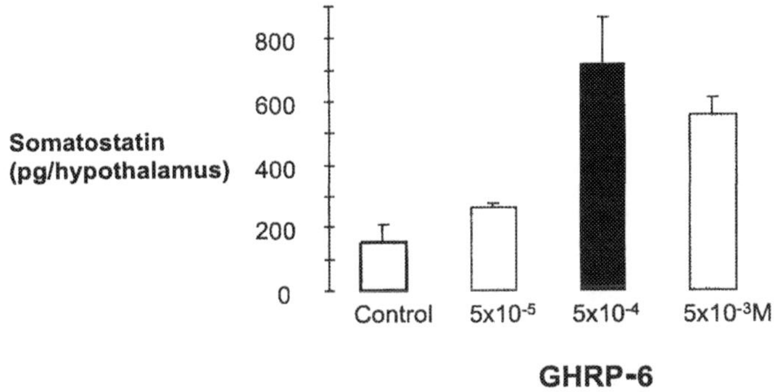

Figure 12. The effect of GHRP-6 on somatostatin release in *in vitro* rat hypothalamic incubations (43). (Figure drawn from data shown in abstract, ref. 43).

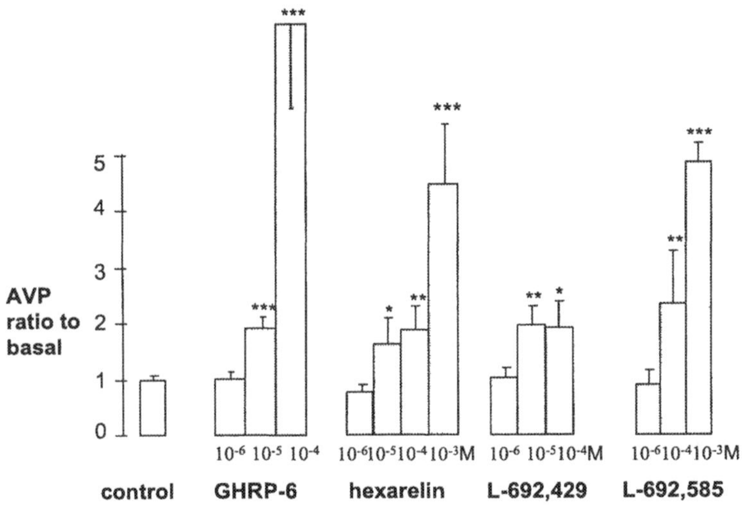

Figure 13. The effect of GHSs on AVP release in rat hypothalamic cultures.

Recently, neuropeptide Y has been implicated as a mediator of the effects of GHSs (44). It has been shown that NPY stimulates SS release (Figure 14) (42,45) and there have been suggestions that it might inhibit GHRH; however, this has never been shown directly. We found that NPY inhibited GHRH release from the hypothalamus *in vitro* (Figure 15). NPY also augmented potassium-induced AVP release, suggesting that it might be a possible mediator of the effects of GHSs in the hypothalamus. The effect of NPY would stimulate the HPA axis, in accordance with *in vivo* data, but the effect on the GH axis is opposite to what would be expected if it was the mediator of the principal effects of GHSs. However, certain *in vivo* studies show compatible results with NPY mediating the effects of GHSs. In a recent study it has been shown that hypophysial portal plasma concentrations of GHRH in sheep did not show a coincident release of GHRH after the intravenous bolus administration of GHRP-6; infusion of GHRP-6 caused no change in GHRH pulse

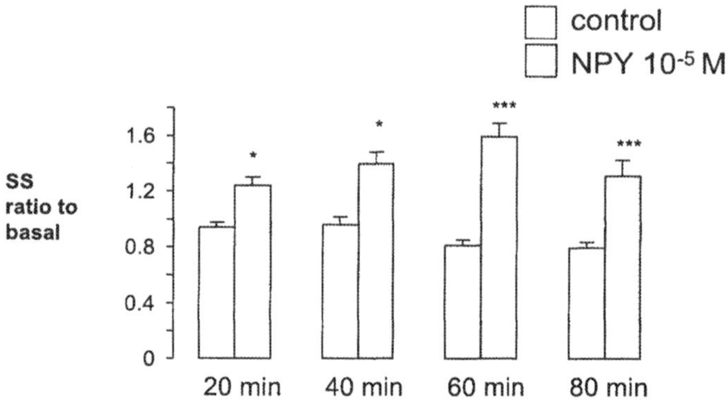

Figure 14. The effect of NPY 10^{-5} on SS release in rat hypothalamic cultures.

amplitude but a 50 percent rise in pulse frequency, suggesting an effect on the frequency of the pulsatile discharge of GHRH (30). Furthermore, the results of experiments of intracerebroventricular administration of GH- releasing peptides on GH levels have failed to show consistent results. While in guinea pigs the expected rise was shown (32), in rats a paradoxical *decrease* in circulating GH levels was observed (31). The dose administered via the intracerebroventricular catheter in the latter study corresponds to a very high concentration of GHRP-6 in solution (2.3×10^{-3} M). Our finding, that the release of GHRH is inhibited by the presence of similarly high doses of GHS analogues, is compatible with these results in the rat, while species-specific differences might explain the conflicting data in guinea pig and rat studies. The regulation of the GH axis in the rat is different from that of other species; for example, stress or hypoglycaemia causes GH inhibition in rats (46), while in guinea pigs or humans these tests cause stimulation of the GH axis. Thus, certain effects observed in rat experiments cannot be readily extrapolated to other species. Other studies indicated that GHSs inhibit the effect of SS (32,47). Our results suggest that certain of the endocrine effects of GHSs may be mediated by NPY, but it remains unclear as to whether the major GH-releasing effect in the hypothalamus involves direct activation of GHRH. Bowers and colleagues have long suggested the effects of GHSs on GH could not be

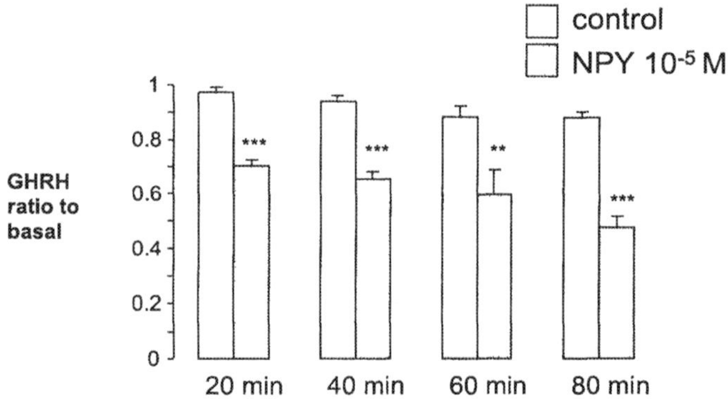

Figure 15. The effect of NPY 10^{-5} on GHRH release in rat hypothalamic cultures.

explained purely in terms of the modulation of GHRH and SS, but required the presence of an unknown endogenous factor to fully explain their complex effects (27). Our results are compatible with this speculation.

CONCLUSIONS

In summary, synthetic GH secretagogues act on a specific receptor which is present in normal and abnormal pituitary tissue and in the hypothalamus. They act via the protein kinase C / phosphatidyl inositol / Ca^{2+} pathway in somatotrophs and augment the effect of GHRH on cAMP release. The effect in the hypothalamus is controversial: somatostatin release is slightly stimulated by high doses of GHSs from the hypothalamus as has now been shown by three separate groups. There are a number of arguments suggesting a direct stimulation of GHRH release; however, we were not able to show this in rat hypothalamic tissue culture. In contrast, we found an opposite effect, and the apparent GH inhibitory effect of intracerebroventricular GHRP-6 is in line with our findings. Clearly, the complex hypothalamic mechanism of effect of GHSs needs further investigation.

REFERENCES

1. Howard, A.D., Feighner, S.D., Cully, D.F., et al. (1996) A receptor in pituitary and hypothalamus that functions in growth hormone release. Science 273, 974–977.
2. deKeyzer, Y., Lenne, F., Bertagna, X. (1997) Widespread transcription of the growth hormone-releasing peptide receptor gene in neuroendocrine human tumors. Eur. J. Endocrinol. 137, 715–718.
3. Adams, E.F., Huang, B., Buchfelder, M., et al. (1998) Presence of growth hormone secretagogue receptor messenger ribonucleic acid in human pituitary tumors and rat GH(3) cells. J. Clin. Endocrinol. Metab. 83, 638–642.
4. Korbonits, M., Jacobs, R.A., Aylwin, S.J.B., Burrin, J.M., Dahia, P.L.M., Monson, J.P., Trainer, P.J., Chew, S.L., Besser, G.M., and Grossman, A.B. (1998) Expression of the growth hormone secretagogue receptor in pituitary adenomas and other neuronendocrine tumors. J. Clin. Endocrinol. Metab. 83, 3624–3630.
5. Shimon, I., Yan, X., Melmed, S. (1998) Human fetal pituitary expresses functional growth hormone-releasing peptide receptors. J. Clin. Endocrinol. Metab. 83, 174–178.
6. Jansson, J., Svensson, J., Bengtsson, B., et al. (1998) Acromegaly and Cushing's syndrome due to ectopic production of GHRH and ACTH by a thymic carcinoid tumour: *in vitro* responses to GHRH and GHRP-6. Clin. Endocrinol. (Oxf.) 48, 243–250.
7. Renner, U., Brockmeier, S., Strasburger, C.J., et al. (1994) Growth-hormone (GH)-releasing peptide stimulation of GH release from human somatotroph adenoma cells — interaction with GH-releasing hormone, thyrotropin-releasing-hormone, and octreotide. J. Clin. Endocrinol. Metab. 78, 1090–1096.
8. Bresson-Bepoldin, L., Odessa, M.F., Dufy-Barbe, L. (1994) GHRP-6 stimulates calcium increase and growth hormone release in human somatotrophs *in vivo*. Endocrine 2, 793–803.
9. Adams, E.F., Lei, T., Bucherfelder, M., Bowers, C.Y., Fahlbush, R. (1996) Protein kinase C-dependent growth hormone releasing peptides stimulate cyclic adenosine 3′, 5′-monophosphate production by humna pituitary somatotropinomas expressing gsp oncogenes: evidence for cross-talk between transduction pathways. Mol. Endocrinol. 10, 432–438.
10. Cheng, K., Chan, W.W., Butler, B., Barreto, A., Jr., Smith, R.G. (1991) Evidence for a role of protein kinase-C in His-D-Trp-Ala-Trp-D-Phe-Lys-NH_2-induced growth hormone release from rat primary pituitary cells. Endocrinology 129, 3337–3342.

11. Herrington, J., Hille, B. (1994) Growth hormone-releasing hexapeptide elevates intracellular calcium in rat somatotropes by 2 mechanisms. Endocrinology 135, 1100–1108.
12. Chen, C., Vincent, J.D., Clarke, I.J. (1994) Ion channels and the signal transduction pathways in the regulation of growth hormone secretion. Trends Endocrinol. Metab. 6, 227–233.
13. Lei, T., Bucherfelder, M., Fahlbush, R., Adams, E.F. (1995) Growth hormone-releasing peptide (GHRP-6) stimulates phosphatidylinositol (PI) turnover in human pituitary somatotroph cells. J. Mol. Endocrinol. 14, 135–138.
14. Adams, E.F., Petersen, B., Lei, T., Bucherfelder, M., Fahlbush, R. (1995) The growth hormone secretagogue, L-692,429, induces phosphatidylinositol hydrolysis and hormone secretion by human pituitary tumours. Biochem. Biophys. Res. Commun. 208, 555–561.
15. Adams, E.F., Lei, T., Buchfelder, M., Petersen, B., Fahlbusch, R. (1995) Biochemical characteristics of human pituitary somatotrophinomas with or without *gsp* mutations. J. Clin. Endocrinol. Metab. 80, 2077–2081.
16. Adams, E.F., Buchfelder, M., Lei, T., et al. (1996) *In vitro* responses of GH-secreting tumours with and without *gsp* oncogenes to octreotide, GHRH and growth hormone-releasing peptide. In: Pituitary Adenomas: From basic research to diagnosis and therapy. K. von Werder and R. Fahlbusch (eds). Elsevier, Amsterdam, pp. 43–47.
17. Emoto, N., Ohmura, E., Isozaki, O., Tsushima, T., Shizume, K., Demura, H. (1991) Phorbol ester, not growth hormone-releasing factor, consistently stimulates growth hormone release from somatotroph adenomas in culture. Clin. Endocrinol. (Oxf.) 34, 377–382.
18. Cheng, K., Chan, W.W., Barreto, A., Jr., Convey, E.M., Smith, R.G. (1989) The synergistic effects of His-D-Trp-Ala-Trp-D-Phe-Lys-NH_2 on growth hormone(GH)-releasing factor-stimulated GH release and intracellular adenosin 3′,5′-monophosphate accumulation in rat pituitary cell culture. Endocrinology 124, 2791–2798.
19. Houslay, M.D. (1998) Crosstalk: a pivotal role for the protein kinase C in modulating relationships between signal transduction pathways. J. Biol. Chem. 264, 8802–8810.
20. Ciccarelli, E., Grottoli, S., Razzore, P., et al. (1996) Hexarelin, a growth hormone-releasing peptide, stimulates prolactin release in acromegalic but not in hyperprolactinaemic patients. Clin. Endocrinol. (Oxf.) 44, 67–71.
21. Popovic, V., Simic, M., Ilic, L., et al. (1998) Growth hormone secretion elicited by GHRH, GHRP-6 or GHRH plus GHRP-6 in patients with microprolactinoma and macroprolactinoma before and after bromocriptine therapy. Clin. Endocrinol. (Oxf.) 48, 103–108.
22. Friedman, E., Adams, E.F., Hoog, A., et al. (1994) Normal structural dopamine type 2 receptor gene in prolactin-secreting and other pituitary tumours. J. Clin. Endocrinol. Metab. 78, 568–574.
23. Barlier, A., Grino, M., Zamora, A.J., et al. (1998) Expression of growth hormone secretagogue-receptors (GHS-R) in human pituitary adenomas. European Congress of Endocrinology, Seville, Spain, P2-341.
24. Lania, A., Ballaré, E., Corbetta, S., Filopanti, M., Persani, L., Spada, A. (1998) Mechanism of action of growth hormone-releasing hexapeptide (GHRP-6) in cultured cells from functioning and non-functioning human pituitary adenomas. European Congress of Endocrinology, Seville, Spain, P2-117.
25. Loche, S., Colao, A., Cappa, M., et al. (1997) Adrenocorticotropin- and cortisol-releasing effect of hexarelin, a synthetic growth hormone-releasing peptide, in normal subjects and patients with Cushing's syndrome. J. Clin. Endocrinol. Metab. 82, 861–864.
26. Dickson, S.L., Leng, G., Dyball, R.E.J., Smith, R.G. (1995) Central actions of peptide and non-peptide growth hormone secretagogues in the rat. Neuroendocrinol. 61, 36–43.
27. Bowers, C.Y., Sartor, A.O., Reynolds, G.A., Badger, T.M. (1991) On the actions of the growth hormone-releasing hexapeptide, GHRP. Endocrinology 128, 2027–2035.
28. Guillaume, V., Magnan, E., Cataldi, M., et al. (1994) GH-releasing hormone secretion is stimulated by a new GH-releasing hexapeptide in sheep. Endocrinology 135, 1073–1076.
29. Tannenbaum, G.S., Lapointe, M., Beaudet, A., and Howard, A.D. (1998) Expression of growth hormone secretagogue receptors by growth hormone-releasing hormone neurons in the mediobasal hypothalamus. Endocrinology 139, 4420–4423.

30. Fletcher, T.P., Thomas, G.B., Clarke, I.J. (1996) Growth hormone-releasing hormone and somatostatin concentrations in the hypophysial portal blood of conscious sheep during the infusion of growth hormone-releasing peptide-6. Domest. Animal Endocrinol. 13, 251–258.
31. Yagi, H., Kaji, H., Sato, M., Okimura, Y., Chihara, K. (1996) Effects of intravenous or intracerebroventricular injections of His-*D*-Trp-Ala-Trp-*D*-Phe-Lys-NH_2 on GH release in conscious, freely moving male rats. Neuroendocrinol. 63, 198–206.
32. Fairhall, K.M., Mynett, A., Robinson, I.C.A.F. (1995) Central effects of growth hormone-releasing hexapeptide (GHRP-6) on growth-hormone release are inhibited by central somatostatin action. J. Endocrinol. 144, 555–560.
33. Bowers, C.Y., Momany, F.A., Reynolds, G.A., Hong, A. (1984) On the *in vitro* and *in vivo* activity of a new synthetic heptatpeptide that acts on the pituitary to specifically release growth hormone. Endocrinology 114, 1537–1545.
34. Ghigo, E., Arvat, E., Gianotti, L., et al. (1994) Growth hormone-releasing activity of Hexarelin, a new synthetic hexapeptide, after intravenous, subcutaneous, and oral administration in man. J. Clin. Endocrinol. Metab. 78, 693–698.
35. Gertz, B.J., Barrett, J.S., Eisenhandler, R. et al. (1993) Growth hormone response in man to L-692,429, a novel nonpeptide mimic of growth hormone-releasing peptide-6. J. Clin. Endocrinol. Metab. 77, 1393–1397.
36. Korbonits, M., Trainer, P.J., Besser, G.M. (1995) The effect of an opiate antagonist on the hormonal changes induced by hexarelin. Clin. Endocrinol. (Oxf.) 43, 365–371.
37. Thomas, G.B., Fairhall, K.M., Robinson, I.C.A.F. (1997) Activation of the hypothalamo-pituitary-adrenal axis by the growth hormone (GH) secretagogue, GH-releasing peptide-6, in rats. Endocrinology 138, 1585–1591.
38. Korbonits, M., Grossman, A.B. (1995) Growth hormone-releasing peptide and its analogues; novel stimuli to growth hormone release. Trends Endocrinol. Metab. 6, 43–49.
39. Massoud, A.F., Hindmarsh, P.C., Brook, C.G.D. (1997) Interaction of the growth hormone releasing peptide hexarelin with somatostatin. Clin. Endocrinol. (Oxf.) 47, 537–547.
40. Cheng, K., Chan, W.W.S., Butler, B., Wei, L., Smith, R.G. (1993) A novel non-peptidyl growth hormone secretagogue. Horm. Res. 40, 109–115.
41. Korbonits, M., Little, J.A., Camacho-Hübner, C., Trainer, P.J., Besser, G.M., Grossman, A.B. (1996) Insulin-like growth factor-I and -II in combination inhibit the release of growth hormone-releasing hormone from the rat hypothalamus *in vitro*. Growth Regul. 6, 110–120.
42. Korbonits, M., Little, J.A., Forsling, M.L. et al. (1997) The effect of growth hormone secretagogues on the release of growth hormone-releasing hormone, somatostatin, vasopressin and corticotrophin-releasing hormone from the rat hypothalamus *in vitro*. In: Growth Hormone Secretagogues. B.B. Bercu and R.F. Walker (eds). Merkel, pp. 231–249.
43. Hao, E.H., Malozowski, S., Ren, S.G., Marin, G., Southers, J., Merriam, G.R. (1988) A comparison of the effect of GH-releasing hormone (GHRH) and GH-releasing peptide (GHRP) on GH and somatostatin (SRIF) release. 70th Annual Meeting of the Endocrine Society, New Orleans, LA.
44. Dickson, S.L., Luckman, S.M. (1997) Induction of *c-fos* messenger ribonucleic acid in neuropeptide Y and growth hormone (GH)-releasing factor neurones in the rat arcuate nucleus following systemic injection of the GH secretagogue, GH-releasing peptide-6. Endocrinology 138, 771–777.
45. Rettori, V., Milenkovic, L., Aguila, M.C., McCann, S.M. (1990) Physiologically significant effect of neuropeptide Y to suppress growth hormone release by stimulating somatostatin discharge. Endocrinology 126, 2296–2301.
46. Lengyel, A.J., Grossman, A.B., Niewenhuyzen-Kruseman, A.C., Ackland, J., Rees, L.H., Besser, G.M. (1984) Glucose modulation of somatostatin and LHRH release from rat hypothalamic fragments *in vitro*. Neuroendocrinol. 39, 31–38.
47. Clark, R.G., Carlsson, L., Trojnar, J., Robinson, I.C.A.F. (1989) The effect of growth hormone-releasing peptide and growth hormone-releasing factor on conscious and anaesthetized rats. J. Neuroendocrinol. 1, 249–255.

Growth Hormone Secretagogues
Edited by E. Ghigo, M. Boghen, F.F. Casanueva and C. Dieguez

Chapter 8

Hypothalamic Site and Mechanism of Action of Growth Hormone Secretagogues

SUZANNE L. DICKSON

Department of Physiology, University of Cambridge, Downing Street, Cambridge, U.K.

The pulsatile pattern of growth hormone (GH) secretion from the anterior pituitary gland reflects a changing balance in the output of two hypothalamic systems, the GH-releasing hormone (GHRH) neurones that stimulate GH secretion (1,2) and the inhibitory somatostatin neurones (3). The GH secretagogues stimulate GH secretion by a direct pituitary action and also by a central mechanism that includes effects on the hypothalamic GHRH-somatostatin pulse generating system.

GROWTH HORMONE SECRETAGOGUES ARE CENTRALLY ACTIVE COMPOUNDS

In the early 1980s C.Y. Bowers and colleagues identified the first member of a class of synthetic growth hormone (GH)-releasing compounds, GH-releasing peptide-6 (GHRP-6), whose actions were potent and selective for GH secretion (4). Direct GH-releasing activity at the pituitary level was observed *in vitro*, providing evidence for a direct pituitary action (4) and the GH-releasing mechanism was clearly different to that of GHRH (5,6). That GHRP-6 may also affect the hypothalamic regulation of GH secretion was first suggested by Clark and Robinson (7). They proposed that part of the GH-releasing mechanism of GHRP-6 is likely to include increased GHRH release, since the GHRP-6-induced GH response was attenuated in rats passively immunized with GHRH antiserum. Furthermore they suggested that GHRP-6 may alter somatostatin secretion, since it disrupted the cyclic changes in GH release following regular injections of GHRH (a response which has been attributed to cyclic changes in somatostatin secretion) (8). Certainly, it emerged that the GH-releasing actions of GH secretagogues could not be explained solely by a pituitary action since, when administered together, GHRP-6 and GHRH had merely additive effects on GH secretion *in vitro* (in pituitary perfusion experiments and in cell culture) but produced an enormous synergistic effect *in vivo* (9). Rather it was suggested that the central actions of GHRP-6 may include the release of an unknown hypothalamic-releasing factor (a "U-factor") into the portal blood; according to this hypothesis, the U-factor would act together with GHRH to stimulate GH secretion from the pituitary (9).

The first direct evidence that the GH secretagogues are centrally active compounds was provided by studies demonstrating increased activity of a sub-population of cells in the hypothalamic arcuate nucleus (6). In this study and in subsequent studies, cells activated following systemic GH secretagogue injection were detected using two complementary approaches: (1) by the immunocytochemical detection of Fos protein (the product of the immediate early gene, *c-fos*, which is expressed in many neuronal systems following activation) and (2) by changes in electrical activity of arcuate neurones recorded in anaesthetised rats (6,10).

INDUCTION OF FOS PROTEIN FOLLOWING GH SECRETAGOGUE ADMINISTRATION

Following systemic injection of GHRP-6, nuclear staining for Fos was detected in a sub-population of cells in the arcuate nucleus; most Fos-positive nuclei were located in the most ventromedial aspects of this nucleus, although there were also a few scattered more ventro-laterally (Figure 1) (3). No increase in Fos expression was detected in any other forebrain structure studied. The selectivity of the GH secretagogue response is quite remarkable, considering the distribution of the GH secretagogue receptor. This receptor was cloned in 1997 by Howard and colleagues and subsequently shown to be present in a number of CNS structures (11). In addition to the arcuate nucleus, GH secretagogue receptor mRNA was detected in several other hypothalamic nuclei (including the ventromedial nucleus, preoptic nucleus, anterior hypothalamic area, suprachiasmatic nucleus) and in discrete regions of many non-hypothalamic structures (including the hippocampus, thalamus and brainstem) (12,13). Thus, assuming that GH secretagogues have access to CNS regions other than the arcuate nucleus, it would appear that GH secretagogue action is not coupled with expression of Fos in these regions. Consistent with this hypothesis, no additional forebrain structures expressed Fos when GHRP-6 was administered directly into the brain ventricles; both the distribution and the number of cells activated was identical to that described for systemic injection of this compound (10).

In a recent study, the effects of GH secretagogues on Fos expression within the CNS have been extended to include the brainstem. Interestingly, systemic GHRP-6 injection induced Fos expression in the area postrema and in the nucleus tractus solitarii (Figure 1) (14). The physiological consequences of activation of these cells by GH secretagogues is not known. The area postrema is a circumventricular organ, that is, an area of the CNS where the blood–brain barrier is incomplete and where information about circulating levels of peripheral factors is monitored. It is closely interconnected with the nucleus tractus solitarii and together they form an important role in relaying such information to the hypothalamus, for the regulation of pituitary hormone secretion. It is possible that the GH secretagogue-responsive cells in the brainstem form part of a projection to the hypothalamus and that this

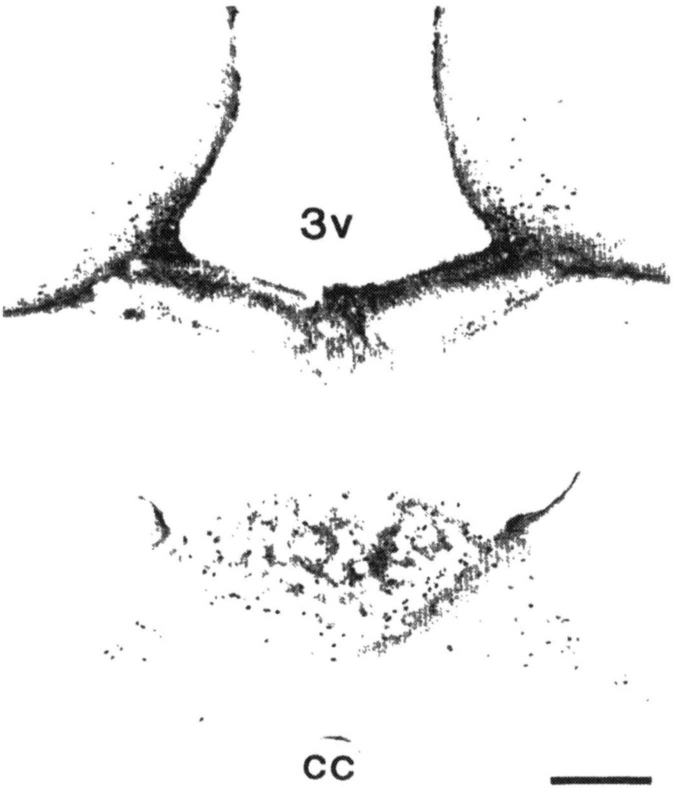

Figure 1. Distribution of cells expressing nuclear immunostaining for Fos protein following systemic injection of 50 μg GHRP-6 in the arcuate nucleus (top) and area postrema and nucleus tractus solitarius of the brainstem (below). Rats were killed at 90 min following GHRP-6 injection. Scale bar = 0.2 mm. 3V (third ventricle). CC (central Canal).

plays some role in regulating hypothalamic activity. It is evident, however, that the GH secretagogues also have more direct actions within the hypothalamus. That the GH secretagogues act directly within the arcuate nucleus to activate cells has been demonstrated in electrophysiological studies *in vitro*: GHRP-6 was shown to activate a sub-population of cells in a hypothalamic slice preparation in which all but closely adjacent inputs to the arcuate nucleus had been severed (15).

GH SECRETAGOGUES INDUCE ELECTRICAL ACTIVATION OF CELLS IN THE ARCUATE NUCLEUS RECORDED *IN VIVO*.

The first recordings of changes in electrical activity following GH secretagogue administration were incorporated into a series of experiments in which attempts were being made to record from identified GHRH neurones in the arcuate nucleus (6). In addition to the GHRH population, the arcuate nucleus contains a number of different cell types, including other neuroendocrine cells (e.g. the tuberoinfundibular dopamine neurones) and many non-neuroendocrine cells. Thus, it was first necessary to determine whether the cells activated by the GH secretagogues are neuroendocrine cells. This can be achieved using *in vivo* electrophysiological recording techniques in which arcuate cells projecting to the median eminence are identified by antidromic identification (Figure 2). Since GHRH neurones are neuroendocrine cells that project from the arcuate nucleus to the median eminence, the antidromically identified cells will include GHRH-neurones. To distinguish GHRH neurones from other neuroendocrine arcuate cells at the time of recording, several criteria were established based upon how GHRH neurones would be expected to behave following electrical stimulation of discrete brain structures. Thus, for example, stimulation

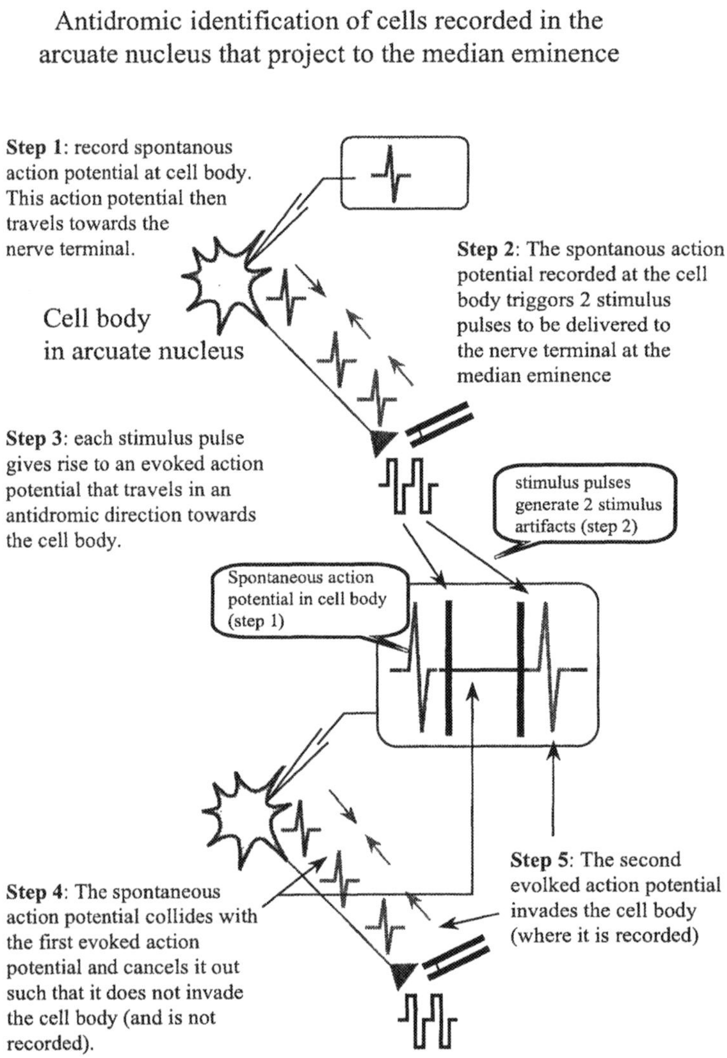

Figure 2. Schematic representation of the test of antidromic identification, a procedure used in electrophysiological studies *in vivo* to identify cells that are likely to be neuroendocrine cells in the arcuate nucleus of anaesthetised rats.

of the periventricular nucleus (a region known to contain the somatostatin neurones that form part of the hypothalamic GH pulse generating mechanism) caused an inhibitory response during electrical stimulation followed by excitation following the end of stimulation (6,16). This inhibition/excitation response may account for the inhibition/rebound release of GH secretion that occurs during/following either a somatostatin infusion (17) or, indeed, during/after electrical stimulation of the periventricular nucleus (18). Another criterion used in these studies was electrical stimulation of the basolateral amygdala, previously shown to stimulate GH secretion without increasing the release of any other pituitary hormone (19). Thus, in summary, putative GHRH neurones were identified in these studies that (1) project to the median eminence (and are therefore likely to be neuroendocrine cells) (2) show an inhibition/excitation response during/after electrical stimulation of the periventricular nucleus, and (3) were excited during electrical stimulation of the basolateral amygdala. Cells fulfilling all of these criteria for identification as putative GHRH neurones showed excitatory responses following systemic injection of GHRP-6 (6), providing the first indication that the GH secretagogues appear to be activating the GHRH neurones.

GH SECRETAGOGUES ACTIVATE NEUROENDOCRINE CELLS IN THE ARCUATE NUCLEUS

If, indeed, the GH secretagogues are targeting the GHRH neurones we might predict that their actions will be fairly selective for a sub-population of the neuroendocrine cells in the arcuate nucleus. Thus, in subsequent electrophysiological studies, we set out to characterize the effects of systemic GHRP-6 injection on the firing rate of both antidromically identified cells (the majority of which are likely to be neuroendocrine cells) and cells that did not fulfil the criteria for antidromic identification (largely non-neuroendocrine cells). The predominant response recorded at the cell bodies of the neuroendocrine cells was excitatory. By contrast, of the cells recorded that did not fulfil the criteria for antidromic identification the predominant response was inhibitory (10). Thus, the excitatory actions of GHRP-6 within the arcuate nucleus do indeed appear to be fairly selective for the neuroendocrine cells, consistent with the idea that the target population of cells for GH secretagogue actions include the GHRH population.

That the majority of cells activated by GH secretagogues are neuroendocrine cells was confirmed in neuroanatomical studies in which GHRP-6-induction of Fos was detected in rats given an intravenous injection five days previously with the retrograde tracer Fluorogold. When injected by this route, Fluorogold does not cross the blood–brain barrier and is not transported synaptically, rather it is taken up by neurones that project to areas supplied by fenestrated capillaries or to the periphery (20). Thus, retrogradely-labelled cells were identified in all hypothalamic nuclei that contain neuroendocrine cells. In the arcuate nucleus, the majority of cells expressing Fos protein following GHRP-6 injection (68–82%) were retrogradely labelled with Fluorogold (21). Thus, consistent with the electrophysiological data, the majority (but not all) arcuate cells activated by GHRP-6 appear to be neuroendocrine cells.

CENTRAL SOMATOSTATIN PATHWAYS INFLUENCE GH SECRETAGOGUE ACTION

If our hypothesis is correct that the neuroendocrine cells activated by GH secretagogues include the GHRH-containing cells, we might expect that these cells would also be subject to inhibitory control by central somatostatin pathways. Somatostatin receptors are present on GHRH-containing cells in this region (22) and we have demonstrated that electrical stimulation in the region of the somatostatin cell bodies in the periventricular nucleus suppresses the activity of putative GHRH-containing cells (6). Indeed, in electrophysiological studies *in vitro* we found that the GHRP-6-responsive cells recorded from the arcuate nucleus of a hypothalamic slice preparation were also inhibited by bath application of somatostatin (23). Also, the induction of Fos protein by GH secretagogues is attenuated by systemic or central injection of Sandostatin, a long acting somatostatin analogue (24,25). This suggests that the central actions of the GH secretagogues are subject to inhibition by central somatostatin pathways, possibly reflecting an inhibitory effect of somatostatin on GHRH producing cells. Interestingly this was not seen in mice with disrupted somatostatin type 2 receptor, indicating that the inhibitory effect of central somatostatin pathways on GH secretagogue-induced Fos protein expression is mediated by type 2 receptors (25).

NEUROCHEMICAL IDENTIFICATION OF THE TARGET CELLS FOR GH SECRETAGOGUES IN THE ARCUATE NUCLEUS

From both the heterogeneity of the electrophysiological responses to GHRP-6 and the demonstration that most (but not all) of the cells activated are neuroendocrine cells in the arcuate nucleus, it seems clear that the GH secretagogues are not targeting a single population of cells in the arcuate nucleus. Direct evidence suggesting that the target cells for GH secretagogue action include the GHRH neurones was provided by a study in which we investigated the neurochemical identity of the arcuate cells expressing c-*fos* mRNA following systemic GHRP-6 injection (26). Of the cells detected that express c-*fos* mRNA following systemic GHRP-6 injection, almost 25% could be identified as expressing GHRH mRNA on the consecutive section, indicating that a sub-population of the GHRH cells are activated by the GH secretagogues. Consistent with this finding the GH secretagogue, hexarelin, has been shown to stimulate GHRH release into portal blood of sheep (27). However, the central GH-releasing activity of the GH secretagogues cannot be explained solely by increased GHRH release since this would not explain the large synergy when GHRH and GHRP-6 are administered concomitantly (5). Rather the central GH-releasing actions of the GH secretagogues must include GHRH-independent effects.

It emerged that the GHRH neurones were not the only population of cells in the arcuate nucleus to be activated following systemic GHRP-6 injection. In this study we determined whether the cells expressing c-*fos* mRNA following GHRP-6 injection also express mRNAs for neurochemical markers for substances known to be expressed in discrete populations of cells within the arcuate nucleus; these include neuropeptide Y (NPY), pro-opiomelanocortin (POMC), tyrosine hydroxylase (present in dopamine neurones) and somatostatin

(26). The only other population of cells to show considerable activation following GH secretagogue administration in this study were the NPY-containing cells; approximately 50% of the cells expressing c-*fos* mRNA could be identified as containing NPY mRNA on the consecutive section. Importantly, not all NPY cells were activated by the GH secretagogues since approximately 30% of the cells expressing NPY were identified as c-*fos*-positive on the consecutive section.

This discovery that the GH secretagogues activate NPY neurones led us to consider what is known about these neurones that might help explain what role they play in mediating the central actions of the GH secretagogues. NPY cells are scattered throughout the arcuate nucleus, notably in the ventromedial arcuate nucleus (28,29). Although NPY is present in portal blood in concentrations which are higher than in the peripheral circulation (30) it is not clear whether the arcuate NPY cells are true neurosecretory adenohypophysiotropic cells which project to the external zone of the median eminence (31,32). Certainly, at least a small number of NPY neurones in the arcuate nucleus must be neuroendocrine cells since about 30% of GHRH neurones appear to co-express NPY (33). However, it seems unlikely that the GH secretagogues are causing significant release of NPY into portal blood since, in contrast to the GH secretagogues, NPY exerts an inhibitory influence on GH secretion (34). The NPY cells in the arcuate nucleus have an established role in the regulation of feeding behaviour. Multiple daily injections of NPY into the PVN increases daily food intake and body weight (35). It seems possible therefore that activation of these cells by the GH secretagogues is responsible for the acute effects on feeding behaviour when these compounds are administered directly into the brain ventricles (36,37). The arcuate NPY cells give rise to a major intrahypothalamic pathway; they project to the anterior hypothalamus, the pre-optic area, the paraventricular nucleus (PVN) as well as to the ventromedial and dorsomedial hypothalamus (38,39). The arcuate-PVN projection is believed to be important for the effects on feeding behaviour and also for activation of the hypothalamo–pituitary adrenal axis. Indeed, activation of NPY cells by GH secretagogues may also explain the small cortisol response following GH secretagogue administration (40,41).

In a recent study, we sought to determine whether the arcuate cells projecting to the PVN (a sub-population of which are likely to be NPY-containing) also project to the median eminence (indicating that they are likely to be neuroendocrine cells) and whether these cells respond to systemic GHRP-6 injection (42). Recordings were made of the electrical activity of individual arcuate neurones, and for each cell, we determined whether it could be antidromically activated from the median eminence and also, from the PVN. Of 43 cells that were identified as projecting to the PVN, only one cell also projected to the median eminence. Ten of the arcuate cells that project to the PVN were tested with systemic GHRP-6 injection and 3 of these responded with an excitation. We concluded that the arcuate cells that project to the PVN (including NPY cells) are unlikely to be neuroendocrine cells and that about 30% of these cells respond to GHRP-6. Taken together with our previous finding that most of the cells in the arcuate nucleus that are activated by GHRP-6 are neuroendocrine cells (10,21), it seems likely that the arcuate nucleus NPY cells activated by GH secretagogue are a subpopulation of neuroendocrine NPY that do not project to the PVN.

Detection of Fos protein following GHRP-6 in injection in arcuate nucleus neurones retrogradely labelled from the PVN

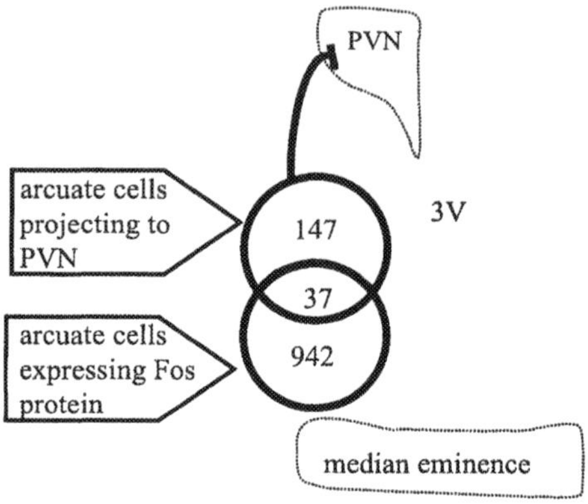

Figure 3. Schematic representation of the results from a study investigating whether the arcuate cells activated by GH secretagogues project to the paraventricular nucleus (PVN) of the hypothalamus (42). The retrograde tracer Fluorogold was administered to the PVN; arcuate cells projecting to that site were identified by the presence of Fluorogold in the cell cytoplasm. Cells activated by GHRP-6 were identified that express Fos protein. Approximately 20% (37/184) of the arcuate cells projecting to the PVN were activated by GHRP-6. However, the majority of cells activated by GHRP-6 (942/979) do not project to the PVN.

In retrograde labelling studies we confirmed that a very small proportion of the arcuate cells expressing Fos protein following GHRP-6 injection project to the PVN. In these studies, the retrograde tracer Fluorogold was administered to the PVN via a microdialysis probe such that arcuate cells projecting to the PVN could be detected (42). Of the total number of arcuate cells detected that were retrogradely labelled as projecting to the PVN (n = 184), approximately 20% expressed Fos protein in response to systemic GHRP-6 injection (Figure 3). These cells, however, only account for a very small proportion (3.8%) of the total number of Fos-positive cells detected (n = 979). Taken together with electrophysiological data, we can conclude that approximately 20–30% of arcuate cells projecting to the PVN are activated by GHRP-6. However, of the total number of cells activated by secretagogues, these comprise a very minor population. It would appear that the NPY population activated by the GH secretagogues do not form an important part of the arcuate-PVN NPY projection previously shown to be important for feeding and for stimulation of the hypothalamo-pituitary–adrenal axis.

Thus, we conclude that the GH secretagogues are centrally active compounds that act directly within the arcuate nucleus to activate cells in this region. The target cells for GH secretagogue action are a heterogenous population including both neuroendocrine cells (that are mainly excited by GH secretagogues) and non-neuroendocrine cells (the majority of which show an inhibitory response). The target cells for GH secretagogue action appear to include both GHRH and NPY cells in this region. Consistent with this hypothesis, GH secretagogue receptors have recently been demonstrated on a sub-population of GHRH neurones (43) and on NPY neurones (M. Wilsen and J. Rømer, Novo Nordisk, Copenhagen, personal communication). Clearly, however, GH secretagogue receptors are present in other parts of the CNS where no effect of these compounds on Fos protein expression has been detected. It will be interesting to discover what physiological role the GH secretagogues have at these sites.

ACKNOWLEDGEMENTS

We thank colleagues who participated in the research described in this chapter: Prof Gareth Leng, Dr Alex R.T. Bailey, Dr Kazu Honda, Dr Odile Viltart, Dr Richard E.J. Dyball, Dr Simon M Luckman, Mrs Lynn P Macdonald. Also, thanks to Prof Gareth Leng and Dr Adrian K. Hewson for reading and commenting on the manuscript.

REFERENCES

1. Guillemin, R., Brazeau, P., Bohlen, P., Esch, F., Ling, N., Wehrenberg, W.B. (1982) Growth hormone-releasing factor from a human pancreatic tumor that caused acromegaly. Science 218, 585–587.
2. Rivier, J., Spiess, J., Thorner, M.O., Vale, W.W. (1982) Characterization of a growth hormone-releasing factor from a human pancreatic islet tumour. Nature 300, 276–278.
3. Brazeau, P., Vale, W., Burgus, R., Ling, N., Butcher, M., Rivier, J., Guillemin, R. (1973) Hypothalamic polypeptide that inhibits the secretion of immunoreactive pituitary growth hormone. Science 179, 77–79.
4. Bowers, C.Y., Momany, F.A., Reynolds, G.A., Hong, A. (1984) On the *in vitro* and *in vivo* activity of a new synthetic hexapeptide that acts on the pituitary to specifically release growth hormone. Endocrinology 114, 1537–1545.
5. McCormick, G.F., Millard, W.J., Badger, T.M., Bowers, C.Y., Martin, J.B. (1985) Dose–response characteristics of various peptides with growth hormone-releasing activity in the unanesthetized male rat. Endocrinology 117, 97–105.
6. Dickson, S.L., Leng, G., Robinson, I.C.A.F. (1993) Systemic administration of growth hormone-releasing peptide activates hypothalamic arcuate neurons. Neuroscience 53, 303–306.
7. Clark, R.G., Carlsson, L.M.S., Trojnar, J., Robinson, I.C.A.F. (1989) The effects of a growth hormone-releasing peptide and growth hormone-releasing factor in conscious and anaesthetized rats. J. Neuroendocrinol. 1, 249–255.
8. Clark, R.G., Robinson, I.C.A.F. (1985) Growth hormone responses to multiple injections of a fragment of human growth hormone-releasing factor in conscious male and female rats. J. Endocrinol. 106, 281–289.
9. Bowers, C.Y., Sartor, A.O., Reynolds, G.A., Badger, T.M. (1991) On the actions of the growth hormone-releasing hexapeptide, GHRP. Endocrinology 128, 2027–2035.
10. Dickson, S.L., Leng, G., Dyball, R.E.J., Smith, R.G. (1995) Central actions of peptide and non-peptide growth hormone secretagogues in the rat. Neuroendocrinology 61, 36–43.
11. Howard, A.D., Feighner, S.D., Cully, D.F., Arena, J.P., Liberator, P.A., Rosenblum, C.I., Hamelin, M., Hreniuk, D.L., Palyha, O.C., Anderson, J., Paress, P.S., Diaz, C., Chou, M., Liu, K.K., McKee, K.K., Pong, S.S., Chaung, L.Y., Elbrecht, A., Dashkevicz, M., Heavens, R., Rigby, M., Sirinathsinghji, D.J.S., Dean, D.C., Melillo, D.G., Patchett, A.A., Nargund, R., Griffin, P.R., DeMartino, J.A., Gupta, S.K., Schaeffer, J.M., Smith, R.G., van-der-Ploeg L.H.T. (1996) A receptor in pituitary and hypothalamus that functions in growth hormone release. Science 273, 974–977.
12. Guan, X.M., Yu, H., Palyha, O.C., McKee, K.K., Feighner, S.D., Sirinathsinghji, D.J., Smith, R.G., van-der-Ploeg, L.H.T., Howard, A.D. (1997) Distribution of mRNA encoding the growth hormone secretagogue receptor in brain and peripheral tissues. Brain Res. Mol. Brain Res. 8, 23–29.

13. Bennett, P.A., Thomas, G.B., Howard, A.D., Feighner, S.D., van-der-Ploeg, L.H.T., Smith, R.G., Robinson, I.C.A.F. (1997) Hypothalamic growth hormone secretagogue-receptor (GHS-R) expression is regulated by growth hormone in the rat. Endocrinology 138, 4552–4557.
14. Bailey, A.R.T., Smith, L.C., Brown, C.H., Smith, R.G., Dickson, S.L. and Leng, G. (1998) Effects of chronic central growth hormone secretagogue infusion on the activation of arcuate nucleus neurones in the conscious male rat. J. Physiol. 506, 142P.
15. Dickson, S.L., Doutrelant-Viltart, O., McKenzie, D.N., Dyball, R.E.J. (1996) Growth hormone-releasing peptide activates rat arcuate neurones recorded *in vitro*. 10th International Congress of Endocrinology, San Francisco, USA. pp. 1–680.
16. Dickson, S.L., Leng, G., Robinson, I.C.A.F. (1994) Electrical stimulation of the rat periventricular nucleus influences the activity of hypothalamic arcuate neurones. J. Neuroendocrinol. 6, 359–367.
17. Clark, R.G., Carlsson, L.M.S., Rafferty, B., Robinson, I.C.A.F. (1988) The rebound release of growth hormone (GH) following somatostatin infusion in rats involves hypothalamic GH-releasing factor release. J. Endocrinol. 119, 397–404.
18. Okada, K., Wakabayashi, I., Sugihara, H., Minami, S., Kitamura, T., Yamada, J. (1991) Electrical stimulation of hypothalamic periventricular nucleus is followed by a large rebound secretion of growth hormone in unanesthetized rats. Neuroendocrinology 53, 306–312.
19. Koibuchi, N., Kato, M., Kakegawa, T., Suzuki, M. (1986) Growth hormone release induced by electrical stimulation of the basolateral amygdala, observed in pentobarbital anesthetized rats. Brain Res. 382, 104–108.
20. Merchenthaler, I. (1991) Neurons with access to the general circulation in the central nervous system of the rat: a retrograde tracing study with Fluorogold. Neuroscience 44, 655–662.
21. Dickson, S.L., Doutrelant-Viltart, O., Dyball, R.E.J., Leng, G. (1996) Retrogradely labelled neurosecretory neurones of the rat hypothalamic arcuate nucleus express Fos protein following systemic injection of GH-releasing peptide-6. J. Endocrinol. 151, 323–331.
22. Epelbaum, J., Moyse, E., Tannenbaum, G.S., Kordon, C., Beaudet, A. (1989) Combined autoradiographic and immunohistochemical evidence for an association of somatostatin binding sites with growth hormone-releasing factor nerve cell bodies in the rat arcuate nucleus. J. Neuroendocrinol. 1, 109–115.
23. Dickson, S.L., Doutrelant-Viltart, O., McKenzie, D.N., Dyball, R.E.J. (1996) Somatostatin inhibits arcuate neurones excited by GH-releasing peptide (GHRP-6) in rat hypothalamic slices. J. Physiol. 495P:109P.
24. Dickson, S.L., Viltart, O., Bailey, A.R.T., Leng, G. (1997) Attenuation of the growth hormone secretagogue induction of Fos protein in the rat arcuate nucleus by central somatostatin action. Neuroendocrinology 66, 188–194.
25. Zheng, H., Bailey, A.R.T., Jiang, M.H., Honda, K., Chen, H.Y., Trumbauer, M.E., van der Ploeg L.H., Schaeffer, J.M., Leng, G., Smith, R.G. (1997) Somatostatin receptor subtype 2 knockout mice are refractory to growth hormone-negative feedback on arcuate neurons. Mol. Endocrinol. 11, 1709–1717.
26. Dickson, S.L., Luckman, S.M. (1997) Induction of c-*fos* messenger ribonucleic acid in neuropeptide Y and growth hormone (GH)-releasing factor neurons in the rat arcuate nucleus following systemic injection of the GH secretagogue, GH-releasing peptide-6. Endocrinology 138, 771–777.
27. Guillaume, V., Magnan, E., Cataldi, M., Dutour, A., Sauze, N., Renard, M., Razafindraibe, H., Conte, D.B., Deghenghi, R., Lenaerts, V., Oliver, C. (1994) Growth hormone (GH)-releasing hormone secretion is stimulated by a new GH-releasing hexapeptide in sheep. Endocrinology 135, 1073–1076.
28. Meister, B. (1993) Gene expression and chemical diversity in hypothalamic neurosecretory neurons. Mol. Neurobiol. 7, 87–110.
29. Everitt, B.J., Meister, B., Hökfelt, T., Melander, T., Terenius, L., Rökaeus, Å., Theodorsson-Norheim, E., Dockray, G., Edwardson, J., Cuello, C., Elde, R., Goldstein, M., Hemmings, H., Ouimet, C., Wallas, I., Greengard, P., Vale, W., Weber, E., Wu, J.-Y., Chang, K.-J. (1986) The

hypothalamic arcuate nucleus-median eminence complex: immunohistochemistry of transmitters, peptides and DARPP-32 with special reference to co-existence in dopamine neurones. Brain Res. Rev. 11, 97–155.
30. McDonald, J.K., Lumpkin, M.D., Samson, W.K., McCann, S.M. (1985) Neuropeptide Y affects secretion of luteinizing hormone and growth hormone in ovariectomized rats. Proc. Natl. Acad. Sci. 82, 561–564.
31. Li, B.H., Xu, B., Rowland, N.E., Kalra, S.P. (1994) C-*fos* expression in the rat brain following central administration of neuropeptide Y and effects of food consumption. Brain Res. 65, 277–284.
32. Clarke, I., Jessop, D., Millar, R., Morris, M., Bloom, S., Lightman, S., Coen, C.W., Lew, R., Smith, I. (1993) Many peptides that are present in the external zone of the median eminence are not secreted into the hypophysial portal blood of sheep. Neuroendocrinology 57, 765–775.
33. Ciofi, P., Tramu, G., Bloch, B. (1990) Comparative immunohistochemical study of the distribution of neuropeptide Y, growth hormone-releasing factor and the carboxyterminus of precursor protein GHRF in the human hypothalamic infundibular area. Neuroendocrinology 51, 429–436.
34. Pierroz, D.D., Catzeflis, C., Aebi, A.C., Rivier, J.E., Aubert, M.L. (1996) Chronic administration of neuropeptide Y into the lateral ventricle inhibits both the pituitary–testicular axis and growth hormone and insulin-like growth factor I secretion in intact adult male rats. Endocrinology 137, 3–12.
35. Stanley, B.G., Kyrkouli, S.E., Lampert, S., Leibowitz, S.F. (1986) Neuropeptide Y chronically injected into the hypothalamus: a powerful neurochemical inducer of hyperphagia and obesity. Peptides 7, 1189–1192.
36. Locke, W., Kirgis, H.D., Bowers, C.Y., Abdoh, A.A. (1995) Intracerebroventricular growth-hormone-releasing peptide-6 stimulates eating without affecting plasma growth hormone responses in rats. Life Sci. 56, 1347–1352.
37. Okada, K., Ishii, S., Minami, S., Sugihara, H., Shibasaki, T., Wakabayashi, I. (1996) Intracerebroventricular administration of the growth hormone-releasing peptide KP-102 increases food intake in free-feeding rats. Endocrinology 137, 5155–5158.
38. Baker, R.A., Herkenham, M. (1995) Arcuate nucleus neurons that project to the hypothalamic paraventricular nucleus: neuropeptidergic identity and consequences of adrenalectomy on mRNA levels in the rat. J. Comp. Neurol. 358, 518–530.
39. Bai, F.L., Yamano, M., Shiotani, Y., Emson, P.C., Smith, A.D., Powell, J.F., Tohyama, M. (1985) An arcuato-paraventricular and -dorsomedial hypothalamic neuropeptide Y-containing system which lacks noradrenaline in the rat. Brain Res. 331, 172–175.
40. Thomas, G.B., Fairhall, K.M., Robinson, I.C.A.F. (1997) Activation of the hypothalamo–pituitary–adrenal axis by the growth hormone (GH) secretagogue, GH-releasing peptide-6, in rats. Endocrinology 138, 1585–1591.
41. Jacks, T., Hickey, G., Judith, F., Taylor, J., Chen, H., Krupa, D., Feeney, W., Schoen, W., Ok, D., Fisher, D., Wyvratt, M., Smith, R.G. (1994) Effects of acute and repeated intravenous administration of L-692,585, a novel non-peptidyl growth hormone secretagogue, on plasma growth hormone, IGF-1, ACTH, cortisol, prolactin, insulin, and thyroxine levels in beagles. J. Endocrinol. 143, 399–406.
42. Honda, K., Bailey, A.R.T., Bull, P.M., Macdonald, L.P., Dickson, S.L., Leng, G. (1999) An electrophysiological and morphological investigation of the projections of GHRP-6-responsive neurons in the rat arcuate nucleus to the median eminence and to the paraventricular nucleus. Neuroscience 90, 875–883.
43. Tannenbaum, G.S., Lapointe, M., Beaudet, A., Howard, A.D. (1998) Expression of GH secretagogue-receptors by GH-releasing hormone neurones in the mediobasál hypothalamus. Endocrinology 139, 4420–4423.

Growth Hormone Secretagogues
Edited by E. Ghigo, M. Boghen, F.F. Casanueva and C. Dieguez

Chapter 9

Mechanisms of Actions of Growth Hormone-Releasing Peptides and their Analogues In Vivo

CHARLES OLIVER[1,2], FRÉDÉRIC DADOUN[1,2], NATHALIE BRIARD[2], VIVIANE GUILLAUME[1,2], NICOLE SAUZE[2], MICHEL GRINO[2] and ANNE DUTOUR[1,2]

[1]*Service d'Endocrinologie, Maladies Métaboliques et de la Nutrition, Hôpital Nord, 13915 Marseille Cedex 20, France*
[2]*Laboratoire de Neuroendocrinologie Expérimentale, INSERM U 501, Faculté de Médecine, 13916 Marseille Cedex 20, France*

INTRODUCTION

Growth hormone (GH) secretion is mainly controlled by two hypothalamic peptides: GH releasing hormone (GHRH), synthesized in neurons of the arcuate nucleus, and somatostatin (somatotropin release-inhibiting hormone, SRIH) mainly synthesized in neurons of the periventricular nucleus, which respectively stimulate and inhibit GH secretion (1). More recently, a class of synthetic molecules, denominated growth hormone secretagogues (GHS), was shown to stimulate GH release. GHS include a group of synthetic oligopeptides termed GH releasing peptides or GHRPs ($GHRP_1$, $GHRP_2$, $GHRP_6$ and its 2-methyl-D-Trp derivative, hexarelin) and non-peptidyl pharmacologic analogues which appeared more recently (L-163,191- L-692,429 - L-692,585, L-700,653 and MK-0677) (2,3). GHS stimulate GH secretion in numerous species including rat, sheep, pig, chicken and human, through a direct action on the pituitary different from that of endogenous GHRH. They bind to a recently cloned specific G-protein coupled receptor and act in synergism with GHRH as demonstrated both *in vivo* and *in vitro*. This receptor is expressed in the pituitary and the hypothalamus supporting the hypothesis that both structures are involved in GHS stimulation of GH release. Indeed, GH response to GHS appears to be complex and depends upon effects at both pituitary and hypothalamic sites. These compounds have been extensively investigated and have been proven to be of clinical interest (2,3).

In the following review, we will discuss the sites and mechanisms of action of GHS on GH release *in vivo*. Their actions on other hormonal systems will also be reviewed. Most studies were performed using GHRP-6 and hexarelin; similar results were shown with other GHRPs and non peptidyl analogues. Such studies are noteworthy with respect to the potential therapeutic applications of these substances. Indeed, a proper understanding of GHS effects and mechanisms of action, especially at the hypothalamic level, is relevant to clinical practice not only for appropriate therapeutic indications in patients with hypothalamic-pituitary diseases, but also in order to be aware of potential side-effects on other endocrine systems and to be able to prevent them.

ACUTE EFFECTS OF GH SECRETAGOGUES ON GH AXIS

Effect of GHS on GH secretion

Administration of GHS induces a rapid increase in plasma GH levels in a broad range of animal species i.e. rat, monkey, sheep, pig, chick, steer as well as in human. *In vivo* activities of GHRPs and GHRH have been compared. However comparison is intricate and depends upon many variables, such as structural form of GHS used, route of administration, age, sex, experimental procedure (anaesthesia, stress...), and animal specie (4). However, GHRP-6 was found more efficient in primates than in rat, dog or farm animals. Walker et al. demonstrated that GHRP-6 was 5–20 times more effective in monkey (*Macaca fasicularis*) than in rat or dog (5); in another strain of monkeys (*Cynomolgus macaque*) GHRP-6 potency was lesser ($ED_{50} > 90$ µg/kg body weight, b.w.) than that of GHRH (ED_{50}: 3.6 µg/kg); GHRP-6 and GHRH dose–response curves showed no parallelism, but unlike in the rat, GHRP-6 was able to evoke a much higher GH peak response than GHRH (> 55 vs 12 µg/l) (6). In human, GHRP-6 is more potent than in other species; an i.v. bolus injection of 1 µg/kg b.w. GHRP-6 induced a greater GH release than GHRH, using the same dose and route of administration (7).

The initial comparative analysis of GHRH and GHRPs effects on GH secretion suggested that their mechanisms of action were both different and complementary. Evidence for different mechanisms of action derives from the ability of GHRPs and GHRH to increase GH release beyond the maximum capacity of the other (7). Besides, homologous but not heterologous desensitization is observed after continuous infusion of GHRH or GHRP-6, followed by acute administration of one of these peptides (8). Different mechanisms of action were confirmed through identification of different receptors and signalling for GHS and GHRH. In human, complementary effects of GHS and GHRH were found more striking *in vivo* than *in vitro*. Using maximally or submaximally effective doses of GHRP-6 and GHRH, GH secretory responses *in vivo* were potentiated rather than additive (7). Results obtained *in vitro* with the association of both substances are conflicting, showing either merely additive effects on rats (9) and ovine pituitary cells (10), or direct synergism with GHRH on rat pituitary cells for both GHRP-6 (11) and L-682,429 (12). As discussed later on, the hypothalamus has been involved in the synergism between GHS and GHRH.

Conversely, a functional hypothalamic–pituitary GHRH system is needed for the GH stimulating release of GHRP. Indeed, passive immunization against GHRH was shown to inhibit GHRP-6 induced GH release (13). Functionally intact GHRH receptors are required for pituitary action of GHRP-6, as shown by the lack of GHRP-6 evoked GH stimulation in GH deficient dwarf lit/lit mouse, whose GHRH receptor bears a point mutation in the N-terminal ligand-binding domain (14). In nine healthy 20–30 year-old men, Pandya et al. recently showed that the administration of a specific GHRH antagonist 20 minutes before i.v. injection of GHRP-6 severely blunts GH response to GHRP-6 (area under the curve: 376 ± 113 vs 1701 ± 278 μg/l/min when saline was injected instead of GHRH antagonist); these data show that endogenous GHRH is necessary for most of GH response to GHRP-6 in human (15). These results diverge from those reported earlier by the same group in human using a different paradigm; a model of complete pituitary desensitization was designed to suppress involvement of endogenous GHRH; after a short term infusion of GHRH resulting in complete pituitary desensitization to a maximally effective dose of GHRH (1 μg/kg), GH rise in response to a bolus dose of GHRP-6 was fully preserved (16). In agreement with the findings of Pandya et al., acute GHRP administration had only limited efficiency in individuals with GHRH deficiency (17). Since cultured pituitary cells respond to GHRP-6, it was first considered that GHS act primarily as a direct GH secretagogue at the pituitary level (18). The same group later demonstrated that GH response to GHRP-6 was higher in an hypothalamic–pituitary incubate than in an isolated pituitary incubate; these data support the concept that GHS have a hypothalamic as well as a direct pituitary site of action (19). Furthermore, GHS show equal potency with GHRH on GH release *in vitro* whereas *in vivo*, GH secretagogues are more efficient than GHRH in elevating plasma GH; this suggests that the pituitary gland is not the sole site of action of GHS (20). GH response to intracerebroventricular (i.c.v.) GHRP-6 injection was higher than that observed after systemic administration of the same dose (21), supporting the assumption of hypothalamic action of GHS. Anaesthesia reduced the amplitude of GH response to GHS providing another evidence for extrapituitary action of GHS; indeed, GH response to GHRP-6 is much smaller in urethane-anaesthetized than in conscious rat (13). Nevertheless, acute i.v. injection of GHRP-6 was still able to evoke GH release in hypothalamopituitary disconnected (HPD) sheep (wethers and ewes) indicating a pituitary site of action for this peptide; as expected, GHRP-6 was less potent than GHRH; indeed, response to GHRP-6 was 5-fold smaller in intact animals and 15-fold smaller in HPD animals; this difference may be explained by a stimulating effect of GHRP-6 on GHRH neurons and suggests that a component of GHRP-6 action is mediated through the hypothalamus (22). Similar findings have been reported with L-692,585 in rat (23) and pig (24). In a group of 12 patients with hypothalamopituitary disconnection, GHRP-6 induced GH release was 15-fold lower than in control subjects; in these patients, GH response to GHRH was similar to that obtained in controls; the authors concluded that the potency of GHRP-6 action at the pituitary level is minimal and that its main action is mediated by hypothalamic structures (25). Similar data have been reported by Hayashi et al. (26).

Probable GHRH mediated GHS effects upon the hypothalamic–pituitary system raise two questions:

- What is the target of GHS in the hypothalamus: GHRH neurons, somatostatin neurons or other groups of neurons?
- What is the relative importance of pituitary and hypothalamic actions of GHS?

Several experimental approaches have been used to address these questions. We will refer mostly to animal studies although human studies providing useful information on the mechanisms of action of GHS will also be cited.

Action of GHS at the level of GHRH neurons

Several studies support this hypothesis. In rats, Dickson et al. showed that systemic administration of GHRP-6 (100 µg i.v.) activates a subpopulation of neurons in the arcuate nucleus of the hypothalamus which contains most of the GHRH neurons; increased Fos-like immunoreactivity (a marker of neuronal activation) was detected 90 min after GHRP-6 injection in many cells throughout the ventrolateral regions of the arcuate nucleus; this effect is highly specific since only a slight increase in Fos-like immunoreactivity was observed in the supraoptic nucleus and no change was seen in other hypothalamic nuclei studied (ventromedial, periventricular and paraventricular nuclei); acute i.v. injection of GHRP-6 using the same dose, stimulated the firing of putative GHRH neurons in the arcuate nucleus; this excitation began one minute after GHRP administration and lasted for at least 10 min (27). This response does not reflect a feedback effect involving increased GH release or increased plasma level of GHRH, since administration of a high dose of GHRH had no effect on c-fos expression in the arcuate nucleus (28). In the lit/lit dwarf mouse which lacks functional pituitary GHRH receptors, administration of GHRP-6 evoked arcuate nucleus neurons activation, demonstrating that central actions of GHRP-6 are not mediated by GHRH, GH or IGF-1 (14). Dense Fos nuclear immunostaining was induced throughout the ventral part of this nucleus when GHRP-6 was given i.c.v., using a much lower dose (0.1 µg/rat) than the dose required to stimulate GH secretion when i.v. injection is used. Similar results were obtained with L-692,585 and L-682,429 (28). Although GHRH neurons constitute a major subgroup in this area, the arcuate nucleus is heterogeneous. Indeed, it contains several groups of neurons which project either to the median eminence and portal primary plexus (neuroendocrine cells) or to other hypothalamic or extrahypothalamic brain structures. Besides GHRH, several peptides and monoamines have been detected, neuropeptide Y (NPY), proopiomelanocortin (POMC) and dopamine being quantitatively the most important ones (29). Effect of intravenous GHRP-6 on electrical activity of arcuate neurons was different in two subpopulations of cells: predominantly excitatory for putative neuroendocrine cells and inhibitory for the remaining unidentified cells (28). Further characterization of the subpopulation of arcuate neurons stimulated by GHRP-6 has been performed using the retrograde tracer Fluorogold which identifies neurosecretory neurons. Between 68% and 82% of the arcuate neurons expressing c-fos protein following the i.v. injection of GHRP-6 are presumably neurosecretory neurons. The majority of these cells were not identified as tyrosine hydroxylase positive (involved in dopamine biosynthesis) or β-endorphin-containing cells (30). In another study, neurochemically identifiable cells expressing c-fos mRNA were shown to coexpress NPY mRNA (51 ± 4%), GHRH mRNA (23 ± 1%) tyrosine hydroxylase mRNA (11 ± 3%), POMC mRNA (11 ± 2%) or somatostatin mRNA (4 ± 1%) (31).

I.c.v. and i.v. injection of MK-0677 induced Fos-like immunoreactivity within the ventromedial region of the arcuate nucleus in conscious male rats. Neurons activated by MK-0677 were confined close to the wall of the third ventricule whereas GHRP-6 induced Fos-like immunoreactivity in the same area as well as in more dorsal and lateral regions of this nucleus. Therefore, GHRP-6 may activate a broader variety of hypothalamic neurons than MK-0677; this observation might explain increased food intake observed with

GHRP-6, but not with MK-0677. In urethane-anaesthetized rats, systemic injection of MK-0677 increased the electrical activity of the same population of arcuate neuroendocrine cells than GHRP-6, this effect being inhibited by the administration of SRIH (32). In addition, it was shown that GHRP-6-induced activation of arcuate nucleus neurons is blunted by prior central administration of a SRIH analog (33). These data confirm GHS action on arcuate neurons involved in the regulation of GH release.

In sheep, we were able to demonstrate GHRH release *in vivo* after GHRPs injection; acute i.v. injection of hexarelin (1 mg) to adult rams induced a significant 2.5-fold enhancement of GHRH release into hypophysial portal blood (HPB) which lasted 45 min; in these animals, a 2.3-fold increase in plasma GH was observed and this GH rise was still detected 60 min after hexarelin injection; SRIH levels in HPB did not change throughout the study (Figure 1); the magnitude of GHRH increase after acute hexarelin administration was similar to that observed after other pharmacological stimulations of GH release, suggesting that the GHRH rise after hexarelin injection may be sufficient to account for GH stimulation (34). In another study, Fletcher et al. gave to conscious ewes a GHRP-6 bolus

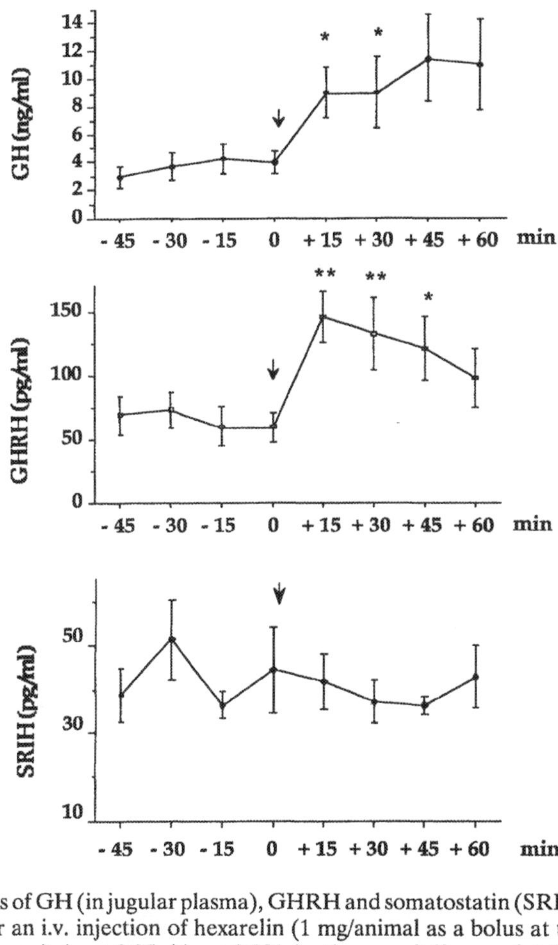

Figure 1. Acute responses of GH (in jugular plasma), GHRH and somatostatin (SRIH) in hypophysial portal plasma in five rams after an i.v. injection of hexarelin (1 mg/animal as a bolus at the time indicated by an arrow). $^{*}p < 0.05$; $^{**}p < 0.001$ (vs time zero) (from ref. 34).

injection (10 μg/kg) followed by a 2-hour GHRP-6 infusion (0.1 μg/kg/hr) and measured GHRH and SRIH secretion in HPB; a 5.3-fold increase in plasma GH levels was observed 5–10 min after the GHRP-6 bolus injection, without a significant coincident release of GHRH; during the infusion period, there was a significant 50% increase in GHRH pulse frequency without any change in GHRH pulse amplitude; mean portal SRIH concentrations, pulse frequency and amplitude were unchanged; the authors concluded that GHRP-6 acts at the hypothalamic level or higher centres of the brain; however, under these experimental conditions, GH secretory response to GHRP-6 injection does not appear to be the result of GHRP-6 action on GHRH or SRIH hypothalamic neurons (35). The difference between Guillaume's and Fletcher's studies may be explained by the sex of the animals (rams are more responsive to GHRPs than ewes) and by the greater potency of hexarelin. Using *in situ* hybridization, prominent expression of GHS receptor in the rat arcuate and ventromedial nucleus was demonstrated, supporting a direct action of GHS at the level of GHRH neurons in the hypothalamic nucleus; abundance of GHS receptors was higher in the hypothalamus than in the anterior pituitary gland (36,37).

Action of GHS on SRIH neurons

From the studies mentioned above, a general consensus emerged that GHRH is integrally involved in GHS mechanisms of action. By contrast, studies questioning GHS putative influence on SRIH neurons are far less conclusive. Indeed, no change in SRIH release into HPB has been observed in both studies performed in sheep (34,35). GHS receptor expression was either barely (36) or not (37) detectable in neurons of the periventricular nucleus, the major source of SRIH released into HPB and no increase in Fos immunoreactivity was detected in these neurons following GHRP-6 injection (27).

Other studies performed *in vivo* suggest that GHS may inhibit SRIH release. Two observations issued from the extensive work of Clark et al. (13) in conscious male rats suggest an effect of GHRPs on endogenous SRIH secretion:

- GHRP-6 infusion induced increased GH release associated with a disruption of normal GH pulsatility which did not resume for at least 2 h after stopping GHRP-6 infusion. This effect may be explained by an alteration of endogenous SRIH release which is thought to underlie rhythmic GH pulsatility in male rat.
- GHRP-6 infusion abolished the cyclic refractoriness to repetitive GHRH injections. A similar abolition of intermittent responsiveness to GHRH has been obtained when rats were passively immunized against SRIH.

Involvement of SRIH has also been addressed in two studies performed in rats pretreated with anti-SRIH antiserum. It was assumed that if GHS release GH through inhibition of SRIH secretion or action, SRIH antiserum pretreatment would not further increase the GH response to GHS. In one study, an augmented response to i.v. GHRP-6 was demonstrated following immunoneutralization of SRIH suggesting that SRIH is not involved in GHRP stimulation of GH release (19). Opposite results were obtained in another study; indeed, in freely moving rats, GHRP-6 induced GH release was not further increased by previous administration of anti-SRIH antiserum. These last data suggest that

GHRP-6 suppresses the somatostatinergic tone. These authors proposed that the discrepancy between both results is related to the use of different paradigms: stress-free conditions in conscious rats in their study as opposed to stressful conditions known to increase somatostatinergic tone in the former study, where therefore, SRIH immunoneutralization might not have been sufficient (38).

Altogether, these studies bring no definite proof of an effect of GHS on GH secretion through decreased SRIH secretion into HPB. Nevertheless, an antagonistic action of GHS on SRIH inhibitory influence at the hypothalamic level was demonstrated; i.c.v. pretreatment with octreotide blocked GH-stimulating effect of i.c.v. (but not i.v.) administration of GHRP-6 (21). these findings support an action of GHS on SRIH neurons set in the intrahypothalamic neuronal network involved in GH neuroregulation.

Action of GHS on the secretion of another putative hypothalamic hormone

Several experimental and clinical studies suggest that GHS may influence the release into HBP of another hypothalamic hormone than GHRH or SRIH. In a study by Bowers et al. (18), effects of GHRP-6 and GHRH on rat pituitary cell cultures were additive with only a small synergistic effect. *In vivo*, a 2.5–3-fold synergism was observed with GHRP plus GHRH which cannot be simply explained by an interaction of these peptides at the pituitary level; since GHRP-6 and GHRH synergistic action on GH release was not explained by inhibition of SRIH or stimulation of GHRH (indeed, synergism was observed between GHRP-6 and supramaximal doses of GHRH), these authors suggested that GHRP-6 releases an unknown hypothalamic factor (U-factor) which may be an endogenous GHS receptor ligand interacting with GHRH on pituitary cells to release GH synergistically. A study performed in monkey by Malozowki et al. (6) supports the same hypothesis; indeed, GHS produced a greater maximal response than GHRH suggesting that a factor other than GHRH may be involved in its action either at the pituitary or hypothalamic level; pretreatment with propranolol, which is assumed to inhibit SRIH release, enhanced GH response to GHRP-6 suggesting that GHRP-6 does not affect SRIH tone; this interpretation conforms with previously discussed experiments performed in sheep, showing no change in portal SRIH levels following hexarelin or GHRP-6 administration (34,35). In human, GHRP-6 and hexarelin were able to potentiate GH release in response to a maximally stimulating dose of GHRH (7,39); stimulation of hypothalamic U-factor secretion may account for this phenomenon. Other clinical findings support this hypothesis. In human, the GH-releasing activity of hexarelin or GHRP-6 is partly refractory to the infusion of SRIH (39) or to pharmacological manipulations (administration of muscarinic antagonist or β-2 adrenergic agonist) which are thought to inhibit GH secretion through SRIH release (40). SRIH and GHS may act as mutual functional antagonists at the pituitary and/or hypothalamic level (21) and involve the U-factor or an endogenous GHS receptor ligand.

Interaction between GHS and brain neurotransmitters

Activity of hypothalamic neurons involved in the control of GH secretion is influenced by several neurotransmitters. Among them, cholinergic and adrenergic pathways have been shown to play a major role. GH response to GHRPs was potentiated by pyridostigmine, a cholinesterase inhibitor, in dog (41), as well as propranolol, a β-adrenoreceptor antagonist, in monkey (6). It is assumed that both drugs stimulate GH secretion through inhibition of somatostatinergic tone although controversy remains in the mechanism of action of cholinergic drugs (42). These data conform with a hypothalamic site of action of GHRPs although they do not identify which neuronal group(s) mediate(s) stimulation of GH release.

Although GHRPs derive from met-enkephalin, their effects on GH release are not mediated by opioid receptors since they are not altered by naloxone and they are synergistical with dermorphin and met-enkephalin analogues (18).

CHRONIC EFFECTS OF GH SECRETAGOGUES ON GH AXIS

Previous *in vitro* and *in vivo* studies in rat demonstrated that continuous exposure to GHRPs results in progressive attenuation of GH response (4). During GHRP-6 infusion in rats, GH remained elevated above spontaneous baseline and the normal GH pulsatile secretory pattern was disrupted; at the end of GHRP-6 infusion, plasma GH levels fell without resumption of normal pulsatile GH (13). Similar results were obtained in human. Number, duration and height of GH pulses, incremental pulse amplitude, interpeak valley concentration and individual pulse areas were significantly greater during GHRP-6 infusion than during saline administration; as a consequence, plasma IGF-1 increased significantly; at the end of the infusion, GH response to a subsequent GHRP-6 bolus injection was significantly reduced; this attenuation of GH response was not caused by depletion of pituitary GH stores, since the response to GHRH bolus was enhanced by prior infusion of GHRP-6 (8). Daily oral administration of MK-0677 for 1 week increased circulating IGF-1 levels together with an enhancement of GH pulse frequency, but without detectably increased GH secretion (43).

Effects of chronic administration of GHRP on hypothalamic structures received little attention. In dwarf (dw/dw) female rats treated with GHRP-6 (1 mg/kg per 24 h) continuously for 14 days, a significant selective increase of GHRH mRNA in the posterior arcuate nucleus was seen, but no significant effect was observed in neurons of the anterior or ventromedial parts of the same nucleus. In addition, SRIH mRNA levels in the posterior periventricular nucleus were decreased (44).

OTHER ENDOCRINE EFFECTS OF GH SECRETAGOGUES

It was early recognized in clinical studies that GHRP-6 also stimulates both prolactin and cortisol release. Bowers et al. observed a 2-fold rise in serum cortisol and prolactin levels after i.v. injection of GHRP-6, using a dose of 1μg/kg; this increase was moderate as compared with the 120-fold increase in serum GH levels, but significant and unlikely related to stress effects; no change in LH and TSH was observed (7). Massoud et al. reported in human the dose–response curves for GH, prolactin and cortisol following i.v. injection of increasing doses (0.125–1 μg/kg) of hexarelin; GH dose–response curve reached a plateau with a dose of 1 μg/kg of hexarelin; GH maximal increase was approximately 100-fold; prolactin dose–response curve paralleled that of GH but prolactin increase was much smaller (approximately 80% increase); serum cortisol concentrations also increased moderately, but significantly; this increase of cortisol was not observed in all subjects; a maximal increase of approximately 40% in cortisol levels occurred with a dose of 0.5 μg/kg; no change in TSH, insulin or blood sugar levels was observed in this study (45). In monkey, no change in prolactin and TSH plasma levels was observed after GHRP-6 injection (6). Small, but consistent elevations in ACTH and cortisol secretion were seen in other human studies as well as in animal models. Acute administration of L-692,585 stimulated ACTH and cortisol secretion in beagles; however, increment of both hormones levels was far less than that of GH level; indeed, increase in cortisol was 2- to 3-fold as compared to 10- to 20-fold for GH; therefore, ACTH and cortisol stimulation following administration of L-682,585 did not induce a maximal adrenal response but rather approximated an endogenous pulse (46); increase in cortisol secretion was similar in magnitude using L-692,429, another non-peptidic GHS (47). GHS stimulatory effect on the hypothalamic–pituitary–adrenal axis seems however transient and may not constitute an important side-effect during chronic treatment with these molecules (43).

GHRP-6 mechanisms of action on the pituitary–adrenal axis have been investigated in animal studies. Several lines of evidence suggest a hypothalamic site of action for GHRPs. GHRPs do not directly stimulate glucocorticoid release from the adrenal glands and ACTH secretion from the pituitary gland. They do not synergize with GHRH to release more ACTH *in vivo* as they do to release GH. Furthermore, ACTH response to GHRP-6 injection was abolished in rats with transected pituitary stalk (48). Therefore, GHRPs and their analogues probably interact with the hypothalamic peptidergic systems controlling ACTH release such as corticotropin releasing hormone (CRH) and arginine vasopressin (AVP). Thomas et al. indirectly tested this hypothesis in rat and measured plasma ACTH levels after GHRP-6, CRH or AVP, alone or combined; GHRP-6 given together with CRH did not increase ACTH levels beyond its response to CRH alone whereas association of GHRP-6 and AVP markedly increased ACTH levels as compared with the effects of AVP alone (× 2.4); these data suggest that GHRP-6 acts on the hypothalamus to stimulate ACTH release; this effect is probably mediated at least partly by release of CRH. GHS effect on ACTH is regulated by glucocorticoids. Indeed, Thomas et al. also found that GHRP-6 induced ACTH release was higher in animals with the lowest basal ACTH and corticosterone output; these findings are probably related to decreased glucocorticoid feedback on hypothalamic CRH neurons (48). However, in human, hexarelin showed no synergistic effect with either AVP or CRH, suggesting that the ACTH-releasing activity of GHS may be, at least partly independent of both CRH and AVP (49). In our laboratory, using the sheep model, a rise in CRH levels in HPB after GHRP-6 i.v. injection (2 mg/animal) was recently confirmed; in four animals, we observed a 2-fold increase in CRH levels; AVP release into HPB showed a 1.6-fold increase (G. Thomas, V. Guillaume, I. Robinson, C. Oliver, manuscript in preparation), suggesting AVP involvement in sheep, in contrast to what is assumed in rat (48). CRH and AVP are both synthesized in neurons of the paraventricular nucleus (PVN), where GHS receptors have recently been identified using *in situ* hybridization (36,37). However, expression of GHS receptor is much higher in the arcuate nucleus; many of the arcuate neurons are NPY positive and it is known that PVN receives major projections from arcuate NPY neurons; it is therefore conceivable that GHRPs action on CRH and AVP neurons is indirectly driven, through NPY arcuate neurons.

ROLE OF A PUTATIVE ENDOGENOUS GHS SYSTEM IN GH REGULATION

In mammals, GH secretion is pulsatile and influenced by various conditions, including stress, feeding and pharmacological manipulations, which involve central neurotransmitters. Based on experiments performed in male rat, it is generally believed that both GH secretion pulsatile pattern and GH response to physiological or pharmacological stimuli depend upon the exquisite interrelationship between GHRH and SRIH secretion and pituitary action (1). However, inter-species differences in GH secretion and neuroregulation have been demonstrated. Indeed, in sheep no obvious correlation was found between most GH peaks and simultaneous increase in GHRH release and decrease in SRIH release into HPB. Besides, in several species, a supramaximal dose of GHRH and several GH stimulating factors (e.g. cholinergic drugs, clonidine) have synergistic effects on GH stimulation. It is assumed that a reduction of SRIH release mediates cholinergic drugs and clonidine effects on GH secretion; however, no change in SRIH levels was detected in sheep HPB, following administration of these substances (42). It is tempting to speculate that a natural ligand for GHS receptors is involved in GH regulation together with both GHRH and SRIH.

The presence of such an endogenous ligand was recently reported in portal plasma of ovariectomized ewes using an *in vitro* assay (based on intracellular calcium responses by HEK-293 AEO cells expressing the recombinant porcine GHS receptor); biological activity of this putative natural ligand in portal plasma was found to be correlated with GH peaks (50). GHS may potentiate GH release and contribute to GH neuroregulation through another mechanism; indeed, Kamegai et al. (51) suggested that GHS-induced GHRH secretion may stimulate GHS receptor expression (51).

GHS receptor expression in the rat arcuate and ventromedial nucleus is highly sensitive to GH, being markedly increased in the dw/dw dwarf strain and decreased after chronic (6 days) treatment with bovine GH. Expression of GHS receptor was unaltered by continuous s.c. infusion of GHRP-6. These data suggest that hypothalamic GHS receptor is involved in feedback regulation of GH, adding some evidence for the participation of an endogenous GHS receptor ligand in the regulation of GH secretion (37).

CONCLUSION

There is now clear evidence that GHS stimulate GH release through a dual action on the pituitary and the hypothalamus. GHRH neurons are presumably the main targets of GHS. However, participation of SRIH and of a putative endogenous GHS receptor ligand cannot be excluded. The relative contribution of the pituitary and the hypothalamus in the GHS-induced GH release is still unknown.

ACKNOWLEDGEMENT

The authors thank Ms Patricia Braccini for her skilled editorial assistance.

REFERENCES

1. Müller, E.E. (1987) Neural control of somatotropic function. Physiol. Rev. 67, 962–1053.
2. Ghigo, E., Arvat, E., Muccioli, G., Camanni, F. (1997) Growth hormone-releasing peptides. Eur. J. Endocrinol. 136, 445–460.
3. Smith, R.G., Van Der Ploeg, L.H.T., Howard, A.D. et al. (1997) Peptidomimetic regulation of growth hormone secretion. Endocr. Rev. 18, 621–645.
4. Sartor, O., Bowers, C.Y., Reynolds, G.A., Momany, F.A. (1985) Variables determining the growth hormone response of His-D-Trp-Ala-Trp-D-Phe-Lys-NH_2 in the rat. Endocrinology 117, 1441–1447.
5. Walker, R.F., Codd, E.E., Barone, F.C., Nelson, A.H., Goodwin, T., Campbell, S.A. (1990) Oral activity of the growth hormone releasing peptide His-D-Trp-Ala-Trp-D-Phe-Lys-NH_2 in rats, dogs and monkeys. Life Sci. 47, 29–36.
6. Malozowski, S., Hao, E.H., Ren, S.G. et al. (1991) Growth hormone (GH) responses to the hexapeptide GH-releasing peptide and GH-releasing hormone (GHRH) in the cynomolgus macaque: evidence for non-GHRH-mediated responses. J. Clin. Endocrinol. Metab. 73, 314–317.
7. Bowers, C.Y., Reynolds, G.A., Durham, D., Barrera, C.M., Pezzoli, S.S., Thorner, M.O. (1990) Growth hormone (GH)-releasing peptide stimulates GH release in normal men and acts synergistically with GH-releasing hormone. J. Clin. Endocrinol. Metab. 70, 975–982.
8. Huhn, W.C., Hartman, M.L., Pezzoli, S.S., Thorner, M.O. (1993) Twenty-four-hour growth hormone (GH)-releasing peptide (GHRP) infusion enhances pulsatile GH secretion and specifically attenuates the response to a subsequent GHRP bolus. J. Clin. Endocrinol. Metab. 76, 1202–1208.
9. Sartor, A.O., Bowers, C.Y., Chang, D. (1985) Parallel studies of His-D Trp-Ala-Trp-D Phe-Lys-NH_2 and human pancreatic growth hormone releasing factor 44-NH_2 in rat pituitary cell monolayer culture. Endocrinology 116, 952–957.
10. Wu, D., Chen, C., Zhang, J., Katoh, K., Clarke, I.J. (1994) Effects *in vitro* of new growth hormone releasing peptide (GHRP-1) on growth hormone secretion from ovine pituitary cells in primary culture. J. Neuroendocrinol. 6, 185–190.
11. Cheng, K., Chan, W.W., Barreto, A. Jr., Convey, E.M., Smith, R.G. (1989) The synergism effects of His-D-Trp-Ala-Trp-D-Phe-Lys-NH_2 on growth hormone (GH)-releasing factor-stimulated GH release and intracellular adenosine 3′,5′-monophosphate accumulation in rat pituitary cell culture. Endocrinology 124, 2791–2798.
12. Cheng, K., Chan, W.W., Butler, B. (1993) Stimulation of growth hormone release from rat primary pituitary cells by L-692,429, a novel non peptidyl GH secretagogue. Endocrinology 132, 2729–2731.
13. Clark, R.G., Carlsson, L.M.S., Trojnar, J., Robinson, I.C.A.F. (1989) The effects of a growth hormone-releasing peptide and growth hormone-releasing factor in conscious and anesthetized rats. J. Neuroendocrinol. 1, 249–255.
14. Dickson, S.L., Doutrelant-Viltart, O., Leng, G. (1995) GH-deficient dw/dw rats and lit/lit mice show increased Fos expression in the hypothalamic arcuate nucleus following systemic injection of GH-releasing peptide-6. J. Endocrinol. 146, 519–526.
15. Pandya, N., De Mott-Friberg, R., Bowers, C.Y., Barkan, A.L., Jaffe, C.A. (1998) Growth hormone (GH)-releasing peptide-6 requires endogenous hypothalamic GH-releasing hormone for maximal GH stimulation. J. Clin. Endocrinol. Metab. 83, 1186–1189.
16. Robinson, B.M., De Mott-Friberg, R., Bowers, C.Y., Barkan, A.L. (1992) Acute growth hormone (GH) response to GH-releasing hexapeptide in humans is independent of endogenous GH-releasing hormone. J. Clin. Endocrinol. Metab. 75, 1121–1124.

17. Bowers, C.Y., Alster, D.K., Frentz, J.M. (1992) The GH-releasing activity of a synthetic hexapeptide in normal men and short stature children after oral administration. J. Clin. Endocrinol. Metab. 74, 292–298.
18. Bowers, C.Y., Momany, F., Reynolds, G.A., Hong, A. (1984) On the *in vitro* and *in vivo* activity of a new synthetic hexapeptide that acts on the pituitary to specifically release growth hormone. Endocrinology 114, 1537–1545.
19. Bowers, C.Y., Sartor, A.O., Reynolds, G.A., Badger, T.M. (1991) On the actions of the growth hormone-releasing hexapeptide, GHRP. Endocrinology 128, 2027–2035.
20. Smith, R.G., Cheng, K., Pong, S.S. et al. (1996) Mechanism of action of GHRP-6 and nonpeptidyl growth hormone secretagogues. In: Growth Hormone Secretagogues. B.B. Bercu and R.F. Walker (eds). Springer-Verlag, New York, pp. 147–163.
21. Fairhall, K.M., Mynett, A., Robinson, I.C.A.F. (1995) Central effects of growth hormone-releasing hexapeptide (GHRP-6) on growth hormone release are inhibited by central somatostatin action. J. Endocrinol. 144, 555–560.
22. Fletcher, T.P., Thomas, G.B., Willoughby, J.O., Clarke, I.J. (1994) Constitutive growth hormone secretion in sheep after hypothalamopituitary disconnection and the direct *in vivo* pituitary effect of Growth Hormone-Releasing Peptide 6. Neuroendocrinology 60, 76–86.
23. Mallo, F., Alvarez, C.V. Benitez, L. et al. (1993) Regulation of His-d Trp-Ala-Trp-d Phe-Lys NH_2 (GHRP-6)-induced GH secretion in the rat. Neuroendocrinology 57, 247–256.
24. Hickey, G.J., Drisko, J., Faidley, T. et al. (1996) Mediation by the central nervous system is critical to the *in vivo* activity of the GH secretagogue L-692585. J. Endocrinol. 148, 371–380.
25. Popovic, V., Damjanovic, S., Micic, D., Djurovic, M., Dieguez, C., Casanueva, F.F. (1995) Blocked growth hormone-releasing peptide (GHRP-6)-induced GH secretion and absence of the synergic action of GHRP-6 plus GH-releasing hormone in patients with hypothalamo-pituitary disconnection: envidence that GHRP-6 main action is exerted at the hypothalamic level. J. Clin. Endocrinol. Metab. 80, 942–947.
26. Hayashi, S., Kaji, H., Ohashi, S., Abe, H., Chihara, K. (1993) Effect of intravenous administration of growth hormone-releasing peptide on plasma growth hormone in patients with short stature. Clin. Pediatr. Endocrinol. 2(suppl 2), 69–74.
27. Dickson, S.L., Leng, G., Robinson, I.C.A.F. (1993) Systemic administration of growth hormone-releasing peptide activates hypothalamic arcuate neurons. Neuroscience 53, 303–306.
28. Dickson, S.L., Leng, G., Dyball, R.E.J., Smith, R.G. (1995) Central actions of peptide and non-peptide growth hormone secretagogues in the rat. Neuroendocrinology 61, 36–43.
29. Chronwal, B.M. (1985) Anatomy and physiology of the neuroendocrine arcuate nucleus. Peptides 6 (supp 2), 1–11.
30. Dickson, S.L., Doutrelant-Viltard, O., Dyball, R.E.J., Leng, G. (1996) Retrogradely labelled neurosecretory neurones of the rat hypothalamic arcuate nucleus express Fos protein following systemic injection of GH-releasing peptide-6. J. Endocrinol. 151, 323–331.
31. Dickson, S.L., Luckman, S.M. (1997) Induction of c-fos messenger ribonucleic acid in neuropeptide Y and growth hormone (GH)-releasing factor neurones in the rat arcuate nucleus following systemic injection of the GH secretagogue, GH-releasing peptide-6. Endocrinology 138, 771–777.
32. Bailey, A.R.T., Smith, R.G., Leng, G. (1998) The nonpeptide growth hormone secretagogue MK-0677 activates hypothalamic arcuate nucleus neurons *in vivo*. J. Neuroendocrinol. 10, 111–118.
33. Dickson, S.L., Viltart, O., Bailey, A.R.T., Leng, G. (1997) Attenuation of the growth hormone secretagogue induction of Fos protein in the rat arcuate nucleus by central somatostatin action. Neuroendocrinology 66, 188–194.
34. Guillaume, V., Magnan, E., Cataldi, M. et al. (1994) Growth hormone (GH)-releasing hormone secretion is stimulated by a new GH-releasing hexapeptide in sheep. Endocrinology 135, 1073–1076.
35. Fletcher, T.P., Thomas, G.B., Clarke, I.J. (1996) Growth hormone-releasing peptide and somatostatin concentrations in the hypophysial portal blood of conscious sheep during the infusion of Growth Hormone-Releasing Peptide-6. Domestic Anim. Endocrinol. 13, 251–258.

36. Guan, X.-M., Yu, H., Palyha, O.C. et al. (1997) Distribution of mRNA encoding the growth hormone secretagogue receptor in brain and peripheral tissues. Mol. Brain Res. 48, 23–29.
37. Bennett, P.A., Thomas, G.B., Howard, A.D. et al. (1997) Hypothalamic growth hormone secretagogue-receptor (GHS-R) expression is regulated by growth hormone in the rat. Endocrinology 138, 4552–4557.
38. Conley, L.K., Telk, J.A., Deghenghi, R. et al. (1995) Mechanism of action of hexarelin and GHRP-6: analysis of the involvement of GHRH and somatostatin in the rat. Neuroendocrinology 61, 44–50.
39. Massoud, A.F., Hindrmarsh, P.C., Brook, C.G.D. (1997) Interaction of the growth hormone releasing peptide hexarelin with somatostatin. Clin. Endocrinol. 47, 537–547.
40. Peñalva, A., Carballo, A., Pombo, M., Casanueva, F.F., Dieguez, C. (1993) Effect of growth hormone (GH)-releasing hormone (GHRH), atropine, pyridostigmine or hypoglycemia on GHRP-6 induced GH secretion in man. J. Clin. Endocrinol. Metab. 76, 168–171.
41. Muruais, J., Peñalva, A., Dieguez, C., Casanueva, F.F. (1993) Influence of endogenous cholinergic tone and alpha-adrenergic pathways on growth hormone responses to His-D-Trp-Ala-Trp-D-Phe-Lys-NH_2 in the dog. J. Endocrinol. 138, 211–218.
42. Dutour, A., Briard, A., Guillaume, V. et al. (1997) Another view of GH neuroregulation: lessons from the sheep. Eur. J. Endocrinol. 136, 553–565.
43. Copinschi, G., Van Onderbergen, A., L'Hermite-Balériaux, M. et al. (1996) Effects of 7-day treatment with a novel orally active, growth hormone (GH) secretagogue, MK-677, on 24-hour GH profiles, insulin-like growth factor I, and adrenocortical function in normal young men. J. Clin. Endocrinol. Metab. 81, 2276–2282.
44. Chowen, J.A., Novak, J., Eschen, C., Gonzalez-Parra, S., Garcia-Segura, L.M., Argente, J. (1995) Growth hormone-releasing peptide-6 (GHRP-6) modulates specific populations of growth hormone releasing hormone (GHRH) and somatostatin (SS) neurons in dwarf rats. Horm. Res. 44, A74.
45. Massoud, A.F., Hindmarsh, P.C., Brook, C.G.D. (1996) Hexarelin-induced growth hormone, cortisol and prolactin release: a dose–response study. Endocrinology 81, 4338–4341.
46. Jacks, T., Hickey, G., Judith, F. et al. (1994) Effects of acute and repeated intravenous administration of L-692,585, a novel non-peptidyl growth hormone secretagogue, on plasma growth hormone, IGF-1, ACTH, cortisol, prolactin, insulin and thyroxine levels in beagles. J. Endocrinol. 143, 399–406.
47. Hickey, G., Jacks, T.M., Judith, F.R. et al. (1994) Efficacy and specificity of L-692-429, a novel non-peptidyl growth hormone secretagogue, in beagles. Endocrinology 134, 695–701.
48. Thomas, G.B., Fairhall, K.M., Robinson, I.C.A.F. (1997) Activation of the hypothalamo-pituitary–adrenal axis by the growth hormone (GH) secretagogue, GH-releasing peptide-6 in rats. Endocrinology 138, 1585–1591.
49. Arvat, E., Maccagno, B., Ramunni, J. et al. (1997) Hexarelin, a synthetic growth-hormone releasing peptide, shows no interaction with corticotropin-releasing hormone and vasopressin on adrenocorticotropin and cortisol secretion in humans. Neuroendocrinology 66, 432–438.
50. Leong, D.A., Pomes, A., Veldhuis, J.D., Clarke, I.J. (1998) A novel hypothalamic hormone measured in hypophysal portal plasma drives rapid bursts of GH secretion. Proc. 70th Meeting of the Endocrine Society, New Orleans, p. 64.
51. Kamegai, J., Wakabayashi, I., Unterman, T.G., Frohman, L.A., Kineman, R.D. (1998) Growth Hormone-Releasing Hormone (GHRH) stimulates pituitary GH-secretagogue receptor (GHS-R) mRNA levels, *in vivo*. Proc. 80th Meeting of the Endocrine Society, New Orleans, p. 63.

Growth Hormone Secretagogues
Edited by E. Ghigo, M. Boghen, F.F. Casanueva and C. Dieguez

Chapter 10

Animal Models of Growth Hormone Deficiency as Tools to Study Growth Hormone Releasing Mechanisms

LAWRENCE A. FROHMAN and RHONDA D. KINEMAN

Department of Medicine, University of Illinois at Chicago, Chicago, Illinois, U.S.A.

A large number of animal models with GH deficiency have become available for study during the past decade and have provided invaluable resources for the investigation of the growth hormone (GH) secretory process and its regulation. The models can be divided into two groups: naturally occurring (genetic) and experimentally generated (transgenic or knockout). Their importance is underscored by the fact that for many, human counterparts have been identified that are associated with clinical disorders of impaired GH secretion.

The GH secretory process is a complex mechanism. It is triggered by an increase in cytosolic Ca^{++}, resulting in the fusion of the plasma membrane with that of the GH secretory granule and exocytosis of GH into the extracellular space. Two separate pathways are currently recognized as being capable of producing this change: one stimulated by GH-releasing hormone (GHRH) and the other by a yet identified ligand for which synthetic analogs, collectively known as GH secretagogues, exist.

GHRH signal transduction is initiated by its binding to a G-protein-coupled, seven transmembrane-spanning receptor. Receptor activation leads to dissociation of the heterotrimeric $G_S\alpha$ subunit from its $\beta\gamma$ subunits, stimulation of adenylyl cyclase, generation of cyclic AMP, and phosphorylation and dissociation of the catalytic subunit of protein kinase A (PKA). Activated PKA initiates the phosphorylation of plasma membrane monovalent ion channels that results in membrane depolarization and entry of extracellular Ca^{++} into the cytoplasm. The rise in intracellular Ca^{++} culminates in the extrusion of GH-containing secretory granules (1).

GH secretagogues bind to a separate G-protein-coupled receptor only recently identified (2). This receptor is linked through the heterotrimeric G_{q11} protein to phospholipase C, resulting in phosphoinositol hydrolysis and stimulation of protein kinase C. Activation of this system leads to liberation of intracellular Ca^{++} stores and a rise in intracellular free Ca^{++}, the point at which the two signaling pathways converge.

TABLE 1

GENETIC AND TRANSGENIC MODELS USED TO STUDY GH SECRETORY MECHANISMS

	Model	Designation	Defect
Genetic	Little mouse	*lit/lit*	GHRH-R gene mutation
	Dwarf rat	*dw/dw*	Post-receptor signal transduction defect (undefined)
	Spontaneous dwarf rat (SDR)	*df/df*	GH gene mutation
Transgenic	Tyrosine hydroxylase-hGH mouse	TH-hGH	hGH overproduction in tyrosine hydroxylase-containing neurons

Regulation of the two receptors and post-receptor mechanisms has been a subject of great interest. The advantage of genetic and transgenic models, as an adjunct to the use of normal animals, has been the ability to observe consequences of perturbations of individual components of the hypothalamic–pituitary GH axis on GH secretion and the somatotropic signaling mechanisms. A list of those models that have provided new insights into the GH secretory process is provided in Table 1.

LITTLE MOUSE (*lit/lit*)

The *lit/lit* mouse was first described by Beamer and Eicher (3). This severely growth-retarded animal carries a recessive mutation, with heterozygotes exhibiting a normal phenotype. Pituitary GH content and mRNA levels are markedly decreased (to 5–10% of normal) and the animal is fully responsive to GH (4). Unstimulated release of GH by primary monolayer cultures of dispersed *lit/lit* pituitaries was markedly decreased, as compared to somatotropes from normal mice, though when expressed as a percentage of cellular GH content, was about twice normal (5). Stimulation with GHRH was completely ineffective in increasing GH release or intracellular GH content (Figure 1). Similar findings were observed *in vivo* (6). Measurement of intracellular cyclic AMP after GHRH stimulation also failed to demonstrate any increase, as compared to a 20-fold increase observed in normal somatotropes. However, probes of the signal transduction system with sites of action distal to the GHRH receptor, including cholera toxin (which stimulates $G_S\alpha$), forskolin (which directly stimulates adenylyl cyclase) and cyclic AMP were fully active in *lit/lit* pituitaries. These results led to the prediction of a mutation in the GHRH receptor. A missense mutation was described several years later (7,8), resulting in an Asp→Gly change in the extracellular portion of the receptor that completely abolishes ligand binding.

Exposure of primary cultures of *lit/lit* somatotropes to GHRP-6, the prototype of GH secretagogues, failed to increase GH secretion and injection of GHRP-6 into anesthetized *lit/lit* mice did not stimulate GH release *in vivo* (Figure 2). Whereas these findings were initially difficult to explain, more recent data has suggested that maintenance of somatotrope responsiveness to GH secretagogues requires the presence of an intact GHRH-GHRH receptor signaling system for some, yet undefined, function.

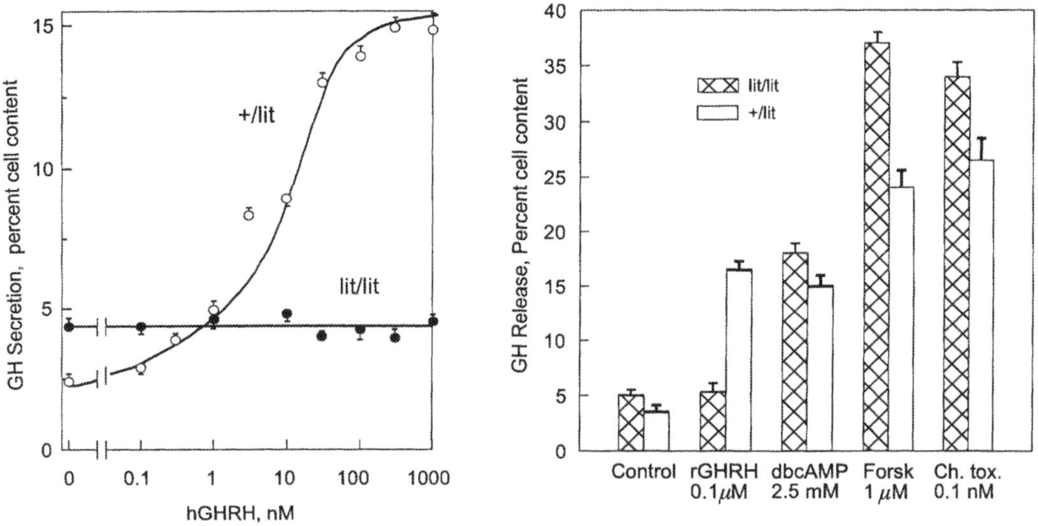

Figure 1. GH secretory responses to GHRH and probes of the GHRH signal transduction system by primary cultures of dispersed pituitary cells from normal and *lit* mice. Left: *Lit* mice pituitaries fail to respond to GHRH. Right: *Lit* mice pituitaries release GH in response to all GHRH signal transduction probes acting distally to the GHRH receptor. [Adapted from Jansson et al. (5).]

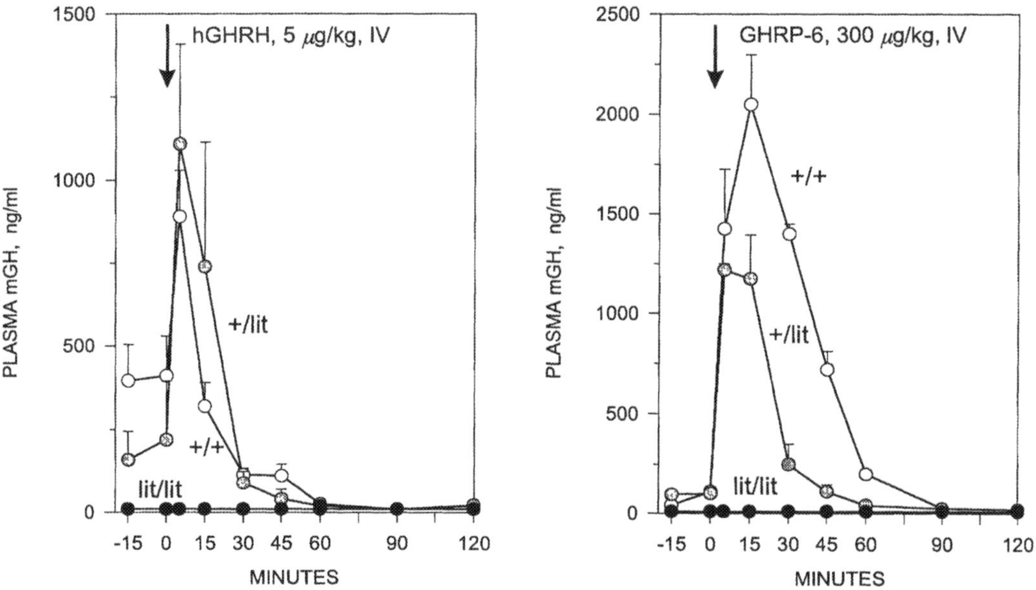

Figure 2. Failure of anesthetized *lit* mice to release GH *in vivo* in response to either GHRH or GHRP-6. Heterozygotic mice (+/–) exhibit a normal response to GHRH but only a partial response to GHRP-6. [Adapted from Jansson et al. (6).]

DWARF RAT (*dw/dw*)

The *dw/dw* rat was initially described by Charlton et al. (9) and has some similarities to the *lit/lit* mouse. It carries an autosomal recessive mutation, exhibits moderate growth retardation, though less severe than in the *lit/lit* mouse, has decreased pituitary GH content, and is fully responsive to GH. Unstimulated GH release from primary pituitary monolayer cultures is reduced but, when expressed as a percentage of cell content, is twice that from normals (10). The *dw/dw* rat differs from the *lit/lit* mouse in that GHRH administration *in vivo* does increase circulating GH levels (9). The GH secretory response to GHRH *in vitro*, however, is only 75% of that in normals, even when expressed as a percentage of cell content of the hormone (Figure 3). Cyclic AMP generation in response to GHRH is markedly impaired, increasing only about 50% over basal levels, in contrast to the 50–100 fold seen in normal rat somatotropes. Because of the considerable difference in sensitivity of the GHRH-induced cyclic AMP response to that of GH (the EC_{50} of GH is about 10 fold less), only a small increase in cyclic AMP appears necessary for a near maximal GH response. Thus, the limited capability of the *dw/dw* somatotropes to increase cyclic AMP is sufficient to permit a partial GH secretory response.

The GHRH signal transduction system of the *dw/dw* rat was explored using probes with direct effects on $G_S\alpha$, adenylyl cyclase, protein kinase A and protein kinase C. Stimulation by cyclic AMP produced a normal GH response and exposure to forskolin resulted in normal cyclic AMP and GH responses, indicating that adenylyl cyclase activity and the downstream pathways were intact. However, direct stimulation of $G_S\alpha$ by cholera toxin and by PGE_1 resulted in markedly reduced cyclic AMP and GH responses, implying impairment

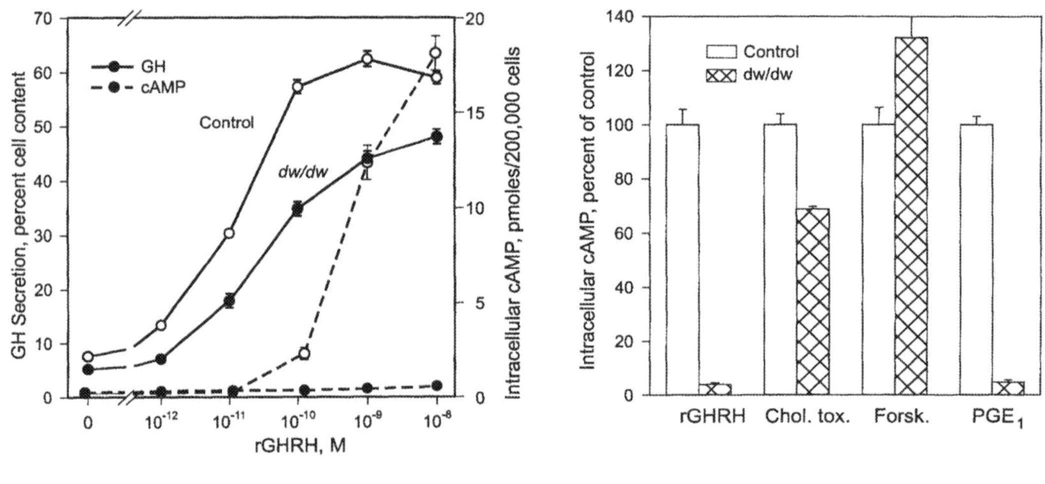

Figure 3. GH and cAMP secretory responses to GHRH and probes of the GHRH signal transduction system by primary cultures of dispersed pituitary cells from normal and *dw* rats. Left: GH secretion is only partially reduced while cAMP generation is almost entirely absent in *dw* pituitaries in response to GHRH. Right: *Dw* rat pituitaries exhibit partial impairment of cAMP generation in response to GHRH signal transduction probes that activate $G_s\alpha$ but normal cAMP generation in response to forskolin, a direct stimulator of adenylyl cyclase. [Adapted from Downs et al. (10).]

in $G_S\alpha$ function. Attempts to define this defect included sequencing of the GHRH receptor and $G_S\alpha$ cDNA, assessment of $G_S\alpha$ protein levels, and measurement of adenosine-diphosphate ribosylation, a measure of $G_S\alpha$ function. However, each of these signaling components was normal (11). Thus, the secretory defect in the *dw/dw* rat remains undefined. Among the possibilities are (1) a defect in the $G_S\beta\gamma$ subunit that impairs its dissociation from the α subunit, (2) a mutation in a pituitary (somatotrope) specific protein corresponding to the β-adrenergic receptor kinase, or (3) a mutation in adenylyl cyclase that affects activation by $G_S\alpha$ but not by forskolin. In addition to their GH secretory impairment, *dw/dw* pituitaries contain proportionally fewer somatotropes (5% of total cells) than do normal pituitaries (40–50%), underscoring the importance of GHRH in somatotrope proliferation (12,13).

Stimulation of primary cultures of *dw/dw* pituitary cells by GHRP-6 revealed no change in the EC_{50} but a 33% reduction in the maximal GH secretory response, when expressed as a percentage of cell hormone content (14). The comparable reduction in secretory responses to GHRH and GHRP-6 argue that the two responses are linked and that normal function of the GHRH receptor is required for an intact response to GHRP (and presumably other GH secretagogues).

TYROSINE HYDROXYLASE–HUMAN GROWTH HORMONE TRANSGENIC MOUSE (TH-hGH)

The TH-hGH mouse carries a transgene composed of 4.8 kb of the tyrosine hydroxylase promoter fused to a human GH reporter gene (15). Although originally designed to investigate tyrosine hydroxylase expression, the model has proven useful for the study of GH feedback regulation (16). The expression of the transgene is restricted to the central nervous system, with the exception of the adrenal medulla, from which secretion does not occur. Within the hypothalamus, GH is expressed primarily in the arcuate nucleus and the periventricular nucleus, and in close proximity to the neurons that secrete GHRH and somatostatin, respectively. As a consequence, a localized feedback occurs that results in increased somatostatin secretion and decreased GHRH secretion. This, in turn, leads to decreased GH secretion and the TH-hGH animals are markedly growth retarded, with decreased circulating GH and IGF-I levels.

TH-hGH mice have reduced hypothalamic GHRH mRNA and GHRH content and therefore, their somatotropes experience reduced GHRH input. The pituitary is decreased in size, though, in contrast to the *dw/dw* rat, the relative abundance of somatotropes in the pituitary of TH-hGH mice is normal (17). Measurement of GHRH receptor mRNA levels in TH-hGH mice revealed a significant reduction (16), though the change can be explained, at least in part, by the reduced number of somatotropes. Although the cellular GH content is reduced, the relative GH secretory response to GHRH is indistinguishable from that in normals (17) indicating that the concentration of GHRH required for maintaining normal GH release is less than that required for normal GH synthesis. To date, no information is available about the secretory response to other GH secretagogues in this model.

SPONTANEOUS DWARF RAT (SDR)

The SDR has a total absence of GH resulting from a point mutation in an intron of the GH gene, which modifies the splice acceptor site of the third exon, creating a premature, in-phase stop codon (18). Pituitary GH mRNA levels are <3% of normal and GH is undetectable both in the pituitary and in circulation. Serum IGF-I levels are markedly reduced (19), body weight is only 25–30% of normal, and the pituitary is proportionally decreased in size (18). Although there are no detectable somatotropes, about 50% of cells are non-immunoreactive for any of the pituitary hormones ("null" cells) and contain a highly developed organelle system, characteristic of somatotropes. Despite the absence of GH synthesis, the GHRH signaling system appears intact since the GH mRNA response to GHRH *in vitro* is proportional to that seen in normal pituitaries.

This model has been useful for examining GH feedback regulation in the total absence of the hormone (19,20), and has provided insights into the regulation of the GHRH and GHS receptors. Although the SDR somatotropes are devoid of GH, GHRH receptor mRNA levels can be measured and receptor function can be assessed measuring intracellular cyclic AMP responses. GHRH mRNA levels are increased in SDR hypothalamus and GHRH receptor mRNA levels are increased in SDR pituitary, providing additional evidence for the role of the ligand in regulating its receptor (Figure 4). Stimulation of dispersed SDR pituitaries with GHRH *in vitro* leads to enhanced intracellular cyclic AMP levels, demonstrating an increase in GHRH receptor function, as well as content.

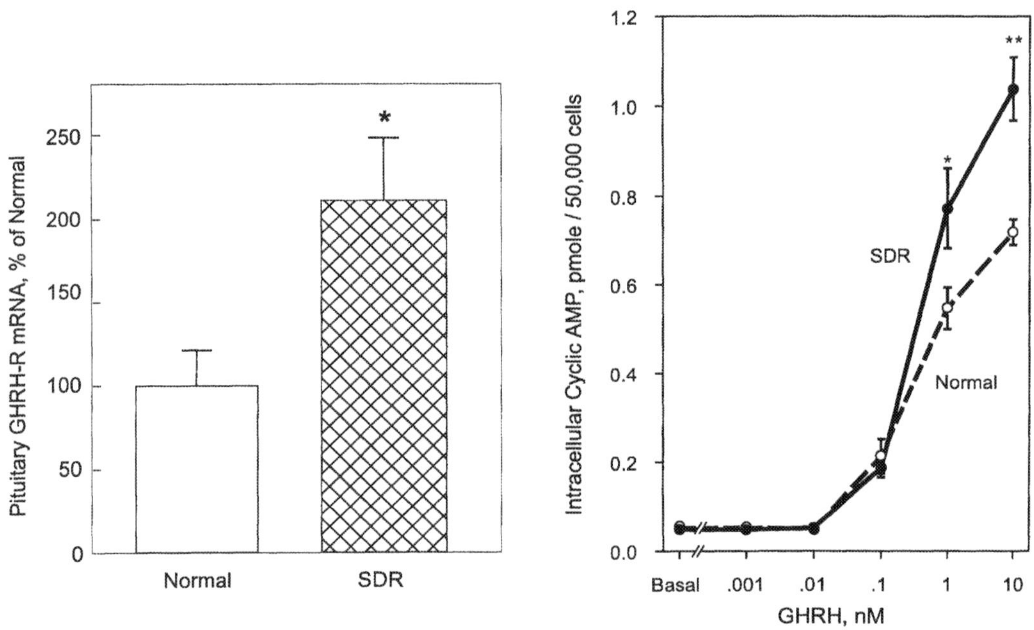

Figure 4. Effect of total deficiency of GH on pituitary GHRH receptor mRNA levels and functional activity. Left: Pituitary GHRH receptor mRNA is significantly increased in SDR, which have a total absence of GH. Right: GHRH receptor function, as assessed by the cAMP response to GHRH, is increased in GH deficient SDR. [Adapted from Kamegai et al. (19).]

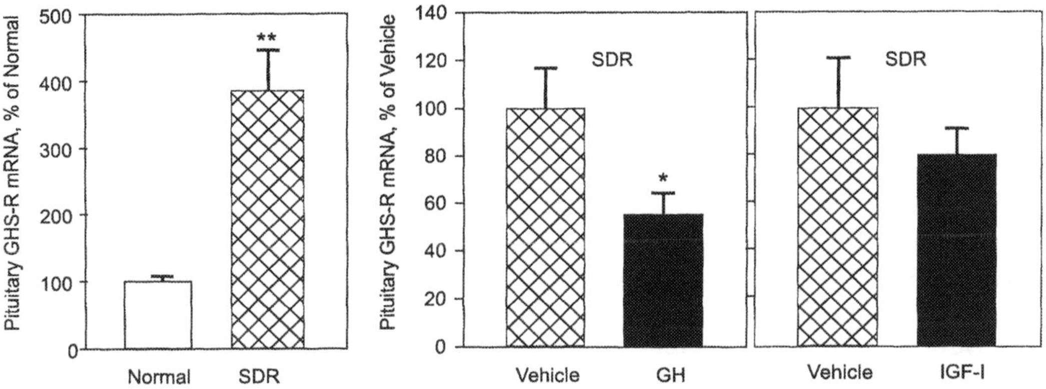

Figure 5. Pituitary GH secretagogue receptor (GHS-R) mRNA levels in SDR (left) and their responses to 3-day infusions of GH or IGF-1 (right). [Adapted from Kamegai et al. (20).]

Treatment of SDRs with GH by subcutaneous infusion for 3 days resulted in a decrease in both hypothalamic GHRH mRNA and pituitary GHRH receptor mRNA levels (19). Treatment of SDRs with IGF-I for the same period of time, however, did not produce any significant change in hypothalamic GHRH mRNA levels. However, a small (30%) but significant decrease in pituitary GHRH receptor mRNA levels was observed. Taken together with our previous observations that GH does not exert acute effects on GHRH receptor mRNA levels *in vitro* (21), these findings suggest a dual role for GH feedback effects on GHRH receptor synthesis: a hypothalamic action on the regulation of GHRH and a direct pituitary or peripheral action by stimulating IGF-I production.

Since the endogenous GHS-GHS-R signaling system is considered to be important in regulation GH secretion, it was postulated that changes in circulating GH levels should directly (or indirectly via IGF-I) influence GHS-R expression. In the total absence of GH feedback, a characteristic of the SDR, pituitary GHS-R mRNA levels were nearly 4-fold greater ($P < 0.01$) than in controls (Figure 5) (20). A 3-day subcutaneous infusion of GH reduced pituitary GHS-R mRNA levels by nearly 50% of vehicle-treated controls ($p < 0.05$). In contrast, treatment with IGF-I for the same period of time did not significantly alter GHS-R mRNA levels, indicating that GH acts independently of systemic IGF-I to regulate GHS-R expression.

The parallel changes in hypothalamic GHRH mRNA and pituitary GHS-R mRNA levels in response to perturbations of the GH status suggested that GHRH might mediate the effects of GH on the GSH-R. A 4 hour intravenous infusion of the GHRH agonist [desNH$_2$Tyr1,D-Ala^2,Ala^{15}]-hGHRH(1–29)NH$_2$ in anesthetized normal rats resulted in a 2.5-fold increase in pituitary GHS-R mRNA levels (22). This confirmed the hypothesis that increased GHRH production mediated the effects of GH deficiency on GHS-R mRNA levels, though it raised another question: does the effect of GHRH occur directly at the pituitary? Exposure of normal rat pituitary cell cultures to GHRH for 4 hours failed to alter GHS-R mRNA levels, suggesting that the effects of GHRH are not mediated through the GHRH-R and may not even occur directly at the level of the pituitary (22). Although the mediation of this effect is presently not known, the observation provides a potential explanation for the requirement of an intact GHRH-GHRH receptor system for the maintenance of pituitary responsiveness to the GH secretagogue.

SUMMARY

Genetic and transgenic models have proven to be of great value in the study of the GHRH-GHS-GH-IGF-1 axis. The variety of models with unique disturbances in the pituitary signaling system associated with GH secretion has provided the means to determine the relative contribution of the multiple pathways that regulate GH secretion and to determine the downstream consequences of altered function of individual components. From studies performed using these models, experimental evidence has linked the GHRH and GHS signaling systems at the level of their respective pituitary receptors. Future studies using these and other models can be expected to further clarify this important physiological system.

ACKNOWLEDGEMENTS

The studies from the authors' laboratory were supported by USPHS Grant DK 30667 and the Bane Foundation.

REFERENCES

1. Frohman, L.A., Kineman, R.D. (1999) Growth hormone-releasing hormone: discovery, regulation, and actions. In: Handbook of Physiology: Hormonal Control of Growth. J. Kostyo (ed). Oxford University Press, New York, pp. 189–221.
2. Howard, A.D., Feighner, S.D., Cully, D.F., Arena, J.P., Liberator, P.A., Rosenblum, C.I., et al. (1996) A receptor in pituitary and hypothalamus that functions in growth hormone release. Science 273, 974–977.
3. Beamer, W.G., Eicher, E.M. (1976) Stimulation of growth in the little mouse. J. Endocrinol. 71, 37–45.
4. Cheng, T.C., Beamer, W.G., Phillips, J.A., III, Bartke, A., Mallonee, R.L., Dowling, A.C. (1983) Etiology of growth hormone deficiency in little, Ames and Snell dwarf mice. Endocrinology 113, 1669–1678.
5. Jansson, J.-O., Downs, T.R., Beamer, W.G., Frohman, L.A. (1986) Receptor-associated resistance to growth hormone-releasing factor in dwarf "little" mice. Science 232, 511–512.
6. Jansson, J.-O., Downs, T.R., Beamer, W.G., Frohman, L.A. (1986) The dwarf "little" (*lit/lit*) mouse is resistant to growth hormone (GH)-releasing peptide (GH-RP-6) as well as to GH-releasing hormone (GRH). Program 68th Ann. Mtg. Endo. Soc. #397 (Abstr).
7. Godfrey, P., Rahal, J.O., Beamer, W.G., Copeland, N.G., Jenkins, N.A., Mayo, K.E. (1993) GHRH receptor of little mice contains a missense mutation in the extracellular domain that disrupts receptor function. Nat. Genet. 4, 227–232.
8. Lin, S.C., Lin, C.R., Gukovsky, I., Lusis, A.J., Sawchenko, P.E., Rosenfeld, M.G. (1993) Molecular basis of the little mouse phenotype and implications for cell type-specific growth. Nature 364, 208–213.

9. Charlton, H.M., Clark, R.G., Robinson, I.C., Porter-Goff, A.E.P., Cox, B.S., Bugnon, C., et al. (1988) Growth hormone-deficient dwarfism in the rat: A new mutation. J. Endocrinol. 119, 51–58.
10. Downs, T.R., Frohman, L.A. (1991) Evidence for a defect in growth hormone-releasing factor signal transduction in the dwarf (*dw/dw*) rat pituitary. Endocrinology 129, 58–67.
11. Zeitler, P.A., Downs, T.R., Frohman, L.A. (1993) Impaired growth hormone releasing-hormone signal transduction in the dwarf (*dw*) rat is independent of a generalized defect in the stimulatory G-protein, Gs-alpha. Endocrinology 133, 2782–2786.
12. Kineman, R.D., Chen, T.T., Frawley, L.S. (1989) A cellular basis for growth hormone deficiency in the dwarf rat: analysis of growth hormone and prolactin release by reverse hemolytic plaque assay. Endocrinology 125, 2035–2040.
13. Zeitler, P.A., Downs, T.R., Frohman, L.A. (1994) Development of pituitary cell types in the spontaneous dwarf (*dw*) rat: evidence for an isolated defect in somatotroph differentiation. Endocrine 2, 729–733.
14. Pinhas-Hamiel, O., Zeitler, P. (1994) Impaired response to GHRP-6 in somatotrophs from dwarf (*dw*) rats. Program 76th Ann. Mtg. Endo. Soc. #663 (Abstr).
15. Banerjee, S.A., Roffler-Tarlov, S., Szabo, M., Frohman, L., Chikaraishi,. D.M. (1994) DNA regulatory sequences of the rat tyrosine hydroxylase gene direct correct catecholaminergic cell type specificity of human growth reporter in the CNS of transgenic mice causing a dwarf phenotype. Mol. Brain Res. 24, 89–106.
16. Szabo, M., Butz, M.R., Banerjee, S.A., Chikaraishi, D.M., Frohman, L.A. (1995) Autofeedback suppression of growth hormone (GH) secretion in transgenic mice expressing a human GH reporter targeted by tyrosine hydroxylase 5′ flanking sequences to the hypothalamus. Endocrinology 136, 4044–4048.
17. Kineman, R.D., Aleppo, G., Frohman, L.A. (1996) The tyrosine hydroxylase-human growth hormone (GH) transgenic mouse as a model of hypothalamic GH deficiency: growth retardation is the result of a selective reduction in somatotrope numbers despite normal somatotrope function. Endocrinology 137, 4630–4636.
18. Takeuchi, T., Suzuki, H., Sakurai, S., Nogami, H., Okuma, S., Ishikawa, H. (1990) Molecular mechanism of growth hormone (GH) deficiency in the spontaneous dwarf rat: Detection of abnormal splicing of GH messenger ribonucleic acid by the polymerase chain reaction. Endocrinology 126, 31–38.
19. Kamegai, J., Unterman, T.G., Frohman, L.A., Kineman, R.D. (1998) Hypothalamic/pituitary axis of the spontaneous dwarf rat: autofeedback regulation of growth hormone (GH) includes suppression of GH releasing-hormone receptor messenger ribonucleic acid. Endocrinology 139, 3554–3560.
20. Kamegai, J., Wakabayashi, I., Miyamoto, K., Unterman, T.G., Kineman, R.D., Frohman, L.A. (1998) Growth hormone (GH)-dependent regulation of pituitary GH secretagogue receptor (GHS-R) mRNA levels in the spontaneous dwarf rat. Neuroendocrinology 68, 312–318.
21. Aleppo, G., Moskal, S.F., II, DeGrandis, P.A., Kineman, R.D., Frohman, L.A. (1997) Homologous down-regulation of growth hormone-releasing hormone receptor mRNA levels. Endocrinology 138, 1058–1065.
22. Kineman, R.D., Kamegai, J., Frohman, L.A. (1999) Growth hormone (GH)-releasing hormone (GHRH) and the GH secretagogue (GHS), L692,585, differentially modulate rat pituitary GHS receptor and GHRH receptor messenger ribonucleic acid levels. Endocrinology 140 (in press).

Growth Hormone Secretagogues
Edited by E. Ghigo, M. Boghen, F.F. Casanueva and C. Dieguez

Chapter 11

Regulation of Growth Hormone (GH) Pulsatility in Humans

ELENI V. DIMARAKI[1] and ARIEL L. BARKAN[2]

[1]*Division of Endocrinology and Metabolism, Department of Internal Medicine, University of Michigan, Ann Arbor, MI, U.S.A.*
[2]*Pituitary and Neuroendocrine Center, University of Michigan, Ann Arbor, MI, U.S.A.*

In all species studied thus far, including man, growth hormone (GH) secretion is pulsatile. Using sensitive GH assays and frequent blood sampling over extended periods of time, it has been shown that in normal humans, plasma GH fluctuates over an approximately 1000 fold range: between 0.01 and 30–40 μg/L (1). These wide fluctuations are not haphazard, but rather tightly organized into discrete secretory pulses and long periods of secretory quiescence. Alterations in GH pulsatility are seen in a variety of physiological and pathological circumstances, including puberty (2), aging (3), menstrual cycle (4), obesity (5), starvation (6), growth delay (7) and acromegaly (8). Even more importantly, the biological efficiency of GH and its tissue specificity depend on the mode of pulsatile GH presentation to the tissues. Administration of GH to hypophysectomized animals in either continuous or pulsatile pattern differentially affects the induction of GH-dependent parameters (9). Moreover, some biological actions of GH are preferentially modulated by the magnitude and duration of the steady, interpulse GH levels, whereas others require frequent and transiently-high GH pulses (10–14). Thus, it is not only the total amount of GH presented to the peripheral tissues, but also the pattern of its pulsatile secretion that is important for the ultimate effect.

Over the past 20 years, several research groups have directed major efforts aimed at identifying the discrete neuroendocrine components responsible for the regulation of GH pulsatility. Crucial for these research endeavors were the identifications of the hypothalamic and the peripheral regulators of GH synthesis and secretion, GH-releasing hormone (GHRH), somatostatin (SRIH) and insulin growth factor-I (IGF-I). More recently the existence of yet another GH stimulant, tentatively labeled GHS (Growth Hormone Secretagogue) has been postulated (15).

The discovery of GHRH and SRIH allowed for the development of physiologic techniques to study their roles in GH secretion. Such techniques, including immunoneutralization of GHRH and SRIH (16–18), and direct sampling of hypophysial-portal blood (19), attributed to GHRH the role of the primary releaser of GH and to SRIH the role of the inhibitor that controls the interpulse GH levels and the amplitude of GH pulses. Recent discovery of hypothalamic and pituitary receptors for a still unidentified GHS peptide (20,21) introduced further complexity to the system. Since this putative compound apparently possesses both GHRH-releasing and anti-SRIH properties, it would be ideally suited to be the primary regulator of GH pulsatility.

Another level of complexity in the study of GH regulation is the heterogeneity of the models. Conclusions derived from one species may not be applicable to another species. Recently, Dutour et al. (22) performed a careful review of the existing data regarding the mechanisms of GH pulsatility. They stressed the unique features of GH neuroregulation in the rat on the one hand and in humans and sheep on the other hand. It is likely that a model different from the rat will be needed in the future to obtain the detailed biochemical and pharmacological data needed to explain GH pulsatility in humans. In this review we shall attempt to present the existing data on the relative roles of GHRH, SRIH and the putative GHS as the hypothalamic regulators of GH pulsatility, with the emphasis on the humans.

GHRH REGULATION OF GH PULSATILITY

A wide variety of *in vivo* animal studies have supported the central role of GHRH in the production of GH pulses.

Animals with experimentally-induced hypothalamic lesions (23) and humans with organic illnesses of the hypothalamus (24) have absent or severely impaired GH secretion. This is especially true with regards to the pulsatile component. Administration of GHRH in these GHRH-deficient models invariably restored GH pulsatility and promoted somatic growth. However, hypothalamic lesioning is too crude to pinpoint the precise neuro-endocrine component of GH dysregulation. More selective and specific models, such as GHRH immunoneutralization (17,18), blockade of GHRH action by a receptor antagonist (25,26), or destruction of the GHRH-containing neuronal bodies in the arcuate nucleus by neonatal monosodium glutamate (MSG) (27), all confirmed the crucial role of GHRH in the generation of GH pulsatility in animal models. Moreover, direct measurement of hypophysial-portal GHRH concentration in sheep has shown that the majority of GH pulses are preceded or accompanied by simultaneous GHRH pulses (28,29). Although SRIH secretory patterns were also pulsatile, there was no correlation between GH and SRIH pulses (28,29). Taken together, these data attribute to GHRH the crucial role of the actual GH pulse generator, with SRIH playing only a secondary role, possibly as a modulator of GH pulse amplitude.

Understandably, the methodologies used in animals such as immunoneutralization of GHRH or direct pituitary-portal sampling are impractical in humans. We have therefore approached this problem pharmacologically, blocking the GHRH receptor to investigate the role of endogenous GHRH in the generation of GH pulsatility in humans. The

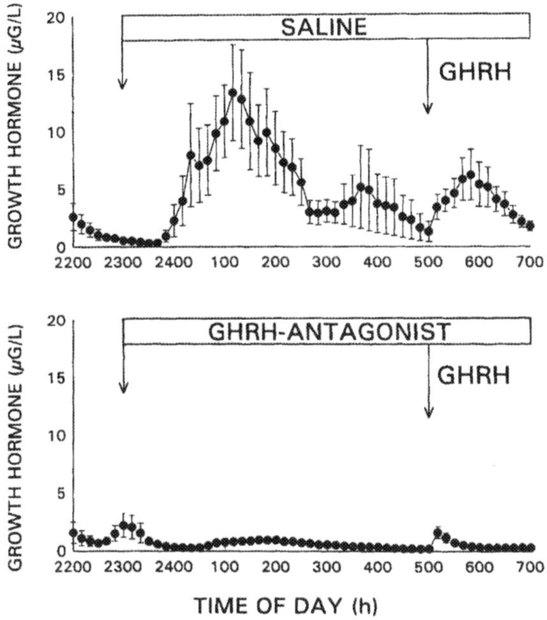

Figure 1. Mean GH concentrations (mean ± SE) during normal saline (upper panel) and GHRH-ant (lower panel) infusion, in healthy young men. The nocturnal GH secretion was suppressed by 89% and the GH response to GHRH bolus was suppressed by 79%. Reproduced from Ocampo-Lim, B., Guo, W., DeMott-Friberg, R., Barkan, A.L., Jaffe, C.A. Nocturnal growth hormone (GH) secretion is eliminated by infusion of GH-releasing hormone antagonist. J. Clin. Endocrinol. Metab. (1996) 81 (12), 4396–4399, by copyright permission of the Endocrine Society.

compound used, [N-Ac-Tyr^1-, D-Arg^2] GHRH (1-29) NH_2, is a specific and selective GHRH antagonist (GHRH-ant) both *in vitro* and *in vivo* (25).

We first showed that a single intravenous (IV) bolus dose of GHRH-ant 400 μg/kg blocked the pituitary response to exogenous GHRH in a time-dependent manner, suppressing the GH response by 95% at 60 minutes and 4% at 24 hours (30). The same bolus dose of GHRH-ant suppressed the nocturnal GH release by 75%. Failure to more completely suppress nocturnal GH secretion could have been due to either (a) a non-GHRH mechanism accounting for some of the nocturnal GH pulsatility, or (b) waning efficacy of GHRH-ant due to rapid clearance. In a subsequent experiment, we administered GHRH-ant as a continuous IV infusion following the loading bolus at 2200 h (31). In this model, nocturnal pulsatile GH secretion in young healthy men was suppressed by almost 90%, confirming the exquisite importance of endogenous GHRH for the generation of pulsatile GH secretion in humans (Figure 1).

The neuroendocrine genesis of pharmacologically-induced GH pulses is unknown. Acute provocative stimuli of GH have been used for more than 30 years as diagnostic tools in investigating potential GH deficiency. Theoretically, these stimuli elicit GH release either through an acute discharge of hypothalamic GHRH or through acute suppression of SRIH secretion. Two pharmacologic paradigms have been traditionally employed to identify the correct possibility. In the first approach, the tested compound is given either alone or in combination with a (presumably) supra-physiological bolus or GHRH. If the combined bolus produces a GH response that is higher than that induced either by GHRH or the compound alone, one is asked to conclude that the compound elicited GH rise through a mechanism different from GHRH, i.e. suppression of SRIH secretion. In the second approach, a continuous infusion of GHRH is used to induce homologous desensitization to a GHRH bolus. If a tested compound is capable of eliciting GH rise at that time, it is assumed to do so through a SRIH suppressive mechanism. Based on these paradigms, L-dopa was postulated to be a pure GHRH releaser (32), whereas pyridostigmine was assigned to a group of "SRIH suppressors" (33). Clonidine and hypoglycemia were variably classified as possessing either or both mechanisms of action.

A study by Magnan et al. (34) cast a serious doubt on the validity of this pharmacologic analysis. First, they showed that combined neostigmine/GHRH bolus in sheep elicited a much greater GH rise than either secretagogue alone. This was fully compatible with the "SRIH inhibitory" action of cholinesterase inhibitors. Yet, neostigmine elicited a marked increase in the hypophysial-portal concentration of GHRH, but had no effect on hypophysial-portal SRIH concentrations. Thus, the seemingly straightforward pharmacologic models may be actually misleading. It is clear that, at least for neostigmine, GHRH is being released. The origin of synergy with GHRH is not yet understood, but implies the existence of a non-GHRH, non-SRIH mechanism.

In humans, we used a model of selective antagonism of GHRH to investigate the neuroendocrine regulation of pharmacologically stimulated GH (35). We first demonstrated that acute cessation of SRIH infusion elicited a small but noticeable GH pulse, and that GHRH-ant could not eliminate this response (Figure 2). This dichotomy provided us with a tool to distinguish between the two potential mechanisms of GH pulse: GHRH increase or SRIH withdrawal. Suppressibility of a secretagogue-induced GH rise by GHRH-ant indicated GHRH involvement whereas its persistence in spite of GHRH-ant administration suggested that the mechanism of secretagogue action was GHRH-independent and, potentially, related to an acute suppression of hypothalamic SRIH secretion. The GH responses to every stimulus tested, including clonidine, L-dopa, arginine, pyridostigmine (Figure 3), and insulin induced hypoglycemia (Figure 4) were markedly attenuated by GHRH-ant. These data suggest that endogenous GHRH plays a critical role in the acute GH responses to all of the above stimuli. Our data have been subsequently reproduced by Hanew et al. (36) who utilized L-dopa and clonidine as GH stimuli.

In summary, it appears that in humans endogenous GHRH is the neurohormone ultimately responsible for the generation of spontaneous GH pulses and pharmacological GH responses. However, our data still do not differentiate between two potential models of GHRH secretion. In one model, GHRH pulsatility is the driving force behind GH pulses. In a second model, GHRH plays a tonic role as a facilitator of GH responses to another stimulus. Based on the animal data, the former possibility is more likely but additional human studies need to be conducted.

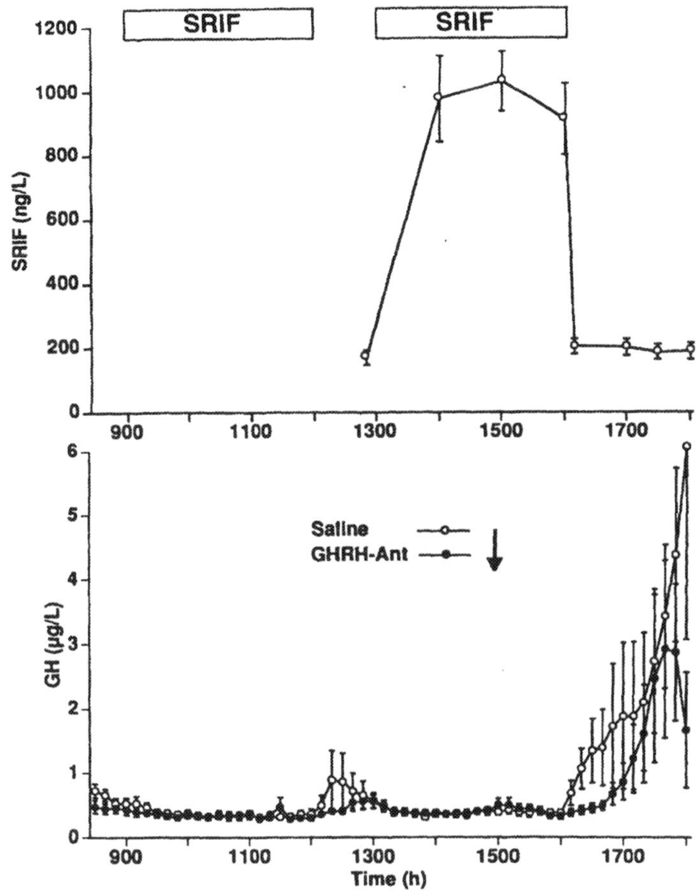

Figure 2. (Upper panel) Plasma SRIH concentration (mean ± SE) in young men during patterned SRIH infusion. (Lower panel) Plasma GH concentrations during the SRIH infusion. Normal saline or GHRH-ant bolus was administered at 1500 h. After the termination of the SRIH infusion at 1600 h, the plasma GH increased in both studies. There were no differences in the magnitude of GH responses during normal saline and GHRH-ant studies. Reproduced from Jaffe, C.A., DeMott-Friberg, R., Barkan, A.L. Endogenous growth hormone (GH)-releasing hormone is required for GH responses to pharmacological stimuli. J. Clin. Invest. (1996) 97(4), 934–940, by copyright permission of the American Society of Clinical Investigation.

SOMATOSTATIN REGULATION OF GH PULSATILITY

A variety of physiologic roles in GH regulation have been attributed to SRIH. Multiple studies suggest that SRIH might allow the accumulation of GH in somatotrophs in an immediately releasable pool. In pituitary cell cultures, SRIH inhibits the GHRH stimulated release of GH but does not alter either basal GH mRNA levels (37) or GH biosynthesis (38). GH release from perifused rat anterior pituitary fragments following SRIH withdrawal was increased by prolonging exposure to SRIH, and GHRH amplified the rebound release that followed SRIH withdrawal (39). Weiss et al. showed that GHRH was capable of inducing GH pulses by itself in perifused rat anterior pituitaries, and that SRIH withdrawal was synergistic with the GHRH effect on GH (40). The data from these *in vitro*

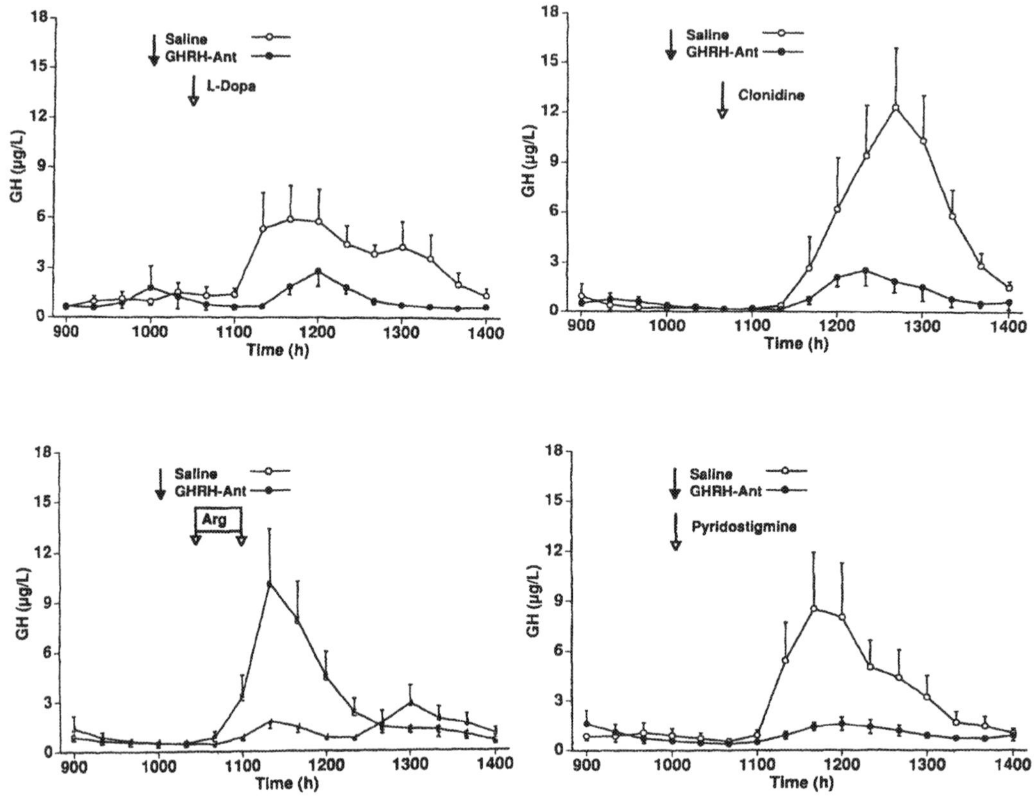

Figure 3. GH responses to oral L-dopa, IV arginine, oral clonidine and pyridostigmine following normal saline or GHRH-ant administration. The GH response ro each stimulus was markedly suppressed by GHRH-ant. Reproduced from Jaffe, C.A., DeMott-Friberg, R., Barkan, A.L. Endogenous growth hormone (GH)-releasing hormone is required for GH responses to pharmacological stimuli. J. Clin. Invest. (1996) 97(4), 934–940, by copyright permission of the American Society of Clinical Investigation.

studies suggest that SRIH withdrawal generates GH release and the magnitude of GH response depends on the pattern of prior exposure to SRIH. Moreover, prior exposure to SRIH may determine the amplitude of the GH response to other hypothalamic factors, including GHRH.

In vivo, these interactions could be more complex, especially since GHRH and SRIH may influence the secretion of each other. There is ample evidence of anatomical association of SRIH and GHRH neurons (41,42). Liposits et al. (43), using immunohistochemistry showed that somatostatinergic axons have synaptic connections with GHRH containing neurons at the arcuate nucleus of the rat. Horvath et al. (44), using electron microscopy and immunohistochemistry demonstrated synaptic connection between SRIH-immunoreactive nerve terminals and GHRH-positive dendrites as well as connections between GHRH-containing nerve terminals and SRIH-containing dendrites. These data support the concept

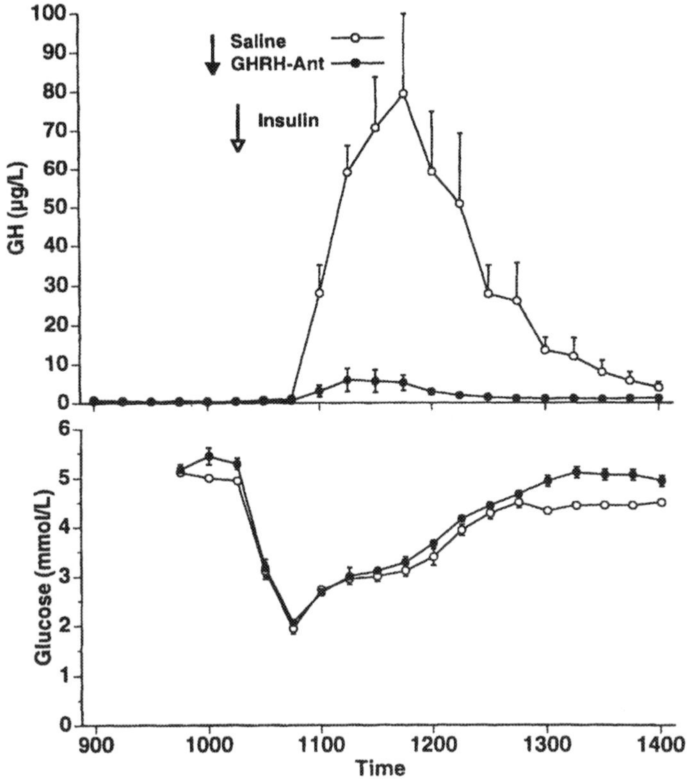

Figure 4. (Upper panel) GH response to hypoglycemia following normal saline or GHRH-ant bolus. (Lower panel) Plasma glucose concentrations after administration of insulin 0.1 U/kg.The GH response to insulin-induced hypoglycemia was markedly suppressed by GHRH-ant. Reproduced from Jaffe, C.A., DeMott-Friberg, R., Barkan, A.L. Endogenous growth hormone (GH)-releasing hormone is required for GH responses to pharmacological stimuli. J. Clin. Invest. (1996) 97(4), 934–940, by copyright permission of the American Society of Clinical Investigation.

that SRIH and GHRH secretions are interrelated. A variety of experiments have demonstrated functional significance of these anatomical observations. In agreement with this concept, *in vivo* intracerebroventricular injection of SRIH to rats released GH (45). The GH release was blunted by GHRH antiserum, indicating that this central stimulating effect of SRIH is mediated through GHRH. On the other hand, intracerebroventricular injection of GHRH inhibited spontaneous GH pulsatility and this effect was blocked by intravenous administration of SRIH antiserum (46). Together these data imply reciprocal central interaction between GHRH and SRIH.

Animal data support the hypothesis that SRIH may play a role in initiating GH pulses. Systemic injection of SRIH antiserum to rats leads to marked increase in serum GH levels, suggesting that acute withdrawal of SRIH is an effective stimulus of GH secretion (47,48). However, much or all of this response may be secondary to GHRH release. Miki et al. (47) showed that immunoneutralization of GHRH inhibited the GH response to SRIH antiserum by approximately 83%. Thomas et al. (48) also found that pretreatment with

GHRH antiserum attenuated antisomatostatin-induced GH release in rats, however the suppression was very incomplete. Similarly, rebound GH secretion following short term SRIH infusion was inhibited by passive immunization against GHRH (49). These results indicate that GH secretion following SRIH withdrawal is caused by hypothalamic release of GHRH. However, failure to completely eliminate the GH response to SRIH withdrawal by GHRH immunoneutralization is consistent with the presence of an additional non-GHRH mechanism of GH release.

In addition to its potential role in GH pulse generation, SRIH might also determine interpulse GH levels. In male rats, immunoneutralization of SRIH increased trough GH levels but had no effect on GH pulses amplitude (19,50). It had, however, a significant effect on the responsiveness of GH to GHRH. GH secretion in male rats is characterized by an endogenous ultradian rhythm with high amplitude GH pulses occurring every 3 hours. The responsiveness to GHRH also follows this ultradian pattern and is low during the trough periods of the GH cycle and high during the periods of expected GH pulses (16). This variability in GHRH sensitivity was abolished by SRIH immunoneutralization. These results led to a hypothesis in which GH is under tonic inhibition by SRIH and the periodic variability of the somatostatinergic tone affects the pituitary responsiveness to GHRH and therefore the amplitude of GH pulses.

Whether these data derived from studies in rats are applicable to other species is unclear. In contrast to the regular GH pulses in male rats, irregular secretory spikes characterize the pattern of GH secretion in ruminants (22). Direct hypophysial-portal blood sampling in sheep showed that hypothalamic SRIH secretion was also pulsatile, however there was no concordance between SRIH and GHRH or SRIH and GH pulses (28,29). Portal GHRH and systemic GH pulses, however, were concordant approximately 2/3 of the time indicating a role of GHRH in initiating GH pulses (28,51). Yet, SRIH does appear to play a role in the regulation of GH secretion, in this species. In underfed sheep decreased hypophysial-portal blood SRIH levels are associated with elevated mean GH levels and increased GH pulse amplitude (51). These findings are consistent with a role of SRIH in regulating pituitary sensitivity to GHRH.

Our understanding of the role of SRIH in humans has come from more indirect experiments. In humans, pulsatile GH secretion persisted during continuous GHRH infusion (52). It was postulated that the preserved rhythmicity of GH in the presence of persistently high GHRH levels was due to periodic withdrawal of tonic inhibition by SRIH. Based on deconvolution analysis, Hartman et al. (53) concluded that GH is secreted in volleys composed of multiple secretory bursts without measurable intervening secretion. They hypothesized that this pattern results from low-frequency episodes of SRIH troughs superimposed on high frequency GHRH secretory events.

However, although acute SRIH withdrawal is an effective stimulus of GH secretion *in vitro* and in some animal models, in humans it is at best a very weak promoter of GH secretion. Hindmarsh et al. (54) showed that the administration of GHRH combined with SRIH withdrawal led to consistent and repeatable GH responses, and therefore suggested

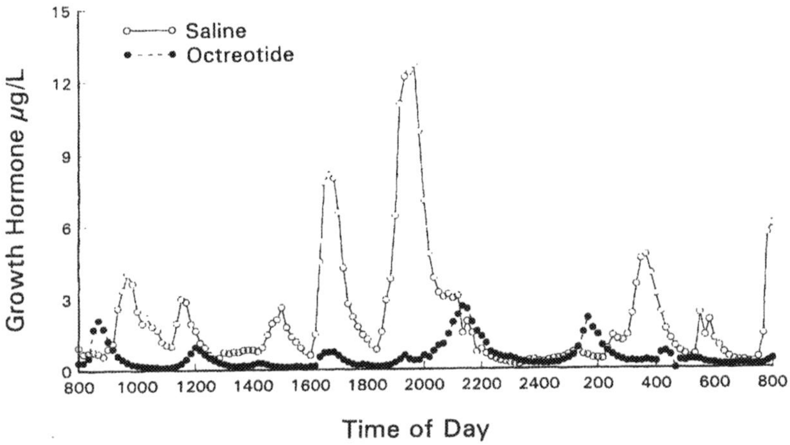

Figure 5. Twenty-four hour plasma GH concentrations profile from a healthy woman during continuous subcutaneous infusion of normal saline or a long-acting SRIH analog, octreotide, at a dose of 400 μg per 24 hours. Continuous infusion of octreotide suppressed the amplitude but not the occurrence of GH pulses.

that SRIH plays a role in determining the ability of somatotrophs to respond to GHRH and that GH pulsatility requires the close interplay of GHRH and SRIH. In this study (54) and in our experience (35,55), the GH response to SRIH withdrawal alone was small. Moreover, in contrast to the studies in animals, in humans the GH response to SRIH withdrawal was not inhibited by GHRH-ant (35). Therefore, in humans SRIH withdrawal is an ineffective promoter of GH pulsatility and this effect is unlikely to be mediated through hypothalamic release of GHRH.

The importance of SRIH withdrawal in the generation of endogenous GH pulsatility in humans has also been investigated through the effects of continuous SRIH administration. In young healthy men continuous intravenous infusion of SRIH decreased the amplitude and frequency of GH pulses, however there were still detectable GH secretory episodes (56). In our experience, continuous subcutaneous infusion of a long-acting SRIH analog, octreotide, at a dose of 400 μg per 24 hours, suppressed the amplitude but not the occurrence of GH pulses (Figure 5). These data suggest that it is unlikely that periodic SRIH declines generate GH pulsatility in humans. Importantly, Sassolas et al. (57) showed that GH pulses persisted during continuous combined infusion of the long acting SRIH analog, BIM 23014, plus GHRH. This finding indicates that there could be additional hypothalamic factors controlling GH secretion besides GHRH and SRIH.

SRIH does, however, likely play a role in modifying the GH response to GHRH and the amplitude of GH pulses in humans. In GH deficient children the magnitude of GH response to equal repeated doses of GHRH varies throughout the day, suggesting fluctuations of SRIH (58). Similarly, in middle-aged men there were fluctuations of the GH response to repetitive stimulation with boluses of GHRH every 2 hours (59). Moreover, the daily pattern of responsiveness to GHRH was parallel to the pattern of spontaneous GH secretion (Figure 6). This suggested a model in which slow changes in SRIH set the pituitary

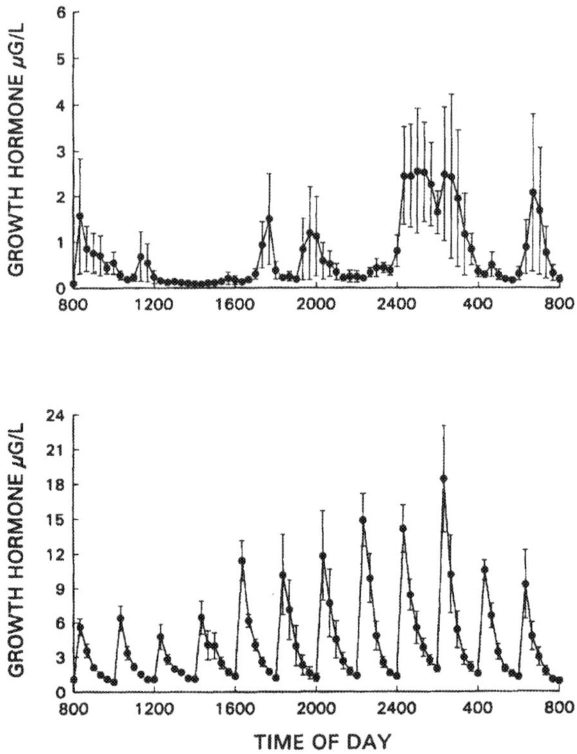

Figure 6. Plasma GH profiles (mean ± SE) from middle-aged men at baseline (upper panel) and during IV bolus administration of GHRH every 2 hours (lower panel). The daily pattern of responsiveness to GHRH is parallel to the pattern of the endogenous GH secretion. Reproduced from Jaffe, C.A., Turgeon, D.K., DeMott-Friberg, R., Watkins, P.B., Barkan, A.L. Nocturnal augmentation of growth hormone (GH) secretion is preserved during repetitive bolus administration of GH-releasing hormone: potential involvement of endogenous somatostatin. J. Clin. Endocrinol. Metab. (1995) 80(11), 3321–3326, by copyright permission of the Endocrine Society.

responsiveness to GHRH. Presumably, nocturnal augmentation of GH secretion is a response, in part, to low SRIH secretion at that time of the day.

In conclusion, in humans there is good evidence that GHRH is the primary driving force of GH pulses. Based on the *in vivo* data in sheep (28,29), SRIH is probably secreted in a pulsatile fashion, however its fluctuations are most likely not responsible for the generation of GH pulses. SRIH levels could be important in regulating trough GH levels and in controlling GH pulse amplitude by influencing the responsiveness of somatotrophs to GHRH. It is also likely involved in the nocturnal augmentation of GH secretion.

ROLE OF GH SECRETAGOGUES (GHS) IN GH PULSATILITY

GH-releasing peptides (GHRPs) and the non-peptidyl GHS are synthetic molecules that stimulate GH release both *in vitro* and *in vivo*, in animals and in humans. For the purposes of this review we will refer to all these molecules with the term GHS. It is now clear that GHS act through distinct receptors and that their action is distinct from the action of GHRH.

The GHS receptor has recently been cloned from swine and human pituitary and hypothalamic tissue (20,21). Therefore, it is likely that an endogenous substance with action similar to GHS exists and plays a significant role in the physiology of GH secretion. The precise role of the endogenous GHS ligand in the regulation of GH pulsatility is currently unknown. Isolation of the natural ligand and creation of models mimicking its selective deficiency are crucial steps for understanding the physiological importance of GHS.

GHS appear to regulate intracellular pathways independent of GHRH and SRIH. GHRH stimulates GH secretion by increasing intracellular cAMP via the protein kinase A pathway. In contrast GHS activate the protein kinase C pathway. They increase the intracellular concentration of inositol triphosphate and Ca^{++}, activate L-type Ca^{++} channels, and inhibit K^+ channels, causing somatotroph depolarization (20,60). Functional antagonism of SRIH by GHS has been proposed based on the inhibition of K^+ channels by GHS. This action is opposite to the effect of SRIH, that causes hyperpolarization of somatotrophs by activating K^+ channels (61,62). Indeed, *in vitro* SRIH antagonizes the stimulatory effects of GHS (63,64).

In vitro, it is clear that GHS and GHRH release GH through different pathways. The GH releasing activity of GHS is weaker than that of GHRH. The maximum effect of GHRP in perifused rat anterior pituitary cells was less than 50% of the maximum effect of GHRH (65). However, combined administration of GHRH and GHS had a synergistic effect (66). GHS alone had no effect on intracellular cAMP, but they amplified the effect of GHRH on cAMP accumulation (66). Also, GHRH-ant did not block the action of GHS, and the action of GHRH was not inhibited by a competitive GHS antagonist (15). Moreover, GHRH and GHS each caused rapid desensitization to its own action, whereas the responsiveness to the opposite peptide remained intact (63,66).

Although *in vitro* data indicate that GHS can release GH alone, other data suggest that an intact hypothalamic-pituitary unit is important for GHS action. In hypothalamic-pituitary incubate GH response to GHS was greater than in pituitary alone incubate, indicating that the presence of intact hypothalamus is needed for the full action of GHS (15). However the mechanisms of interaction of GHS with other hypothalamic factors, especially with GHRH and SRIH is not fully understood.

In contrast to the *in vitro* studies with pituitary cultures, *in vivo* the synergy of GHS and GHRH is much more pronounced (15). Moreover, sustained exposure to GHS *in vivo* amplifies the GH secretion for at least 24 hours (67), whereas somatotrophs are rapidly desensitized to GHS *in vitro* (65). Studies in animals and humans have proposed a variety of potential explanations for the *in vivo* action of GHS: (a) direct action on pituitary GHS receptors, (b) indirect effect through release of hypothalamic GHRH, (c) functional antagonism of SRIH and/or inhibition of hypothalamic SRIH release, (d) mediation through a yet unknown hypothalamic factor (factor U). More than one mechanism is likely involved. It has been suggested that GHS has a dual pituitary and hypothalamic action, with the latter being probably more important.

The above *in vitro* data indicate that GHS can have some GH releasing effect on the pituitary even without hypothalamic input. The difficulties of studying isolated pituitaries *in vivo* are obvious. In one study, in order to eliminate the effect of GHRH and SRIH on GH secretion hypophysectomized rats transplanted with two pituitaries under the renal capsule were studied (68). In these animals there was clear GH response to GHS, demonstrating the ability of GHS to independently stimulate GH secretion *in vivo*.

In vivo animal studies indicate that the full effect of GHS requires the presence of GHRH. For example, GHS do not have GH releasing activity in the *lit/lit* mouse, which lacks functional GHRH receptors (69). In addition, passive immunization of rats with GHRH antiserum almost completely inhibited the response of GH to GHS (15,70,75). In swine, hypothalamic–pituitary stalk transection significantly decreased the GH response to the GHS L-692,585, but the GH response to combined GHRH/GHS bolus was similar to the GH rise in the intact animals (71). Based on these data, mediation of GHS action through hypothalamic release of GHRH has been proposed. In support of this hypothesis increased electrical activity and c-*fos* mRNA expression following the administration of GHS have been observed at the level of arcuate nucleus (72,73). This is the principal location of GHRH neuronal bodies. A critical experiment is the direct measurement of GHRH and SRIH in the hypophysial portal blood following GHS administration. In sheep, GHS do not appear to alter SRIH secretion, however, injection of hexarelin caused a prompt GH response that was preceded by an elevation of GHRH levels in the hypophysial portal blood (74). The authors suggested a dual action of GHS via activation of hypothalamic GHRH neurons and direct action on the pituitary cells. However this is not a uniform finding. Smith et al. did not find consistent GHRH responses in the hypophysial-portal blood of sheep injected with the GHS L-692,585 (20). Clearly, further direct investigations are needed.

Based on the cellular mechanism of GHS and SRIH action, functional antagonism of SRIH has been thought to be an important role of GHS. In male rats, continuous administration of GHS disrupted the 3-hour endogenous GH pulsatility, which is thought to be controlled by the hypothalamic SRIH fluctuations, and abolished the cyclic refractoriness to repetitive boluses of GHRH (75). These effects were remarkably similar to those of SRIH immunoneutralization.

Whether the combined potential effects of GHRH release and SRIH antagonism could account for all the experimental observations on GHS is uncertain. In order to explain the synergistic interaction between GHS and GHRH, Bowers has hypothesized the existence of an unknown hypothalamic factor (U factor) (15). The U factor would be induced by GHS, require GHRH for its action and act synergistically with GHRH at the pituitary level.

In humans continuous intravenous administration of GHS enhances the endogenous GH pulsatile secretion and the GH response to GHRH, suggesting functional antagonism of SRIH by GHS (76,77). Huhn et al. (76) showed that continuous GHS infusion increased GH secretion rate and also increased the number and amplitude of spontaneous GH pulses and the interpulse valley concentration. The response to subsequent bolus of GHS was partially attenuated but the response to GHRH was increased. Similarly, Jaffe et al. (77) found that continuous infusion of GHS augmented spontaneous GH secretion and increased the GH response to GHRH. Moreover it was shown that GHS infusion was least effective in increasing GH secretion during the period corresponding to the natural nocturnal augmentation of GH secretion. This is the time of nadir of somatostatinergic tone (59). This observation is consistent with a role of GHS as functional SRIH antagonists. However, in the same study no effect of GHS on GH pulse frequency or on the interpulse GH concentrations was found. If SRIH is responsible for interpulse trough levels of GH and GHS are functional SRIH antagonists, elevation of the interpulse GH concentrations during GHS infusion would be expected. In these two studies the GH assays that were used (an immunoradiometric assay with limited sensitivity and a radioimmunoassay, respectively) were unable to measure all the trough GH levels, since many of them were at or below the detection limit of the assays. Subsequent studies using more sensitive GH assays have demonstrated an increase in interpulse GH concentrations during continuous GHS administration. Chapman et al. (78), using a sensitive chemiluminescence GH assay also showed that continuous infusion of the GHS L-692,429 caused an increase in mean GH concentrations, with increase in the amplitude of GH pulses but no change in the number of pulses. Similarly, chronic administration of the GHS MK-677 to healthy elderly subjects enhanced GH pulse amplitude, without a change in the number of GH pulses (79). The nadir GH concentrations between pulses were increased. These data support the hypothesis that GHS act through functional antagonism of SRIH.

In humans, as in other animals, GHS and GHRH induce GH secretion through different mechanisms. However, the two peptides administered together have a synergistic effect (80). Several investigators have addressed the question whether the hypothalamic action of GHS in humans is mediated through GHRH release. As reviewed above, animal studies support the hypothesis of GHS action through GHRH release. In contrast, in children with pituitary stalk transection and in adults with hypothalamic–pituitary disconnection the GH response to GHS alone or in combination with GHRH is impaired (81), and the synergistic action of GHRP and GHRH is lost (82). However, in these studies the decreased GH response to GHS and GHRH could be due to decreased responsiveness of somatotrophs because of chronic deprivation of hypothalamic stimulation. Similar to the findings in animals, in humans there is no cross-desensitization between GHS and GHRH. Continuous infusion of GHRH for 4 hours abolished the GH response to a bolus dose of GHRH, but the acute response to a bolus of GHS was fully preserved (83) (Figure 7). From these data in humans it can be concluded that GHS action is not mediated by release of endogenous GHRH alone. However, the GH response to a GHRP-6 bolus was suppressed by almost 80% by GHRH-ant, indicating that the presence of endogenous GHRH is essential for the full *in vivo* effect of GHS (84) (Figure 8). If the role of GHRH as a direct mediator of GHS action is less likely, an essential permissive role of GHRH can be hypothesized. The lack of complete suppression of GH response to GHS by GHRH-ant indicates that other hypothalamic factors, besides GHRH, may play a role in mediating GHS action. Alternatively

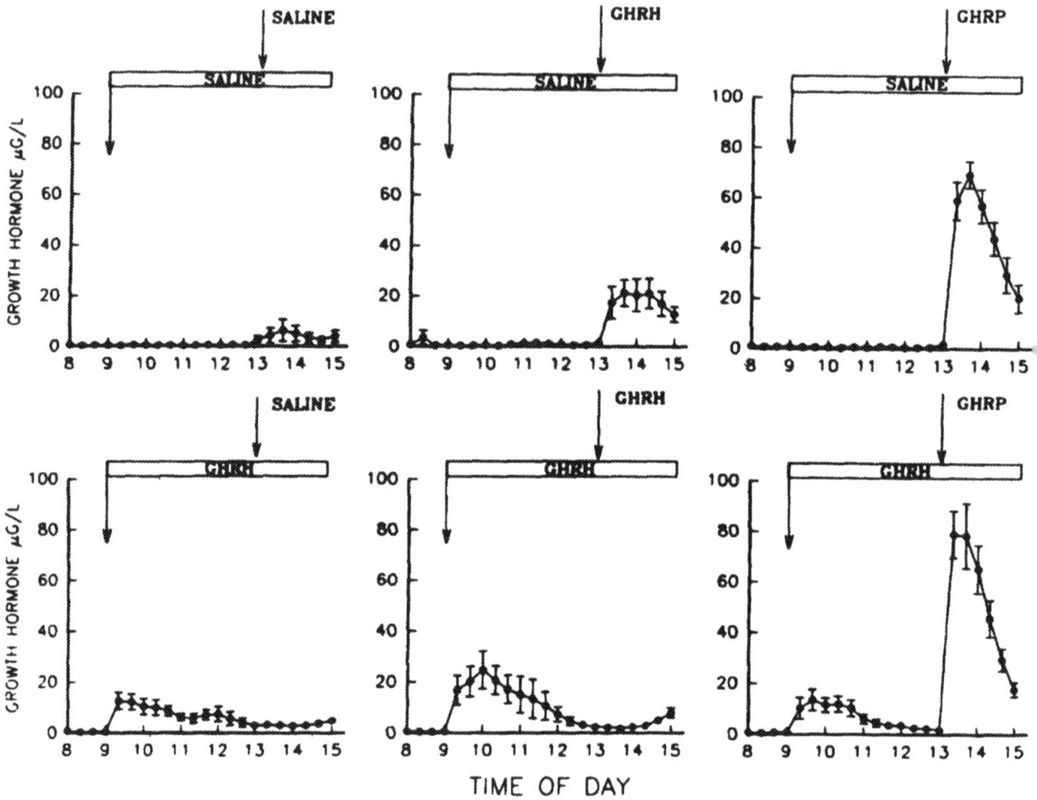

Figure 7. Plasma GH concentrations (mean ± SE) in five normal men. GH response to IV boluses of normal saline, GHRH and GHRP during IV infusion of normal saline (upper panels) or GHRH 1 µg/kg/h/ (lower panels). During continuous infusion of GHRH the GH response to a bolus dose of GHRH was abolished, but the acute response to a bolus of GHS was fully preserved. Reproduced from Robinson, B.M., DeMott-Friberg, R., Bowers, C.Y., Barkan, A.L. Acute growth hormone (GH) response to GH-releasing hexapeptide in humans is independent of endogenous GH-releasing hormone. J. Clin. Endocrinol. Metab. (1992) 75(4), 1121–1124, by copyright permission of the Endocrine Society.

GHS action could be a combined effect of hypothalamic GHRH release and direct GHS pituitary action. It is possible that the direct action of GHS on somatotrophs regulates sensitivity to GHRH and the mechanisms leading to GHRH desensitization.

It is conceivable that pulsatile endogenous GHS release is important in the generation of GH pulses. Recently, Leong et al. (85), using a functional assay for the natural GHS ligand, showed a correlation between natural ligand secretion bursts in the hypothalamic-portal plasma in sheep and GH pulses in the peripheral blood. Their results suggest that an endogenous GHS ligand is an authentic hormone and may play an important role as a pulse generator for GH. Isolation of the endogenous ligand for the GHS receptor will greatly expand our understanding of the role of GHS.

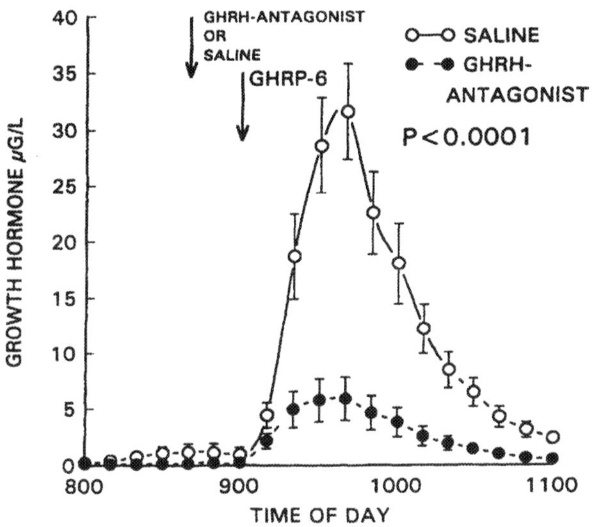

Figure 8. Mean GH responses to the GHS GHRP-6 IV bolus following normal saline or GHRH-ant treatment. The GH response to the GHRP-6 bolus was suppressed by almost 80% by GHRH-ant. Reproduced from Pandya, N., DeMott-Friberg, R., Bowers, C.Y., Barkan, A.L., Jaffe, C.A. Growth hormone (GH)-releasing peptide-6 requires endogenous hypothalamic GH-releasing hormone for maximal GH stimulation. J. Clin. Endocrinol. Metab. (1998) 83(4), 1186–1189, by copyright permission of the Endocrine Society.

NEGATIVE FEEDBACK REGULATION OF GH SECRETION

Similar to other hormonal systems, GH synthesis and secretion are subjected to the negative feedback regulation. In fact, both GH and IGF-I can function in this capacity, providing a two-tier system of feedback regulation. Likely, these levels of regulation fulfill distinct regulatory functions.

IGF-I

The importance of IGF-I in the negative feedback of GH secretion is demonstrated in studies in patients with GH insensitivity syndrome. These patients have elevated GH levels and low IGF-I levels. Administration of recombinant IGF-I leads to a decrease in GH secretion (86,87). The level of the feedback effect by IGF-I is not clear. Early *in vitro* data from Berelowitz et al. (88) showed that IGF-I suppressed GH release by acting directly at the pituitary level and also by augmenting hypothalamic SRIH secretion. In the *dw/dw* rat, systemic IGF-I does not affect hypothalamic GHRH or SRIH mRNA levels (89). However in the same model, intracerebroventricular IGF-I decreases GHRH mRNA and increases SRIH mRNA. Since IGF-I and its receptors are found in the hypothalamus (90,91), and systemic GH treatment increases hypothalamic IGF-I synthesis and secretion (92), a central effect of IGF-I was postulated. This may occur via a direct effect on GHRH neurons or through augmentation of SRIH neuronal activity.

Early approaches to defining the role of IGF-I in the regulation of GH secretion used caloric deprivation as a method of changing IGF-I concentration. In sheep, undernutrition-induced decline in IGF-I was associated with heightened GH secretion and lowered SRIH levels in the pituitary-portal blood (51). In humans, fasting was accompanied by gradual fall in circulating IGF-I and augmentation of GH pulse amplitude (93). However, in the same subjects, GH responsiveness to GHRH or TSH responsiveness to TRH did not change, casting doubt on the potential involvement of SRIH in humans. The model of fasting is, however, multifactorial and likely alters multiple GH regulatory steps.

Availability of recombinant IGF-I allowed more direct studies to be performed. Jaffe et al. (1) administered IGF-I to healthy young men and women. In both sexes IV infusion of IGF-I, 10 μg/kg/h, suppressed plasma GH, but the effect was more prominent in men (Figure 9). In both sexes, inhibition of GH pulse amplitude was the most obvious neuro-endocrine event but in men it was associated with a parallel suppression of the GH

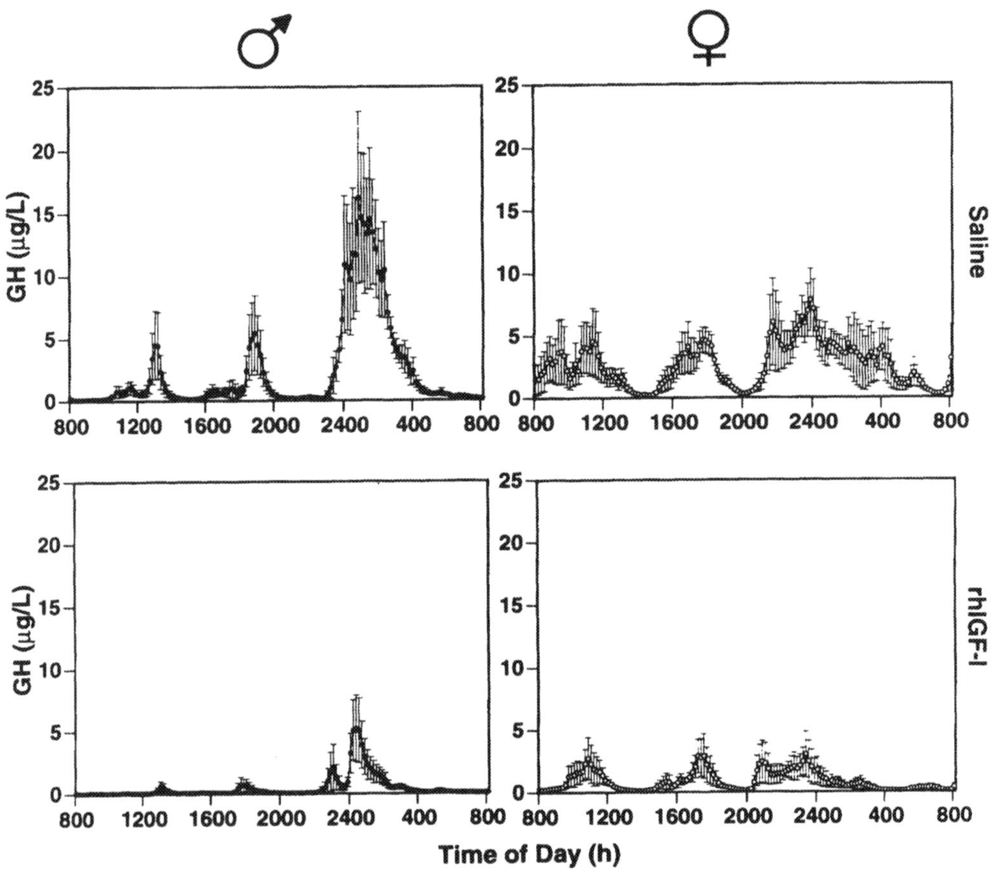

Figure 9. Plasma GH concentration profiles in six men and seven women during intravenous normal saline and recombinant human IGF-I infusions. In both genders IV infusion of IGF-I 10 μg/kg/h, suppressed plasma GH, but the effect was more prominent in men. Reproduced from Jaffe, C.A., Ocampo-Lim, B., Guo, W. et al. Regulatory mechanisms of growth hormone secretion are sexually dimorphic. J. Clin. Invest. (1998) 102(1), 153–164, by copyright permission of the American Society of Clinical Investigation.

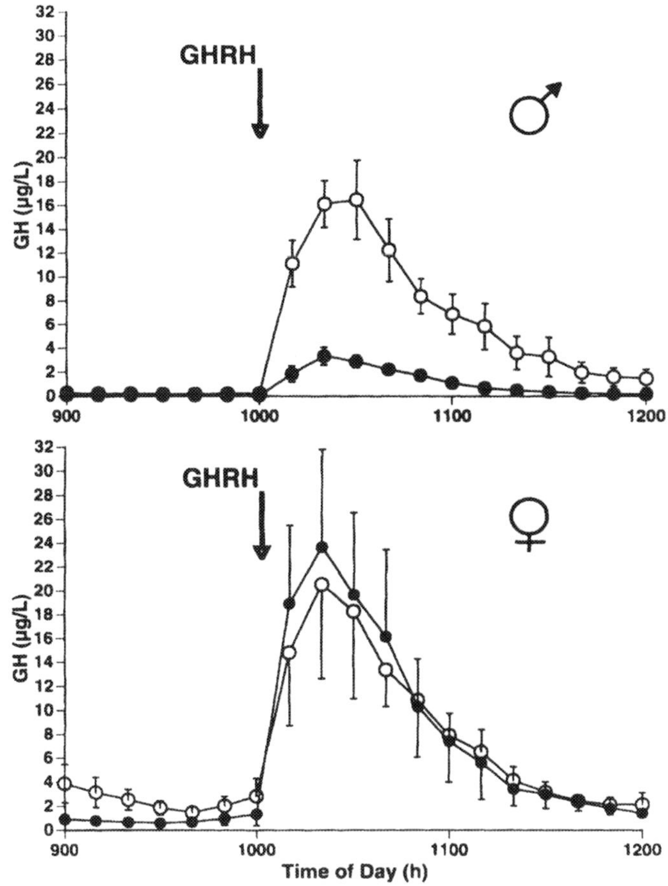

Figure 10. Plasma GH responses (mean ± SE) to IV bolus of GHRH in men (upper panel) and women (lower panel) during normal saline (open circles) and recombinant human IGF-I (closed circles) infusions. Reproduced from Jaffe, C.A., Ocampo-Lim, B., Guo, W. et al. Regulatory mechanisms of growth hormone secretion are sexually dimorphic. J. Clin. Invest. (1998) 102(1), 153–164, by copyright permission of the American Society of Clinical Investigation.

responsiveness to GHRH whereas in women the latter was unaffected (Figure 10). Thus, there appears to exist a sexual dimorphism of the negative IGF-I feedback mechanism. In men, it likely involves augmentation of SRIH release, whereas in women, suppression of GHRH secretion may be more prominent. The potential effects of gonadal hormone milieu are, therefore, of significant interest.

Although the negative feedback effect of IGF-I on GH secretion is clear, it is unlikely to have a role in the generation of endogenous GH pulses. Chapman et al. (94) have shown that the time course of the suppressive effect of free IGF-I upon GH secretion is relatively slow, with the GH changes occurring within one to two hours. Moreover, plasma levels of IGF-I appear to be very stable throughout the day despite periodic discharges of GH. Thus, IGF-I more likely functions as a long-term (hours to days) adjuster of the overall magnitude of GH secretion.

Growth Hormone

GH is likely to reach the hypothalamus via the retrograde pituitary portal blood flow and the GH receptors were found in the arcuate nucleus (95). Indeed, earlier studies in spontaneously dwarf mutant animal models with low GH levels (96) or in hypophysectomized rats (92,96,97) have demonstrated increased hypothalamic GHRH mRNA and low SRIH mRNA levels. Both were brought back to the normal ranges by exogenous GH (92). This approach, however, could not differentiate between the direct effect of GH and the secondary rise in central IGF-I. More recently, animals with centrally-expressed human GH transgene were shown to have low hypothalamic GHRH mRNA (98). Whether the latter effect was caused by human GH acting directly on GHRH neurons or through SRIH or neuropeptide Y interneurons (99) remains uncertain. Attempts were made to measure GH responses to GHRH or to neuroactive drugs in humans treated with exogenous GH. These studies are difficult to interpret because of the problems created by the crossreactivity of recombinant GH with the endogenous hormone. Brain et al. (100) administered GH infusions creating circulating hormone levels within the physiologic range and have shown that the responses to GHRH remained normal in this model. However, levels of GH reaching the hypothalamus via the retrograde pituitary-portal flow are likely to be much higher than the peripheral circulating hormone levels (101). Thus, the negative GH feedback of its own secretion is likely to be present in humans. The elucidation of the neuroendocrine nature of the GH negative feedback would be of major interest, since it is the most logical mechanism whereby the GH pulse is terminated.

SUMMARY

Can the existing information on a multitude of GH regulators be integrated into a coherent mechanism(s) of GH pulsatility? Recently, Wagner et al. (102) attempted to perform mathematical modeling of GH pulses in male rats. According to this model, GH pulse is initiated by an acute GHRH discharge. Subsequent GH elevation suppresses its own secretion through a two-step, SRIH-related mechanism: initially, SRIH terminates GHRH release, and later it acts directly at the pituitary level. Gradual extinction of the SRIH effect allows GHRH to be secreted again, and the cycle repeats itself. Whether a similar mechanism can be invoked in humans is uncertain. Also, this model excludes the potential contribution of the putative endogenous GHS-like ligand. Initial strides in the elucidation of the neuroendocrine network governing GH pulsatility in humans have been made using GHRH-ant. Several other tools, such as the endogenous GHS and its antagonist, as well as SRIH antagonist, appear to be necessary to assemble enough information for a similar mathematical modeling in humans.

REFERENCES

1. Jaffe, C.A., Ocampo-Lim, B., Guo, W., et al. (1998) Regulatory mechanisms of growth hormone secretion are sexually dimorphic. J. Clin. Invest. 102, 153–164.

2. Martha, P.M., Rogol, A.D., Veldhuis, J.D., Kerrigan, J.R., Goodman, D.W., Blizzard, R.M. (1989) Alterations in the pulsatile properties of circulating growth hormone concentrations during puberty in boys. J. Clin. Endocrinol. Metab. 69, 563–570.
3. Ho, K.Y., Evans, W.S., Blizzard, R.M., et al. (1987) Effects of sex and age on the 24-hour profile of growth hormone secretion in man: Importance of endogenous estradiol concentrations. J. Clin. Endocrinol. Metab. 64, 51–58.
4. Faria, A.C.S., Bekenstein, L.W., Booth, R.A., et al. (1992) Pulsatile growth hormone release in normal women during the menstrual cycle. Clin. Endocrinol. 36, 591–596.
5. Iranmanesh, A., Lizarralde, G., Veldhuis, J.D. (1991) Age and relative adiposity are specific negative determinants of the frequency and amplitude of growth hormone (GH) secretory bursts and the half-life of endogenous GH in healthy men. J. Clin. Endocrinol. Metab. 73, 1081–1088.
6. Hartman, M.L., Veldhuis, J.D., Johnson, M.L., et al. (1992) Augmented growth hormone (GH) secretory burst frequency and amplitude mediate enhanced GH secretion during a two-day fast in normal men. J. Clin. Endocrinol. Metab. 74, 757–765.
7. Hizuka, N., Takano, K., Shizume, K., Tanaka, I., Honda, N., Ling, N.C. (1985) Plasma growth hormone (GH) and somatomedin C response to continuous growth hormone releasing factor (GRF) infusion in patients with GH deficiency. Acta Endocrinol. 110, 17–23.
8. Barkan, A.L., Stred, S.E., Reno, K., et al. (1989) Increased growth hormone pulse frequency in acromegaly. J. Clin. Endocrinol. Metab. 69, 1225–1233.
9. Isgaard, J., Carlsson, L., Isaksson, O.G.P., Jansson, J.O. (1988) Pulsatile intravenous growth hormone (GH) infusion to hypophysectomized rats increases insulin-like growth factor I messenger ribonucleic acid in skeletal tissues more effectively than continuous GH infusion. Endocrinology 123, 2605–2610.
10. Clark, R.G., Jansson, J.O., Isaksson, O., Robinson, I.C.A.F. (1985) Intravenous growth hormone: growth responses to patterned infusions in hypophysectomized rats. J. Endocrinol. 104, 53–61.
11. Kashimata, M., Hiramatsu, M., Minami, N. (1989) Differential secretory rhythm of growth hormone controls the number of hepatic epidermal growth factor receptors in the rat. J. Endocrinol. 123, 75–81.
12. Jansson, J.O., Oscarsson, J., Mode, A., Ritzén, E.M. (1989) Plasma growth hormone pattern and androgens influence the levels of corticosteroid-binding globulin in rat serum. J. Endocrinol. 122, 725–732.
13. Oscarsson, J., Olofsson, J., Bondjers, G., Edén, S. (1989) Differential effects of continuous versus intermittent administration of growth hormone to hypophysectomized female rate on serum lipoproteins and their apoproteins. Endocrinology 125, 1638–1649.
14. Shapiro, B.H., MacLeod, J.N., Pampori, N.A., Morrissey, J.J., Lapenson, D.P., Waxman, D.J. (1989) Signaling elements in the ultradian rhythm of circulating growth hormone regulating expression of sex-dependent forms of hepatic cytochrome P450. Endocrinology 125, 2935–2944.
15. Bowers, C.Y., Sartor, A.O., Reynolds, G.A., Badger, T.M. (1991) On the actions of the growth hormone-releasing hexapeptide, GHRP. Endocrinology 128, 2027–2035.
16. Tannenbaum, G.S., Ling, N. (1984) The interrelationship of growth hormone (GH)-releasing factor and somatostatin in generation of the ultradian rhythm of GH secretion. Endocrinology 115, 1952–1957.
17. Wehrenberg, W.B., Brazeau, P., Luben, R., Böhlen, P., Guillemin, R. (1982) Inhibition of the pulsatile secretion of growth hormone by monoclonal antibodies to the hypothalamic growth hormone releasing factor (GRF). Endocrinology 111, 2147–2148.
18. Magnan, E., Mazzocchi, L., Cataldi, M., et al. (1995) Effect of actively immunizing sheep against growth hormone-releasing hormone or somatostatin on spontaneous pulsatile and neostigmine-induced growth hormone secretion. J. Endocrinol. 144, 83–90.
19. Plotsky, P.M., Vale, W. (1985) Patterns of growth hormone-releasing factor and somatostatin secretion into the hypophysial-portal circulation of the rat. Science 230, 461–463.
20. Howard, A.D., Feighner, S.D., Cully, D.F., et al. (1996) A receptor in pituitary and hypothalamus that functions in growth hormone release. Science 273, 974–977.

21. Smith, R.G., Pong, S.S., Hickey, G., et al. (1996) Modulation of pulsatile GH release through a novel receptor in hypothalamus and pituitary gland. Recent Prog. Horm. Res. 51, 261–285.
22. Dutour, A., Briard, N., Guillaume, V. et al. (1997) Another view of GH neuroregulation: lessons from the sheep. Eur. J. Endocrinol. 136, 553–565.
23. Katakami, H., Downs, T.R., Frohman, L.A. (1988) Inhibitory effect of hypothalamic medial preoptic area somatostatin of growth hormone-releasing factor in the rat. Endocrinology 123, 1103–1109.
24. Müller, E.E. (1987) Neural control of somatotropic function. Physiol. Rev. 67, 962–1053.
25. Lumpkin, M.D., Mulroney, S.E., Haramati, A. (1989) Inhibition of pulsatile growth hormone (GH) secretion and somatic growth in immature rats with a synthetic GH-releasing factor antagonist. Endocrinology 124, 1154–1159.
26. Lumpkin, M.D., McDonald, J.K. (1989) Blockade of growth hormone releasing factor (GRF) activity in the pituitary and hypothalamus of the conscious rat with a peptidic GRF antagonist. Endocrinology 124, 1522–1531.
27. Hu, Z.Y., DeMott-Friberg, R., Barkan, A.L. (1993) Ontogeny of GH mRNA and GH secretion in male and female rats: regulation by GH-releasing hormone. Am. J. Physiol. 265, E236–E242.
28. Frohman, L.A., Downs, T.R., Clarke, I.J., Thomas, G.B. (1990) Measurement of growth hormone-releasing hormone and somatostatin in hypothalamic-portal plasma of unanesthetized sheep. J. Clin. Invest. 86, 17–24.
29. Cataldi, M., Magnan, E., Guillaume, V. et al. (1994) Relationship between hypophysial portal GHRH and somatostatin and peripheral GH levels in the conscious sheep. J. Endocrinol. Invest. 17, 717–722.
30. Jaffe ,C.A., DeMott-Friberg, R., Barkan, A.L. (1993) Suppression of growth hormone (GH) secretion by a selective GH-releasing hormone (GHRH) antagonist. J. Clin. Invest. 92, 695–701.
31. Ocampo-Lim, B., Guo, W., DeMott-Friberg, R., Barkan, A.L., Jaffe, C.A. (1996) Nocturnal growth hormone (GH) secretion is eliminated by infusion of GH-releasing hormone antagonist. J. Clin. Endocrinol. Metab. 81, 4396–4399.
32. Masuda, A., Shibasaki, T., Hotta, M., et al. (1990) Insulin induced hypoglycemia, L-dopa and arginine stimulate GH secretion through different mechanisms in man. Regul. Pept. 31, 53–64.
33. Ghigo, E., Bellone, J., Mazza, E. et al. (1990) Arginine potentiates the GHRH-but not the pyridostigmine-induced GH secretion in normal short children, further evidence for a somatostatin suppressing effect of arginine. Clin. Endocrinol. 32, 763–767.
34. Magnan, E., Cataldi, M., Guillaume, V. et al. (1993) Neostigmine stimulates growth hormone-releasing hormone release into hypophysial portal blood of conscious sheep. Endocrinology 132, 1247–1251.
35. Jaffe, C.A., DeMott-Friberg, R., Barkan, A.L. (1996) Endogenous growth hormone (GH)-releasing hormone is required for GH responses to pharmacological stimuli. J. Clin. Invest. 97, 934–940.
36. Hanew, K., Tanaka, A., Utsumi, A., Sugawara, A., Abe, K. (1996) The inhibitory effects of growth hormone-releasing hormone (GHRH)-antagonist on GHRH, L-dopa, and clonidine-induced GH secretion in normal subjects. J. Clin. Endocrinol. Metab. 81, 1952–1955.
37. Namba, H., Morita, S., Melmed, S. (1989) Insulin-like growth factor-I action on growth hormone secretion and messenger ribonucleic acid levels: interaction with somatostatin. Endocrinology 124, 1794–1799.
38. Fukata, J., Diamond, D.J., Martin, J.B. (1985) Effects of rat growth hormone (rGH)-releasing factor and somatostatin on the release and synthesis of rGH in dispersed pituitary cells. Endocrinology. 117:457-467.
39. Stachura, M.E., Tyler, J.M., Farmer, P.K. (1988) Combined effects of human growth hormone (GH)-releasing factor-44 (GRF) and somatostatin (SRIF) on post-SRIF rebound release of GH and prolactin: a model for GRF-SRIF modulation of secretion. Endocrinology 123, 1476–1482.
40. Weiss, J., Cronin, M.J., Thorner, M.O. (1987) Periodic interactions of GH-releasing factor and somatostatin can augment GH release *in vitro*. Am. J. Physiol. 253, E508–E514.

41. Daikoku, S., Hisamo, S., Kawano, H. et al. (1988) Ultrastructural evidence for neuronal regulation of growth hormone secretion. Neuroendocrinology 47, 405–415.
42. Epelbaum, J., Moyse, E., Tannenbaum, G.S., Kordon, C., Beaudet, A. (1989) Combined autoradiographic and immunohistochemical evidence for an association of somatostatin binding sites with growth hormone-releasing factor containing nerve cell bodies in the rat arcuate nucleus. J. Neuroendocrinol. 1, 109–115.
43. Liposits, Z., Merchenthaler, I., Paull, W.K., Flerkó, B. (1988) Synaptic communication between somatostatinergic axons and growth hormone-releasing factor (GRF) synthesizing neurons in the arcuate nucleus of the rat. Histochemistry 89, 247–252.
44. Horvath, S., Palkovits, M., Görcs, T., Arimura, A. (1989) Electron microscopic immunocytochemical evidence for the existence of bidirectional synaptic connections between growth hormone-releasing hormone and somatostatin-containing neurons in the hypothalamus of the rat. Brain Res. 481, 8–15.
45. Murakami, Y., Kato, Y., Kabayama, Y., Inoue, T., Koshiyama, H., Imura, H. (1987) Involvement of hypothalamic growth hormone (GH)-releasing factor in GH secretion induced by intracerebroventricular injection of somatostatin in rats. Endocrinology 120, 311–316.
46. Katakami, H., Arimura, A., Frohman, L.A. (1986) Growth hormone (GH)-releasing factor stimulates hypothalamic somatostatin release: an inhibitory feedback effect on GH secretion. Endocrinology 118, 1872–1877.
47. Miki, N., Ono, M., Shizume, K. (1988) Withdrawal of endogenous somatostatin induces secretion of growth hormone-releasing factor in rats. J. Endocrinol. 117, 245–252.
48. Thomas, C.R., Groot, K., Arimura, A. (1985) Antiserum to rat growth hormone (GH)-releasing factor suppresses but does not abolish antisomatostatin-induced GH release in the rat. Endocrinology 116, 2174–2178.
49. Clark, R.G., Carlsson, M.S., Rafferty, B., Robinson, I.C.A.F. (1988) The rebound release of growth hormone (GH) following somatostatin infusion in rats involves hypothalamic GH-releasing factor release. J. Endocrinol. 119, 397–404.
50. Terry, L.C., Martin, J.B. (1981) The effects of lateral hypothalamic-medial forebrain stimulation and somatostatin antiserum on pulsatile growth hormone secretion in freely behaving rats: evidence for a dual regulatory mechanism. Endocrinology 109, 622–627.
51. Thomas, G.B., Cummins, J.T., Francis, H., Sudbury, A.W., McCloud, P.I., Clarke, I.J. (1991) Effect of restricted feeding on the relationship between hypophysial portal concentrations of growth hormone (GH)-releasing factor and somatostatin, and jugular concentrations of GH in ovariectomized ewes. Endocrinology 128, 1151–1158.
52. Vance, M.L., Kaiser, D.L., Evans, W.S. et al. (1985) Pulsatile growth hormone secretion in normal men during a continuous 24-hour infusion of human growth hormone releasing factor (1–40). J. Clin. Invest. 75, 1584–1590.
53. Hartman, M.L., Faria, A.C.S., Vance, M.L., Johnson, M.L., Thorner, M.O., Veldhuis, J.D. (1991) Temporal structure of *in vivo* growth hormone secretory events in humans. Am. J. Physiol. 260, E101–E110.
54. Hindmarsh, P.C., Brain, C.E., Robinson, I.C.A.F., Matthews, D.R., Brook, C.G.D. (1991) The interaction of growth hormone releasing hormone and somatostatin in the generation of a GH pulse in man. Clin. Endocrinol. 35, 353–360.
55. Ho, P.J., Kletter, G.B., Hopwood, N.J., DeMott-Friberg, R., Barkan, A.L. (1993) Somatostatin withdrawal alone is an ineffective generator of pulsatile growth hormone release in man. Acta Endocrinol. 129, 414–418.
56. Calabresi, E., Ishikawa, E., Bartoloni, L. et al. (1996) Somatostatin infusion suppresses GH secretory burst frequency and mass in normal men. Am. J. Physiol. 270, E975–E979.
57. Sassolas, G., Khalfallah, Y., Chayvialle, J.A. et al. (1989) Effects of the somatostatin analog BIM 23014 on the secretion of growth hormone, thyrotropin, and digestive peptides in normal men. J. Clin. Endocrinol. Metab. 68, 239–246.
58. Martha, P.M., Blizzard, R.M., McDonald, J.A., Thorner, M.O., Rogol, A.D. (1988) A persistent pattern of varying pituitary responsivity to exogenous growth hormone (GH)-releasing hormone

in GH-deficient children: evidence supporting periodic somatostatin secretion. J. Clin. Endocrinol. Metab. 67, 449–454.
59. Jaffe, C.A., Turgeon, D.K., DeMott Friberg, R., Watkins, P.B., Barkan, A.L. (1995) Nocturnal augmentation of growth hormone (GH) secretion is preserved during repetitive bolus administration of GH-releasing hormone: potential involvement of endogenous somatostatin. J. Clin. Endocrinol. Metab. 80, 3321–3326.
60. Smith, R.G., Van der Ploeg, L.H.T., Howard, A.D. et al. (1997) Peptidomimetic regulation of growth hormone secretion. Endocr. Rev. 18, 621–645.
61. Koch, B.D., Blalock, J.B., Schonbrunn, A. (1988) Characterization of the cyclic AMP-independent actions of somatostatin in GH cells. An increase in potassium conductance is responsible for both the hyperpolarization and the decrease in intracellular free calcium produced by somatostatin. J. Biol. Chem. 263, 216–225.
62. Yatani, A., Birnbaumer, L., Brown, A.M. (1990) Direct coupling of the somatostatin receptor to potassium channels by a G protein. Metabolism 39, 91–95.
63. Blake, A.D., Smith, R.G. (1991) Desensitization studies using perifused rat pituitary cells show that growth hormone-releasing hormone and His-D-Trp-Ala-Trp-D-Phe-Lys-NH_2 stimulate growth hormone release through distinct receptor sites. J. Endocrinol. 129, 11–19.
64. Goth, M.I., Lyons, C.E., Canny, B.J., Thorner, M.O. (1992) Pituitary adenylate cyclase activating polypeptide, growth hormone (GH)-releasing peptide and GH-releasing hormone stimulate GH release through distinct pituitary receptors. Endocrinology 130, 939–944.
65. Badger, T.M., Millard, W.J., McCormick, G.F., Bowers, C.Y., Martin, J.B. (1984) The effects of growth hormone (GH)-releasing peptides on GH secretion in perifused pituitary cells of adult male rats. Endocrinology 115, 1432–1438.
66. Cheng, K., Chan, W.W.S., Barreto, A., Convey, E.M., Smith, R.G. (1989) The synergistic effects of His-D-Trp-Ala-Trp-D-Phe-Lys-NH_2 on growth hormone (GH)-releasing factor-stimulated GH release and intracellular adenosine 3′,5′-monophosphate accumulation in rat primary pituitary cell culture. Endocrinology 124, 2791–2798.
67. Jaffe, C.A., Ho, P.J., DeMott-Friberg, R., Bowers, C.Y., Barkan, A.L. (1993) Effects of a prolonged growth hormone (GH)-releasing peptide infusion on the pulsatile GH secretion in normal men. J. Clin. Endocrinol. Metab. 77, 1641–1647.
68. Mallo, F., Alvarez, C.V., Benitez, L. et al. (1993) Regulation of His-dTrp-Ala-Trp-dPhe-Lys-NH_2 (GHRP-6)-induced GH secretion in the rat. Neuroendocrinology 57, 247–256.
69. Dickson, S.L., Doutrelant-Viltart, O., Leng, G. (1995) GH-deficient *dw/dw* rats and *lit/lit* mice show increased Fos expression in the hypothalamic arcuate nucleus following systemic injection of GH-releasing peptide-6. J. Endocrinol. 146, 519–526.
70. Conley, L.K., Teik, J.A., Deghenghi, R. et al. (1995) Mechanism of action of hexarelin and GHRP-6: analysis of the involvement of GHRH and somatostatin in the rat. Neuroendocrinology 61, 44–50.
71. Hickey, G.J., Drisko, J., Faidley, T. et al. (1996) Mediation by the central nervous system is critical to the *in vivo* activity of the GH secretagogue L-692,585. J. Endocrinol. 148, 371–380.
72. Dickson, S.L., Leng, G., Dyball, R.E.J., Smith, R.G. (1995) Central actions of peptide and non-peptide growth hormone secretagogues in the rat. Neuroendocrinology 61, 36–43.
73. Dickson, S.L., Luckman, S.D. (1997) Induction of *c-fos* messenger ribonucleic acid in neuropeptide Y and growth hormone (GH)-releasing factor neurons in the rat arcuate nucleus following systemic injection of the GH secretagogue, GH-releasing peptide-6. Endocrinology 138, 771–777.
74. Guillaume, V., Magnan, E., Cataldi, M. et al. (1994) Growth hormone (GH)-releasing hormone secretion is stimulated by a new GH-releasing hexapeptide in sheep. Endocrinology 135, 1073–1076.
75. Clark, R.G., Carlsson, L.M.S., Trojnar, J., Robinson, I.C.A.F. (1989) The effects of a growth hormone-releasing peptide and growth hormone-releasing factor in conscious and anaesthetized rats. J. Neuroendocrinol. 1, 249–255.

76. Huhn, W.C., Hartman, M.L., Pezzoli, S.S., Thorner, M.O. (1993) Twenty-four-hour growth hormone (GH)-releasing peptide (GHRP) infusion enhances pulsatile GH secretion and specifically attenuates the response to a subsequent GHRP bolus. J. Clin. Endocrinol. Metab. 76, 1202–1208.
77. Jaffe, C.A., Ho, P.J., DeMott-Friberg, R., Bowers, C.Y., Barkan, A.L. (1993) Effects of a prolonged growth hormone (GH)-releasing peptide infusion on pulsatile GH secretion in normal men. J. Clin. Endocrinol. Metab. 77, 1641–1647.
78. Chapman, I.M., Hartman, M.L., Pezzoli, S.S., Thorner, M.O. (1996) Enhancement of pulsatile growth hormone secretion by continuous infusion of a growth hormone-releasing peptide mimetic, L-692,429, in older adults — A Clinical Research Center study. J. Clin. Endocrinol. Metab. 81, 2874–2880.
79. Chapman, I.M., Bach, M.A., Van Cauter, E. et al. (1996) Stimulation of the growth hormone (GH)-insulin-like growth factor I axis by daily oral administration of GH secretagogue (MK-677) in healthy elderly subjects. J. Clin. Endocrinol. Metab. 81, 4249–4257.
80. Bowers, C.Y., Reynolds, G.A., Durham, D., Barrera, C.M., Pezzoli, S.S., Thorner, M.O. (1990) Growth hormone (GH)-releasing peptide stimulates GH release in normal men and acts synergistically with GH-releasing hormone. J. Clin. Endocrinol. Metab. 70, 975–982.
81. Pombo, M., Barrerro, J., Peñalva, A., Peino, R., Dieguez, C., Casanueva, F.F. (1995) Absence of growth hormone (GH) secretion after the administration of either GH-releasing hormone (GHRH), GH-releasing peptide (GHRP-6), or GHRH plus GHRP-6 in children with neonatal pituitary stalk transection. J. Clin. Endocrinol. Metab. 80, 3180–3184.
82. Popovic, V., Damjanovic, S., Micic, D., Djurovic, M., Dieguez, C., Casanueva, F.F. (1995) Blocked growth hormone-releasing peptide (GHRP-6)-induced GH secretion and absence of the synergic action of GHRP-6 plus GH-releasing hormone in patients with hypothalamo-pituitary disconnection: evidence that GHRP-6 main action is exerted at the hypothalamic level. J. Clin. Endocrinol. Metab. 80, 942–947.
83. Robinson, B.M., DeMott-Friberg, R., Bowers, C.Y., Barkan, A.L. (1992) Acute growth hormone (GH) response to GH-releasing hexapeptide in humans is independent of endogenous GH-releasing hormone. J. Clin. Endocrinol. Metab. 75, 1121–1124.
84. Pandya, N., DeMott-Friberg, R., Bowers, C.Y., Barkan, A.L., Jaffe, C.A. (1998) Growth hormone (GH)-releasing peptide-6 requires endogenous hypothalamic GH-releasing hormone for maximal GH stimulation. J. Clin. Endocrinol. Metab. 83, 1186–1189.
85. Leong, D.A., Pomes, A., Veldhuis, J.D., Clarke, I.J. (1998) A novel hypothalamic hormone measured in hypophysial portal plasma drives rapid bursts of GH secretion. Proc. of the 80th Annual Meeting of the Endocrine Society 1998, p.64.
86. Cotterill, A.M., Camacho-Hubner C., Holly, J.M., Savage, M.O. (1993) The effect of recombinant human insulin-like growth factor-I treatment on growth hormone secretion in two subjects with growth hormone insensitivity (Laron syndrome). Clin. Endocrinol. 39, 119–122.
87. Vaccarello, M.A., Diamond, F.B. Jr, Guevara-Aguirre, J. et al. (1993) Hormonal and metabolic effects and pharmacokinetics of recombinant insulin-like growth factor-I in growth hormone receptor deficiency/Laron syndrome. J. Clin. Endocrinol. Metab. 77, 273–280.
88. Berelowitz, M., Szabo, M., Frohman, L.A., Firestone, S., Chu, L., Hintz, R.L. (1981) Somatomedin-C mediates growth hormone negative feedback by effects on both the hypothalamus and the pituitary. Science 212, 1279–1281.
89. Sato, M., Frohman, L.A. (1993) Differential effects of central and peripheral administration of growth hormone (GH) and insulin-like growth factor on hypothalamic GH-releasing hormone and somatostatin gene expression in GH-deficient dwarf rats. Endocrinology 133, 793–799.
90. Olchovsky, D., Song, J., Gelato, M.C. et al. (1993) Pituitary and hypothalamic insulin-like growth factor-I (IGF-I) and IGF-I receptor expression in food-deprived rats. Mol. Cell Endocrinol. 93, 193–198.
91. Aguado, F., Fernandez, T., Martinez-Murillo, R., Rodrigo, J., Cacicedo, L., Sanchez-Franco, F. (1992) Immunocytochemical localization of insulin-like growth factor I in the hypothalamo-hypophyseal system of the adult rat. Neuroendocrinology 56, 856–863.

92. Wood, T.L., Berelowitz, M., Gelato, M.C. et al. (1991) Hormonal regulation of rat hypothalamic neuropeptide mRNAs: effect of hypophysectomy and hormone replacement on growth-hormone-releasing factor, somatostatin and the insulin-like growth factors. Neuroendocrinology 53, 298–305.
93. Ho, P.J., Friberg, R.D., Barkan, A.L. (1992) Regulation of pulsatile growth hormone secretion by fasting in normal subjects and patients with acromegaly. J. Clin. Endocrinol. Metab. 75, 812–819.
94. Chapman, I.M., Hartman, M.L., Pieper, K.S. et al. (1998) Recovery of growth hormone release from suppression by exogenous insulin-like growth factor I (IGF-I): evidence for a suppressive action of free rather than bound IGF-I. J. Clin. Endocrinol. Metab. 83, 2836–2842.
95. Burton, K.A., Kabigting, E.B., Steiner, R.A., Clifton, D.K. (1995) Identification of target cells for growth hormone's action in the arcuate nucleus. Am. J. Physiol. 269, E716–E722.
96. Frohman, M.A., Downs, T.R., Chomczynski, P., Frohman, L.A. (1989) Cloning and characterization of mouse growth hormone-releasing hormone (GRH) complementary DNA: increased GRH messenger RNA levels in the growth hormone-deficient *lit/lit* mouse. Mol. Endocrinol. 3, 1529–1536.
97. Chomczynski, P., Downs, T.R., Frohman, L.A. (1988) Feedback regulation of GHRH gene expression by GH in rat hypothalamus. Mol. Endocrinol. 2, 236–241.
98. Szabo, M., Butz, M.R., Banerjee, S.A., Chikaraishi, D.M., Frohman, L.A. (1995) Autofeedback suppression of growth hormone (GH) secretion in transgenic mice expressing a human GH reporter targeted by tyrosine hydroxylase 5′-flanking sequences to the hypothalamus. Endocrinology 136, 4044–4048.
99. Chan, Y.Y., Steiner, R.A., Clifton, D.K. (1996) Regulation of hypothalamic neuropeptide-Y neurons by growth hormone in the rat. Endocrinology 137, 1319–1325.
100. Brain, C., Thakrar, D.N., Hindmarsh, P.C., Brooks, C.G. (1993) Physiological levels of growth hormone fail to suppress growth hormone releasing hormone (1-29) NH2-stimulated growth hormone secretion in man. J. Endocrinol. Invest. 16, 15–20.
101. Barkan, A.L., Jaffe, C.A., Padmanabhan, V. (1997) *In vivo* investigation of hypothalamic secretory activity. Trends Endocrinol. Metab. 8, 105–111.
102. Wagner, C., Caplan, R., Tannenbaum, G.S. (1998) Genesis of the ultradian rhythm of growth hormone secretion: a new model unifying experimental observations under different conditions. Proc. of the 80th Annual Meeting of the Endocrine Society 1998, p.293.

Growth Hormone Secretagogues
Edited by E. Ghigo, M. Boghen, F.F. Casanueva and C. Dieguez

Chapter 12

Hormonal Activities of Growth Hormone Secretagogues (GHS) across Human Lifespan

EMANUELA ARVAT, FABIO BROGLIO, ROBERTA GIORDANO, GIAMPIERO MUCCIOLI, MAURO MACCARIO, FRANCO CAMANNI and EZIO GHIGO

Division of Endocrinology, Department of Internal Medicine; Department of Pharmacology and Forensic Medicine, University of Turin, Italy

INTRODUCTION

Growth hormone secretagogues (GHS) are synthetic, non-natural peptidyl and non-peptidyl molecules. GH-releasing peptides (GHRPs), the first component of the GHS family, were invented rather than isolated in 1977, and are endowed with strong GH-releasing effect both in animals and in humans (1–4). The first GHRPs were derivatives of the pentapeptide Met-enkephalin but devoid of opiate activity; they showed low activity *in vitro* only (1). The first molecule active *in vitro* and *in vivo* was the hexapeptide GHRP-6, which releases GH in dose-dependent manner in several species and particularly in humans after intravenous, subcutaneous, intranasal and even oral administration (1–4).

Now the GHS family includes both peptidyl analogues, such as GHRP-1, a heptapeptide, GHRP-2 and Hexarelin, two hexapeptides (2–6), Tyr-Ala-Hexarelin, an octapeptide (7), Ipamorelin, a pentapeptide (8), some tetra- and pseudo-tripeptides (6), as well as non-peptidyl GHRP mimetics, such as MK-677, a spiroindoline, whose activity has been extensively studied in humans (9). One of the most important reasons which prompted the research in GHS field was to improve potency and bioavailability of GHS. In fact, peptidyl compounds have the disadvantage that they are less than 1% orally bioavailable and their effect is short lasting (2), while MK-677 shows more than 60% oral bioavailability and seems endowed with long-lasting effect (9). This evidence proposed MK-677 as the best orally active GHS.

Interestingly, the activity of GHS is not fully specific for GH release; in fact, they possess also slight prolactin- (PRL-), adrenocorticotropin hormone- (ACTH-) and cortisol-releasing effect, at least after acute administration (2–4,10). Moreover, GHS also have extra-hormonal activities; in fact, they have been reported to influence food intake and sleep pattern (11–13), while evidence is increasing about the effects of GHS on cardiac function, independently of their endocrine-releasing effects (14,15 and see other chapters in this book).

The endocrine activities of GHS are mediated by specific receptors, at the pituitary and hypothalamic levels (16–19); however, specific GHS receptors are present also in other CNS areas and in peripheral tissues (20,21). The human GHS receptor has already been characterized further pointing to the existence of an endogenous, natural GHS-like ligand, still unknown (22). Now the existence of specific GHS receptor subtypes has also been demonstrated (19,21) and they could mediate different GHS activities at different levels.

Many issues alluded to above are discussed elsewhere in this book. The aim of this chapter is to focus on reviewing data about the endocrine activities of GHS in humans in physiological conditions. Particular attention will be given to the endocrine responses as function of age and sex as well as to the mechanisms underlying these activities of GHS.

GHS RECEPTORS IN HUMANS

As in animals (see other chapters in this book), so also in humans the neuroendocrine effects of GHS are mediated at the pituitary and hypothalamic level, though present also in other CNS areas, such as the cerebral cortex, hippocampus, pons medulla and choroid plexuses (16–20). Animal and human GHS receptors subtypes have been demonstrated (19) and one of them has already been cloned (22). It belongs to the G-protein coupled receptor family and its sequence shows no significant homology with other receptors known so far (22). Interestingly, human pituitary and hypothalamic GHS receptor density does not seem sex-dependent, while advancing age does not affect pituitary but significantly decreases hypothalamic GHS receptor number (our unpublished results).

Notice that our recent studies demonstrate that GHS receptors, apparently distinct from that recently cloned, are also present in peripheral tissues, particularly in adrenal, heart, skeletal muscle, ovary, testis and lung (20,21). In some of these tissues, the GHS binding density is even higher than that recorded in pituitary and hypothalamus (20).

The presence of both central and peripheral specific GHS receptors strengthens hypothetical existence of an unidentified natural ligand.

ENDOCRINE EFFECTS OF GHS

GH-releasing activity

The GH-releasing effect of GHS is dose-related: the GH response to 1 μg/kg i.v. of peptidyl compounds is generally higher than that elicited by 1 μg/kg i.v. GHRH (2–4), its maximal effective dose (23). However, 1 μg/kg is not the maximal effective dose of GHRPs. In fact, dose-range studies indicate that 2.0 μg/kg is the maximal GH-releasing effective dose in humans, being higher than 1.0 μg/kg but similar to 3.0 μg/kg (24). Also non-peptidyl GHS have dose-related effect but their potency is lower than that of peptides (9). Interestingly, the GH response to intravenous (i.v.) GHS shows good intraindividual reproducibility (25), differently from that observed with GHRH (26).

A dose-related effect of GHS has also been shown after subcutaneous (s.c.), intranasal (i.n.) and even oral route of administration (2–4). While peptidyl GHS have low bioavailability after oral administration, MK-677, a spiroindoline non-peptidyl GHS, shows a bioavailability markedly higher than that of peptidyl compounds and, differently from other GHS, is endowed with long lasting GH-releasing effect (9).

The stimulating GHS activity on GH secretion does not depend on gender, at least in adulthood (4,25,27), while it is influenced by BMI. In fact, it is reduced, though still higher than that of GHRH, in obesity (28–31), though it has been reported preserved or even reduced in anorexia nervosa (32,33).

Particularly, the somatotrope responsiveness to GHS undergoes marked age-related variations, which are clearly different from those to GHRH. In agreement with the existence of GHS receptors as early as in human fetal pituitary (34), the GH-releasing activity of Hexarelin is present in newborns when, however, it is lower than that of GHRH (35). The GH response to GHS has been found similar to that to GHRH in prepubertal children by some (36) and higher by other authors (37). Then, it clearly increases at puberty, persists similar in young adults and then decreases with aging, being clearly reduced by the sixth decade of life (4,24,36–45) (Figure 1). Anyway, as well as in pubertal children and adults (25,36,46,47), even in elderly subjects the GH response to GHS is higher than that to GHRH (24,39–41). The age-related reduction of the GH response has been recorded after administration of both peptidyl and non-peptidyl GHS independently of the route of administration (48–51).

At present, the mechanisms underlying the age-related variations in the somatotrope responsiveness to GHS are not definitively clarified.

Some data suggest that the enhanced activity of GHS at puberty could depend on gonadal steroids, in particular on oestrogen increase. In fact, the GH response to Hexarelin is more marked in pubertal girls than in boys, being positively related to oestradiol levels (36). Moreover, the Hexarelin-induced GH rise in prepubertal children is enhanced by testosterone as well as by ethinyl-oestradiol but not by oxandrolone pretreatment (37,52).

While oestradiol could play a role in the increased GH-releasing activity of GHS at puberty, the fall of oestrogen levels in menopause does not seem to play a role in the reduction of the somatotrope response to GHS in this period of life. In fact, treatment with transdermal oestradiol does not modify the reduced GH response to Hexarelin in postmenopausal women (42). These data, together with the evidence that in young adults there

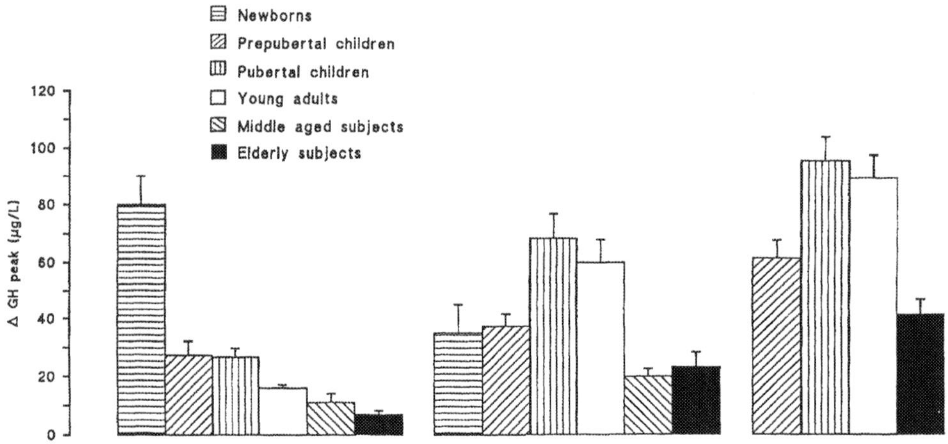

Figure 1. Age-related variations of the GH response to GHRH, Hexarelin and Hexarelin + GHRH in humans.

is no sex-related difference in the GH response to GHS, suggest that also the modulatory effect of gonadal steroids on the GH-releasing activity of GHS may be age-dependent.

As concomitant reduction of GHRH activity and somatostatinergic hyperactivity have been hypothesized to play a role in the reduced GH secretion in aging (53), these alterations may also be involved in the reduction of GH responsiveness to GHS in this period of life. In fact, in elderly subjects the GH response to GHS is enhanced by either GHRH or arginine (24,39,41), which acts via inhibition of hypothalamic somatostatin release (54). However, only the coadministration of these three substances fully restores the GH response to young levels (24). Finally, the existence of an unknown endogenous U factor mediating the effects of GHRPs (55), whose activity could be impaired in aging, has also been hypothesized (56).

Beside age-related variations in the neural control of somatotrope function, it has also been hypothesized that the reduced GH-releasing activity of GHS in aging could, at least partially, depend on impaired receptor or post-receptor mechanisms. In fact, the GH response to Hexarelin is improved, but not restored, by supramaximal doses of the hexapeptide in elderly subjects (24).

Several studies have been performed to clarify the hypothalamo–pituitary mechanisms underlying the GH-releasing activity of GHRPs in humans.

GHS and GHRH have synergistical effect (46,47,57) and even very low GHS doses potentiate the GHRH-induced GH rise in man (46), pointing, in agreement with data in animals (see other chapters in this book), to different mechanisms of action for these substances.

However, also in humans the GH-releasing activity of GHS seems dependent on the functional hypothalamic integrity, including the full activity of GHRH-secreting neurons. In fact, the GH-releasing effect of GHRP-6, either alone or in combination with GHRH, as well as that of Hexarelin, are strongly reduced, though not abolished, in patients with pituitary stalk disconnection (37,58). Moreover, it has recently been shown that the GH responsiveness to GHRP-6 is nearly abolished by pretreatment with a specific GHRH antagonist (59).

The GH-releasing effect of GHS seems enhanced by physiological conditions such as physical exercise and sleep (60,61). On the other hand, the stimulatory effect of GHS on GH secretion is enhanced by insulin-induced hypoglycemia (47) and prazosin, an alpha1 adrenergic antagonist (27) but not modified by pyridostigmine, an indirect cholinergic agonist, arginine, an amino acid, galanin, a neuropeptide, atenolol, a beta1-adrenergic antagonist, and clonidine, an alfa2-adrenergic agonist, which are all known to potentiate the GHRH-induced GH rise (27,57,62,63) by inhibiting hypothalamic somatostatin (54), as well as by naloxone, an opioid antagonist (64) (Figure 2). On the other hand, pirenzepine, a muscarinic receptor antagonist, salbutamol, a beta2 adrenergic agonist, glucose load, and rhGH, which are known to nearly abolish the somatotrope response to GHRH via stimulation of hypothalamic somatostatin (54), only blunt the GHS-induced GH response (57,62,65–67) (Figure 2). A similar effect is shown by atropine, another anticholinergic drug, on the somatotrope response to GHRP-6 + GHRH (47). Also the inhibitory effect of

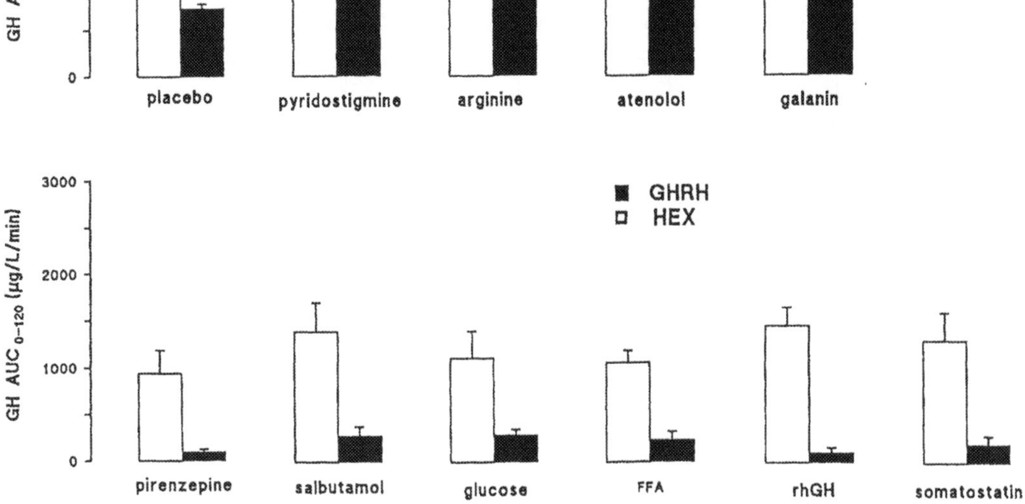

Figure 2. Mean (± SEM) GH AUCs after GHRH or Hexarelin (HEX) alone and combined with pyridostigmine, arginine, atenolol, galanin, pirenzepine, salbutamol, glucose, free fatty acids (FFA), rhGH and somatostatin in young adults.

cyproeptadine and diphenhydramine, serotoninergic and histaminergic antagonists, on GH response to Hexarelin seems related to their anticholinergic activity (63,68); in fact, terphenadine, a more specific H1-antagonist, does not modify this response (68).

Also the increase of circulating free fatty acids, which inhibits GH secretion by acting at the pituitary level (54), only blunts the GH response to GHS (65), which is enhanced by acipimox, a lypolisis inhibitor (69). It is noteworthy that even the infusion of exogenous somatostatin (SS) inhibits but does not abolish the somatotrope responsiveness to Hexarelin (57,70) as well as that to Hexarelin + GHRH, which persisted synergic even during high SS dose infusion (70).

As GHS do not seem to influence hypothalamic somatostatin release (1–3), these findings in man agree with others in animals favouring the hypothesis that GHS antagonize somatostatinergic activity, both at the pituitary and the hypothalamic level (see other chapters in this book). This could also explain the good intraindividual reproducibility of the GH response to GHS (25). These evidences also strengthen the hypothesis that GHS and GHRH act, at least partially, via different mechanisms. Among the central influences, it has also been reported that hCRH, which has been shown to inhibit GHRH-induced GH rise (71), does not affect somatotrope responsiveness to GHS (72), while, interestingly, alprazolam, a benzodiazepine which possesses a slight stimulatory effect on GH secretion (54), significantly blunts the GH-releasing activity of Hexarelin (73). The reason why the interaction between Hexarelin, a strong GH secretagogue, and alprazolam, another GH stimulatory input, results in a reduction of GH secretion is unclear, but these data suggest that GABAergic pathways may play a role in the CNS-mediated GH-stimulatory activity of GHS.

Among the influence of peripheral hormones on the GH-releasing activity of GHS, the effects of gonadal steroids have been mentioned above. The biphasic effect of glucorticoid on GH release is well-known, depending on dose, time and length of exposure; it is probably mediated at both pituitary and hypothalamic level, involving stimulation of somatostatin release (74). In humans, acute dexamethasone pretreatment had no or slight increasing effect on the GH response to GHS depending on timing of glucocorticoid administration (73,75), while prolonged and chronic glucocorticoid treatment as well as endogenous hypercortisolism invariably inhibited the GH response to GHS, which, however, persisted remarkably (76–79).

Concerning thyroid hormone and PRL influences, it has been shown that the GH-stimulating effect of GHS is preserved in hyperthyroidism (80), while it is reduced in hypothyroidism (81) and in patients with both functional and tumoral hyperprolactinemia (82–85), in agreement with a general impairment of somatotroph responsiveness to provocative stimuli in these conditions (54).

It has been reported that GHS induce homologous desensitization (31,50,86–92), while some authors showed even heterologous desensitization to GHRH induced by GHS (32). The homologous desensitization is marked during continuous GHS infusion (86–88), while it is less pronounced during intermittent treatment (89–92). Indeed, significant blunting of somatotroph responsiveness to acute GHS administration has been reported during prolonged and chronic i.n., s.c. and oral treatment by most (31,50,89–92) but not all the authors (40,93). The desensitization to GHS was always reversed by treatment withdrawal (89,91).

In spite of desensitization, prolonged administration of GHS by parenteral (i.v. or s.c.), i.n. and oral administration has been reported to enhance spontaneous GH pulsatility over 24 h and increase IGF-I levels in normal young adults as well as in short children and even in elderly subjects and in critically ill patients (13,40,50,86,88,89,94,95). Thus, it is clear that chronic treatment with GHS, particularly with the non-peptidyl MK-677, augments the activity of GH/IGF-I axis being able to restore it in elderly subjects, obese patients and even in GH-deficient adults (31,50,90). It is noteworthy that prolonged treatment with MK-677 was found able to prevent IGF-I reduction and to improve nitrogen balance in normal subjects on diet-induced catabolism (92). This evidence points to the assumption that prolonged GHS treatment exerts an anabolic effect. The latter probably depends on IGF-I increase which seems easy to obtain by single oral MK administration but not by GHRPs twice daily. In fact, chronic treatment with Hexarelin or GHRP-2 up to 4 months was unable to modify IGF-I and IGF-BP3 levels in both short children and elderly subjects (91,93) as well as body composition and bone markers in the latter group (91).

PRL-releasing activity

Slight but significant and dose-dependent PRL increase has generally been reported in humans after i.v., s.c. and even oral GHS administration (2,4,9). The PRL response to both peptidyl and non-peptidyl GHS generally remains within the normal basal range (2,4,9). Moreover, the PRL response to Hexarelin is markedly lower than that observed after TRH and even after arginine administration (44) which, in turn, show a PRL-releasing activity clearly lower than that of dopaminergic antagonists (54).

The lactotrope responsiveness to GHS is not dependent on gender and, differently from somatotrope responsiveness, is also independent of age. In fact, the PRL response to Hexarelin administration in prepubertal and pubertal children, young adults and elderly subjects is similar (43). Differently from GH, the PRL response to GHS in primary fetal human pituitary cultures is absent (34), suggesting different time onset for the various endocrine activities of GHS.

The mechanisms underlying the PRL-releasing activity of GHS are still unclear. There is evidence suggesting direct stimulatory effect of GHS on lactotroph cells. Though fetal human pituitary cells do not show PRL response to GHS (34), pure PRL- and mixed GH/PRL-secreting pituitary tumors express GHS receptors and release PRL after GHS incubation (96,97). The PRL-releasing effect of GHS has been found preserved in acromegalic patients bearing somatomammotrope pituitary adenoma but not in patients bearing pure micro- or macro-prolactinoma (82,85). Moreover, GHS have been found able to stimulate PRL secretion in puerperal women (83).

However, that the PRL-releasing activity of GHS is CNS-mediated can not be definitively ruled out. There is evidence showing that opioidergic, serotoninergic and histaminergic pathways are not involved in mediating this effect in humans (63,64,68). On the other hand, as Hexarelin and galanin, a PRL-secretagogue, show no interaction on PRL secretion (63), an involvement of galaninergic pathways in the PRL-releasing effect of GHS cannot be ruled out.

ACTH-releasing activity

The stimulatory effect of GHS on the activity of the hypothalamo–pituitary–adrenal (HPA) axis has been clearly demonstrated both in animals and in humans (2–4,9).

An increase in ACTH and cortisol levels has been demonstrated in man after acute administration of both peptidyl and non-peptidyl GHS both in the morning and during the night (2–4,9). Interestingly, the maximal stimulatory dose of GHS on HPA is lower than that needed for maximal GH-releasing effect (98).

The stimulatory effect of GHS on the HPA axis is more remarkable than that on PRL secretion. In fact, the ACTH and cortisol response to acute GHS administration in humans overlaps with that after AVP or naloxone, a CRH-mediated stimulus, and it is even similar to that after hCRH in young adults, at least when mean peaks are considered (44,64,72,79) (Figure 3). The stimulatory effect of GHS on HPA seems, however, an acute neuro-endocrine effect; in fact, during prolonged treatment no variation in 24 h cortisol profile has been found (13,31,40,50,90,92).

The effect of GHS on the HPA axis is gender-independent while it could be affected by BMI; it seems enhanced in obese patients while it is decreased in anorectic patients (33,99).

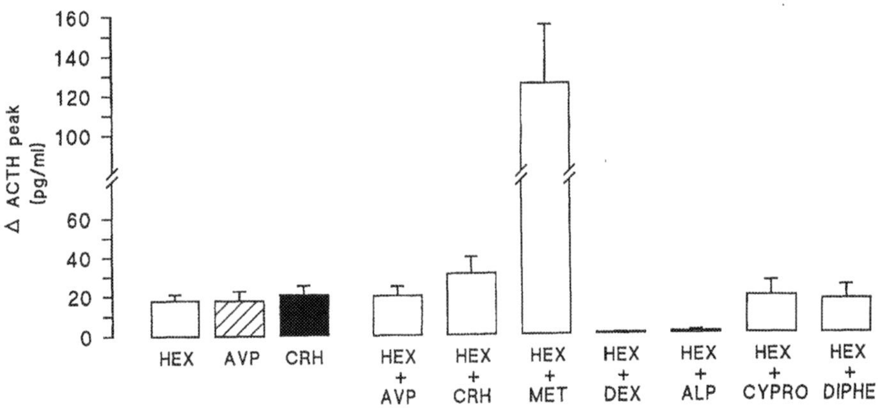

Figure 3. Mean (± SEM) ACTH Δ peak responses to AVP, hCRH and Hexarelin (HEX) alone and combined with AVP, hCRH, metyrapone (MET), dexamethasone (DEX), alprazolam (ALP), cyproeptadine (CYPRO), diphenhydramine (DIPHE) in young adults.

Above all, the HPA response to GHS shows peculiar age-related variations. The ACTH response to Hexarelin is present in prepubertal children, increases at puberty, then shows clear reduction in adulthood and, finally, a new trend toward increase in aging (43); these age-related variations are not shared by cortisol, the response of which does not show significant age-related variations (43). This ACTH response pattern is clearly different from that recorded after hCRH, the ACTH-releasing effect of which is slightly enhanced in aging but is unaffected by puberty (100).

Evidence that the ACTH response to GHS increases in aging when their GH-releasing effect is reduced again strengthens the hypothesis that GHS probably act at different levels and/or on different receptor subtypes to induce different endocrine responses.

Concerning the mechanisms underlying the ACTH/cortisol-releasing effect of GHS, there is evidence showing that, at least in physiological conditions, they are mediated via central actions. In fact, GHS are unable to stimulate ACTH secretion from both animal and human pituitary cells in culture (101,102). Moreover, the ACTH-releasing activity of GHS is abolished after hypothalamo–pituitary disconnection (103).

Data in animals suggested the hypothesis that the ACTH-releasing activity of GHS could be mediated by CRH- and/or AVP-secreting neurons (see other chapters in this book). On the other hand, in young adults the coadministration of Hexarelin and hCRH or naloxone, a CRH-mediated ACTH secretagogue, has an effect less than additive on corticotroph secretion (64,72), while Hexarelin and AVP have no interaction at all (72) (Figure 3). These data could agree with the hypothesis that CRH and/or AVP neurons mediate the stimulatory effect of GHS on the HPA axis. However, taking into account that CRH and AVP show well-known synergistical interaction on ACTH secretion (72,100), were the ACTH-releasing activity of Hexarelin mediated by CRH or by AVP, some clear interaction should have been found between Hexarelin and AVP or CRH, respectively. Thus, data in humans suggest the hypothesis that the stimulatory effect of GHS on HPA could be, at least partially, independent of both CRH and AVP.

The possibility that the ACTH-releasing activity of GHS could be mediated by other neural pathways has also been investigated. Our recent data demonstrate that the ACTH-releasing activity of GHS is not mediated by histaminergic or serotoninergic pathways, which play a critical role in the neural control of the HPA axis (54); in fact, both terphenadine and diphenydramine, two histaminergic antagonists, as well as ciproheptadine, a serotoninergic antagonist, do not modify the ACTH and cortisol response to Hexarelin in young adults (63,68) (Figure 3). On the contrary, alprazolam, a benzodiazepine which possesses an ACTH-inhibiting effect probably mediated by the reduction of CRH release (54), completely abolishes the ACTH and cortisol response to Hexarelin (73) (Figure 3). These findings indicate that GABAergic mechanisms could play a key role in mediating the stimulatory effect of GHS on HPA axis.

Recently, it has also been demonstrated that the majority of hypothalamic neurons activated by GHS administration are NPY positive (104). As the key role of this neuropeptide in the control of HPA axis is widely accepted (54), NPY may be another factor candidate to mediate the ACTH-releasing effect of GHS, whose activity could also mimic that of the putative endogenous GHS-like ligand.

In physiological conditions, the ACTH-releasing activity of GHS is sensitive to the negative glucocorticoid-mediated feed-back mechanism. In fact, the ACTH responsiveness to Hexarelin is enhanced by metyrapone-induced removal of the negative glucocorticoid feed-back, while it is completely abolished by dexamethasone pretreatment (73,105) (Figure 3). In agreement with these data, the ACTH-releasing activity of Hexarelin is absent in patients with Cushing's syndrome due to cortisol-secreting adrenal adenoma (79) but it is enhanced in hypocortisolemic patients with Addison's disease (105). Interestingly, in spite of hypercortisolism, in patients with Cushing's disease due to pituitary ACTH-secreting adenoma the ACTH and cortisol increase after Hexarelin administration is exaggerated and markedly higher than that to hCRH (79,106). These findings strengthen the interest about the ACTH-releasing activity of GHS which represent a new tool to investigate the control of HPA axis. However, it is unlikely that testing with GHS is a useful diagnostic tool for the differential diagnosis of Cushing's syndrome. In fact, our recent studies indicate that the ACTH hyper-responsiveness to Hexarelin is present in patients with micro- but not macro- ACTH-secreting pituitary adenoma (107). Moreover, marked ACTH and cortisol response to Hexarelin have been recently observed even in a patient with ACTH-secreting ectopic tumour (our unpublished results). The mechanisms underlying the exaggerated ACTH-releasing activity of GHS in hypercortisolemic patients with ACTH-secreting tumours is not necessarily mediated by CNS actions. In fact, the existence of specific GHS receptors in human pituitary and ectopic ACTH-secreting tumours (96,108) points toward a direct action of GHS on ACTH-secreting tumours.

EXTRA-ENDOCRINE ACTIVITIES OF GHS

Beside neuroendocrinological activities, GHS possess also pure central actions and have significant influence on both sleep and food intake (see other chapters in this book). In fact, repetitive i.v. GHRP-6 has been found able to increase stage D2 sleep in young adults (12), while prolonged oral MK-677 treatment significantly increased REM sleep and decreased REM latency in elderly subjects (51).

On the other hand, at least in animals, the stimulatory effect of GHS on food intake has been demonstrated (11); in humans, increased appetite has been occasionally reported after both acute and chronic GHS administration (31).

Finally, in agreement with the existence of specific receptors in peripheral animal and human tissues (20,21), it has recently been demonstrated that GHS probably possess peripheral biological activities. Particularly, peptidyl, but not non-peptidyl GHS, have specific receptors in the heart (20,21) and there is evidence that GHRPs have GH-independent cardiotropic activities both in animals and in humans (15 and other chapters of this book). This evidence clearly indicates that the biological activities of the putative endogenous GHS-like ligand(s) could be much less specific for GH than previously thought.

CONCLUSIONS

More than 20 years after their discovery GHS are still a family of synthetic, non-natural molecules; the members of this family have increased decade by decade, but the endogenous GHS-like ligand still has to be identified, even if at least one of the specific GHS receptors has already been cloned. The most important reason prompting the research in this field was the ability of these molecules to release GH after oral administration; indeed, the oral activity of non-peptidyl GH secretagogues, such as MK-677, is impressive and the possibility exists that chronic treatment with these molecules will be useful in clinical practice for diagnostic and therapeutic purposes. However, it is clear that GHS activity is not fully specific for GH and the relevance of other neuroendocrine and extra-endocrine both central and peripheral activities of GHS is now becoming more clear. Actions at different levels and/or on different receptors subtypes could mediate these other than GH-releasing activities which could represent a new era for the understanding of this fascinating field.

ACKNOWLEDGEMENTS

Personal studies included in this review were supported by grants from CNR, MURST, FSMEM and Europeptides. The authors wish to thank Dr Romano Deghenghi and Dr Muni Franklin Boghen as well as Gianluca Aimaretti, Edoardo Bartolotta, Jaele Bellone, Andrea Benso, Lodovico Benso, Giampaolo Ceda, Lidia Di Vito, Corrado Ghè, Mariacristina Ghigo, Laura Gianotti, Silvia Grottoli, Barbara Maccagno, Massimo Procopio, Josefina Ramunni and Guido Rizzi for participating in our studies.

REFERENCES

1. Bowers, C.Y., Veeraragavan, K., Sethumadhavan, K. (1993) Atypical growth hormone releasing peptides. In: Growth hormone II, basic and clinical aspects. B.B. Bercu, R.F. Walker (eds). Springer-Verlag, New York, pp. 203–222.
2. Ghigo, E., Arvat, E., Muccioli, G., Camanni, F. (1997) Growth hormone-releasing peptides. Europ. J. Endocrinol. 136, 445–460.
3. Korbonits, M., Grossman, A.B. (1995) Growth hormone-releasing peptide and its analogues. Novel stimuli to growth hormone release. Trends Endocrinol. Metab. 6, 43–49.
4. Bowers, C.Y. (1993) GH releasing peptides; structure and kinetics. J. Pediatr. Endocrinol. 6, 21–31.
5. Laron, Z. (1995) Growth hormone secretagogues. Drugs 50, 595–601.
6. Deghenghi, R., Boutignon, F., Luoni, M., Grilli, R., Guidi, M., Locatelli, V. (1995) Small peptides as potent releasers of growth hormone. J. Ped. Endocrinol. Metab. 8, 311–313.
7. Arvat, E., Di Vito, L., Lanfranco, F. et al. (1999) Tyr-Ala-Hexarelin, a synthetic octapeptide, possesses the same endocrine activities of hexarelin and GHRP-6 in humans. J. Endocrinol. Invest. 21, 92–97.
8. Rasmussen, M.H., Sogaard, B., Ynddal, L., Groes, L., Helmgaard, L., Nordholm, L. (1998) Ipamorelin — a very potent novel growth hormone secretagogue. Proc. 80th Meeting of the Endocrine Society, New Orleans, p. 159.
9. Smith, R.G., Van der Ploeg, L.X.T., Howard, A.D. et al. (1997) Peptidomimetic regulation of growth hormone secretion. Endocrine Reviews 18(5), 621–645.

10. Ghigo, E., Arvat, E., Camanni, F. (1998) Growth hormone secretagogues as corticotrophin releasing factors. In: 24th International Symposium: GH and growth factors in endocrinology and metabolism. R. Bouillon, G.P. Baumann, G. Johannsson, R. Gunnarsson (eds). Oxford, OCC, pp. 145–158.
11. Locke, W., Kirgis, H.D., Bowers, C.Y., Abdo, A.A. (1995) Intracerebroventricular growth-hormone-releasing peptide-6 stimulates eating without affecting plasma growth hormone responses in rats. Life Science 56, 1347–1352.
12. Frieboes, R.M., Murck, H., Maier, P., Schier, T., Holsboer, F., Steiger, A. (1995) Growth hormone-releasing peptide-6 stimulates sleep, growth hormone, ACTH and cortisol release in normal man. Neuroendocrinology 61, 584–589.
13. Copinschi, G., van Onderbergen, A., L'hermite-Baleriaux, M. et al. (1996) Effects of 7-day treatment with a novel, orally active, growth hormone (GH) secretagogue, MK-677, on 24-hour GH profiles, insulin-like growth factor I, and adrenocortical function in normal young men. J. Clin. Endocrinol. Metab. 81, 2776–2782.
14. De Gennaro Colonna, V., Rossoni, G., Bernareggi, M., Muller, E.E., Berti, F. (1997) Hexarelin, a growth hormone-releasing peptide, discloses protectant activity against cardiovascular damage in rats with isolated growth hormone deficiency. Cardiologia 42(11), 1165–1172.
15. Broglio, F., Valetto, M.R., Podio, V., Del Rio, G., Boghen, M.F., Deghenghi, R., Bisi, G., Ghigo, E. (1998) The acute administration of Hexarelin, a synthetic GHRP, but not of growth hormone, increases left ventricular ejection fraction in normal man. Proc. 80th Meeting of the Endocrine Society, New Orleans, p. 72.
16. Codd, E.E., Shu, A.Y.L., Walker, R.F. (1989) Binding of a growth hormone releasing hexapeptide to specific hypothalamic and pituitary binding sites. Neuropharmacology 28, 1139–1144.
17. Sethumadhavan, K., Veeraragavan, K., Bowers, C.Y. (1991) Demonstration and characterization of the specific binding of growth hormone-releasing peptide to rat anterior pituitary and hypothalamic membranes. Biochem. Biophys. Res. Commun. 178, 31–37.
18. Muccioli, G., Ghè, C., Ghigo, M.C. et al. (1998) Specific receptors for synthetic GH secretagogues in the human brain and pituitary gland. J. Endocrinol. 157, 99–106.
19. Ong, H., McNicoll, N., Escher, E. et al. (1998) Identification of a pituitary GHRP receptor subtype by the photoaffinity labeling approach using ^{125}I p-benzoyl-phenylalanine Hexarelin derivative. Endocrinology. 139, 432–435.
20. Muccioli, G., Ghè, C., Ghigo, M.C. et al. (1997) GHRP receptors in pituitary, central nervous system and peripheral human tissues. J. Endocrinol. Invest. (Suppl 4) 20, 52.
21. Bodart, V., McNicoll, N., Carriere, P. et al. (1998) Identification and characterization of a new GHRP receptor in the hearth. Proc. 80th Meeting of the Endocrine Society, New Orleans, 1998, p. 302.
22. Howard, A.D., Feighner, S.D., Cully, D.F. et al. (1996) A receptor in pituitary and hypothalamus that functions in growth hormone release. Science 273, 974–977.
23. Gelato, M.C., Pescovitz, D.H., Cassorla, F., Loriaux, D.L., Merriam, G.R. (1984) Dose–response relationships for the effects of GH-releasing factor 1-44 NH2 in young adult men and women. J. Clin. Endocrinol. Metab. 59, 197–203.
24. Arvat, E., Ceda, G.P., Di Vito, L. et al. (1998) Age-related variations in the neuroendocrine control, more than impaired receptor sensitivity, cause the reduction in the GH-releasing activity of GHRPs in human aging. Pituitary 1, 51–58.
25. Ghigo, E., Arvat, E., Gianotti, L. et al. (1994) Growth hormone-releasing activity of Hexarelin, a new synthetic hexapeptide, after intravenous, subcutaneous, intranasal and oral administration in man. J. Clin. Endocrinol. Metab. 78, 693-–98.
26. Mazza, E., Ghigo, E., Goffi, S. et al. (1989) Effect of the potentiation of cholinergic activity on the variability in individual GH response to GHRH. J. Endocrinol. Invest. 12, 795–799.
27. Penalva, A., Pombo, M., Carballo, A., Barreiro, J., Casanueva, F.F., Dieguez, C. (1993) Influence of sex, age and adrenergic pathways on the growth hormone response to GHRP-6. Clin. Endocrinol. (Oxf) 38, 87–91.

28. Cordido, F., Penalva, A., Dieguez, C., Casanueva, F.F. (1993) Massive growth hormone (GH) discharge in obese subjects after the combined administration of GH-releasing hormone and GHRP-6: evidence for a marked somatotroph secretory capability in obesity. J. Clin. Endocrinol. Metab. 76, 819–823.
29. Grottoli, S., Maccario, M., Procopio, M. et al. (1996) Selective refractoriness of somatotrope secretion to the inhibitory effect of hyperglycemia in obesity. Eur. J. Endocrinol. 135, 678–682.
30. Kirk, S.E., Gertz, B.J., Schneider, S.H. et al. (1997) Effect of obesity and feeding on the growth hormone (GH) response to the GH secretagogue L-692,429 in young men. J. Clin. Endocrinol. Metab. 82, 1154–1159.
31. Svensson, J., Lonn, L., Jansson, J.O. et al. (1998) Two-month treatment of obese subjects with the oral growth hormone (GH) secretagogue MK-677 increases GH secretion, fat-free mass, and energy expenditure. J. Clin. Endocrinol. Metab. 83, 362–369.
32. Popovic, V., Micic, D., Damjanovic, S. et al. (1997) Absence of desensitization by Hexarelin to subsequent GH releasing hormone-mediated GH secretion in patients with anorexia nervosa. Clin. Endocrinol. 46, 539–543.
33. Giusti, M., Foppiani, L., Ponzani, P. et al. (1997) Hexarelin is a stronger GH-releasing peptide than GHRH in normal cycling women but not in anorexia nervosa. J. Endocrin. Invest. 20, 257–263.
34. Shimon, I., Yan, X., Melmed, S. (1998) Human fetal pituitary expresses functional growth hormone-releasing peptide receptors. J. Clin. Endocrinol. Metab. 83, 174–178.
35. Bartolotta, E., Bellone, J., Aimaretti, G. et al. (1997) The GH-releasing effect of Hexarelin, a synthetic hexapeptide, in newborns is lower than in young adults. J. Pediatr. Endocrinol. Metab. 10, 491–497.
36. Bellone, J., Aimaretti, G., Bartolotta, E. et al. (1995) Growth hormone-releasing activity of Hexarelin, a new synthetic hexapeptide, before and during puberty. J. Clin. Endocrinol. Metab. 80, 1090–1094.
37. Loche, S., Cambiaso, P., Carta, D. et al. (1995) The growth hormone-releasing activity of Hexarelin, a new synthethic hexapeptide, in short normal and obese children and in hypopituitary subjects. J. Clin. Endocrinol. Metab. 80, 674–678.
38. Ghigo, E., Arvat, E., Rizzi, G. et al. (1994) Arginine enhances the growth hormone (GH)-releasing activity of a synthetic hexapeptide (GHRP-6) in elderly but not in young subjects after oral administration. J. Endocrinol. Invest. 17, 157–162.
39. Ghigo, E., Arvat, E., Rizzi, G. et al. (1994) Growth hormone-releasing activity of growth hormone-releasing peptide-6 is maintained after short-term oral pretreatment with the hexapeptide in normal aging. Eur. J. Endocrinol. 131, 499–503.
40. Ghigo, E., Arvat, E., Gianotti, L. et al. (1996) Short term administration of intranasal or oral hexarelin, a synthetic hexapeptide, does not desensitize the GH responsiveness in human aging. Europ. J. Endocrinol. 135, 407–412.
41. Arvat, E., Gianotti, L., Grottoli, S. et al. (1994) Arginine and growth hormone-releasing hormone restore the blunted growth hormone-releasing activity of hexarelin in elderly subjects. J. Clin. Endocrinol. Metab. 79, 1440–1443.
42. Arvat, E., Gianotti, L., Broglio, F. et al. (1997) Oestrogen replacement does not restore the reduced GH-releasing activity of hexarelin, a synthetic hexapeptide, in postmenopausal women. Eur. J. Endocrinol. 136, 483–487.
43. Arvat, E., Ramunni, J., Bellone, J. et al. (1997) The GH, PRL, ACTH and cortisol responses to Hexarelin, a synthetic hexapeptide, undergo different age-related variations. Europ. J. Endocrinol. 137, 635–642.
44. Arvat, E., Di Vito, L., Maccagno, B. et al. (1997) Effects of GHRP-2 and Hexarelin, two synthetic GH-releasing peptides, on GH, prolactin, ACTH and cortisol level in man. Comparison with the effects of GHRH, TRH, and hCRH. Peptides 18, 885–891.
45. Micic, D., Popovic, V., Kendereski, A. et al. (1995) Growth hormone secretion after the administration of GHRP-6 or GHRH combined with GHRP-6 does not decline in late adulthood. Clin. Endocrinol. 42, 191–194.

46. Bowers, C.Y., Reynolds, G.A., Durham, D., Barrera, C.M., Pezzoli, S.S., Thorner, M.O. (1990) Growth hormone (GH)-releasing peptide stimulates GH release in normal men and acts synergistically with GH-releasing hormone. J. Clin. Endocrinol. Metab. 70, 975–982.
47. Penalva, A., Carballo, A., Pombo, M., Casanueva, F.F., Dieguez, C. (1993) Effect of growth hormone (GH)-releasing hormone (GHRH), atropine, pyridostigmine, or hypoglycemia on GHRP-6-induced GH secretion in man. J. Clin. Endocrinol. Metab. 76, 168–171.
48. Gertz, B.J., Barrett, J.S., Eisenhandler, R. et al. (1993) Growth hormone response in man to L-692,429, a novel non peptide mimic of growth hormone-releasing peptide-6. J. Clin. Endocrinol. Metab. 77, 1393–1397.
49. Aloi, J.A., Gertz, B.J., Hartman, M.L. et al. (1994) Neuroendocrine responses to a novel growth hormone secretagogue, L-692,429, in healthy older subjects. J. Clin. Endocrinol. Metab. 79, 943–949.
50. Chapman, I.M., Bach, M.A., Van Cauter, E. et al. (1996) Stimulation of the growth hormone (GH)-insulin-like growth factor I axis by daily oral administration of a GH secretagogue (MK-677) in healthy elderly subjects. J. Clin. Endocrinol. Metab. 81, 4249–4257.
51. Copinschi, G., Leproult, R., Van Onderbergen, A. et al. (1997) Prolonged oral treatment with MK-677, a novel growth hormone secretagogue, improves sleep quality in man. Neuroendocrinology 66, 278–286.
52. Loche, S., Colao, A., Cappa, M. et al. (1997) The growth hormone response to Hexarelin in children: reproducibility and effect of sex steroids. J. Clin. Endocrinol. Metab. 82, 861–864.
53. Ghigo, E., Arvat, E., Gianotti, L. et al. (1996) Human aging and the GH/IGF-I axis. J. Pediatr. Endocrinol. Metab. 9, 271–278.
54. Muller, E.E. and Nisticò, G. (1989) Brain messengers and the pituitary. Academic Press, Chicago.
55. Bowers, C.Y., Sartor, A.O., Reynolds, G.A., Badger, T.M. (1991) On the action of the growth hormone-releasing hexapeptide, GHRP. Endocrinology 128, 2027–2035.
56. Bowers, C.Y., Reynolds, G.A., Field, G. et al. (1996) Independency, dependency and synergism of GHRP-2 and GHRH on GH release in humans. Proc. 10th International Congress of Endocrinology, San Francisco, 1996, p. 770.
57. Arvat, E., Gianotti, L., Di Vito, L. et al. (1995) Modulation of growth hormone-releasing activity of hexarelin in man. Neuroendocrinology 61, 51–56.
58. Popovic, V., Damjanovic, S., Micic, D., Djurovic, M., Dieguez, C., Casanueva, F.F. (1995) Blocked growth hormone-releasing peptide (GHRP-6)-induced GH secretion and absence of the synergic action of GHRP-6 plus GH-releasing hormone in patients with hypothalamo-pituitary disconnection: evidence that GHRP-6 main action is exerted at the hypothalamic level. J. Clin. Endocrinol. Metab. 80, 942–947.
59. Pandya, N., DeMott-Friberg, R., Bowers, C.Y., Barkan, A.L., Jaffe, C.A. (1998) Growth hormone (GH)-releasing peptide-6 requires endogenous hypothalamic GH-releasing hormone for maximal GH stimulation. J. Clin. Endocrinol. Metab. 83, 1186–1189.
60. Nooitgedagt, A., Koppeschaart, H.P.F., de Vries, W.R. et al. (1997) Influence of endogenous cholinergic tone and growth hormone-releasing peptide-6 on exercise induced growth hormone release. Clin. Endocrinol. 46, 195–202.
61. Loche, S., Colao, A., Cappa, M. et al. (1997) Acute administration of hexarelin stimulates GH secretion during day and night in normal men. Clin. Endocrinol. 46, 275–279.
62. Arvat, E., Gianotti, L., Ramunni, J. et al. (1996) Influence of beta-adrenergic agonists and antagonists on the GH-releasing effect of Hexarelin in man. J. Endocrinol. Invest. 19, 25–29.
63. Arvat, E., Maccagno, B., Ramunni, J. et al. (1998) Influence of galanin and serotonin on the endocrine response to hexarelin, a synthetic peptidyl GH-secretagogue, in normal women. J. Endocrinol. Invest. 21, 673–679.
64. Korbonits, M., Trainer, P.J., Besser, G.M. (1995) The effect of an opiate antagonist on the hormonal changes induced by hexarelin. Clin. Endocrinol. 43, 365–371.
65. Maccario, M., Arvat, E., Procopio, M. et al. (1995) Metabolic modulation of the growth-hormone-releasing activity of hexarelin in man. Metabolism 44(1), 134–138.

66. Arvat, E., Di Vito, L., Gianotti, L. et al. (1997) Mechanisms underlying the negative growth hormone (GH) autofeedback on the GH-releasing effect of Hexarelin in man. Metabolism 46, 83–88.
67. Massoud, A.F., Hindmarsh, P.C., Brook, C.G. (1995) Hexarelin induced growth hormone release is influenced by exogenous growth hormone. Clin. Endocrinol. 43, 617–621.
68. Arvat, E., Maccagno, B., Ramunni, J. et al. (1997) Effects of histaminergic antagonists on the GH-releasing activity of GHRH or hexarelin, a synthetic hexapeptide, in man. J. Endocrinol. Invest. 20, 122–127.
69. Peino, R., Cordido, F., Penalva, A., Alvarez, V., Dieguez, C., Casanueva, F.F. (1996) Acipimox-mediated plasma free fatty acid depression per se stimulates growth hormone (GH) secretion in normal subjects and potentates the response to other GH-releasing stimuli. J. Clin. Endocrinol. Metab. 81, 909–913.
70. Massoud, A.F., Hindmarsh P.C., Brook, C.G.D. (1997) Interaction of the growth hormone releasing peptide hexarelin with somatostatin. Clin. Endocrinol. 47, 537–547.
71. Barbarino, A., Corsello, S.M., Della casa, S. et al. (1990) Corticotropin-releasing hormone inhibition of growth hormone-releasing hormone-induced growth hormone release in man. J. Clin. Endocrinol. Metab. 71, 1368–1374.
72. Arvat, E., Maccagno, B., Ramunni, J. et al. (1997) Hexarelin, a synthetic growth-hormone releasing peptide, shows no interaction with corticotropin-releasing hormone and vasopressin on adrenocorticotropin and cortisol secretion in humans. Neuroendocrinology 66, 432–438.
73. Arvat, E., Maccagno, B., Ramunni, J. et al. (1998) Effects of dexamethasone and alprazolam, a benzodiazepine, on the stimulatory effect of Hexarelin, a synthetic GHRP, on ACTH, cortisol and GH secretion in humans. Neuroendocrinology 67, 310–316.
74. Dieguez, C., Mallo, F., Senaris, R. et al. (1996) Role of glucocorticoids in the neuroregulation of growth hormone secretion. J. Pediatr. Endocrinol. Metab. 9, 255–260.
75. Pinto, A.C.R., Finamor, F., Silva, M., Oliveira, J.H.A., Lengyel, A.M.J. (1998) Acute administration of dexamethasone enhances GH response to GHRP-6 in men. Proc. 80th Meeting of the Endocrine Society, New Orleans, p. 159.
76. Gertz, B.J., Sciberras, D.G., Yogendran, L. et al. (1994) L-692,429, a non peptide growth hormone (GH) secretagogue reverses glucocorticoid suppression of GH secretion. J. Clin. Endocrinol. Metab. 79, 745–749.
77. Leal Cerro, A., Pumar, A., Garcia-Garcia, E., Dieguez, C., Casanueva, F.F. (1994) Inhibition of growth hormone release after the combined administration of GHRH and GHRP-6 in patients with Cushing's syndrome. Clin. Endocrinol. 41, 649–654.
78. Giustina, A., Bussi, A.R., Deghenghi, R. et al. (1995) Comparison of the effects of growth hormone-releasing hormone and hexarelin, a novel growth hormone-releasing peptide-6 analog, on growth hormone secretion in humans with or without glucocorticoid excess. J. Endocrinol. 146, 227–232.
79. Ghigo, E., Arvat, E., Ramunni, J. et al. (1997) Adrenocorticotropin- and cortisol-releasing effect of hexarelin, a synthetic growth hormone-releasing peptide, in normal subjects and patients with Cushing's syndrome. J. Clin. Endocrinol. Metab. 82, 2439–2444.
80. Ramos-Dias, J.C., Pimentel Filho, F., Reis, A.F., Lengyel, A.M.J. (1996) Different growth hormone (GH) responses to GH-releasing peptide and GH-releasing hormone in hyperthyroidism. J. Clin. Endocrinol. Metab. 81, 1343–1346.
81. Damjanovic, S., Popovic, V., Petakov, M., Djurovic, M., Dieguez, C., Casanueva, F.F. (1996) Pituitary enlargement due to primary hypothyroidism: growth hormone response to GHRH, GHRP-6 and GHRH plus GHRP-6. J. Pediatr. Endocrinol. Metab. 9, 549–553.
82. Ciccarelli, E., Grottoli, S., Razzore, P. et al. (1996) Hexarelin, a synthetic growth hormone releasing peptide, stimulates prolactin secretion in acromegalic but not in hyperprolactinemia patients. Clin. Endocrinol. 44, 67–71.
83. DeZegher, F., Spitz, B., Van den Berghe, G. et al. (1997) Postpartum hyperprolactinemia and hyporesponsiveness of growth hormone (GH) to GH-releasing peptide. J. Clin. Endocrinol. Metab. 83, 103–106.

84. Grottoli, S., Razzore, P., Arvat, E. et al. (1997) Reduction of the somatotrope responsiveness to GHRH and Hexarelin but not to arginine plus GHRH in hyperprolactinemic patients. J. Endocrinol. Invest. 20, 21–26.
85. Popovic, V., Simic, M., Illic, L.J. et al. (1998) Growth hormone secretion elicited by GHRH, GHRP-6 or GHRH plus GHRP-6 in patients with microprolattinoma and macroprolattinoma before and after bromocriptine therapy. Clin. Endocrinol. 48, 103–108.
86. Jaffe, C.A., Ho, P.J., Demott-Friberg, R., Bowers, C.Y., Barkan, A.L. (1993) Effects of a prolonged growth hormone (GH)-releasing peptide infusion on pulsatile GH secretion in normal men. J. Clin. Endocrinol. Metab. 77, 1641–1647.
87. DeBell, W.K., Pezzoli, S.S., Thorner, M.O. (1991) Growth hormone (GH) secretion during continuous infusion of GH-releasing peptide: partial response attenuation. J. Clin. Endocrinol. Metab. 72, 1312–1316.
88. Huhn, W.C., Hartman, M.L., Pezzoli, S.S., Thorner, M.O. (1993) Twenty-four-hour growth hormone (GH)-releasing peptide (GHRP) infusion enhances pulsatile GH secretion and specifically attenuates the response to a subsequent GHRP bolus. J. Clin. Endocrinol. Metab. 76, 1202–1208.
89. Klinger, B., Silbergeld, A., Deghenghi, R., Frenkel, J., Laron, Z. (1996) Desensitization from long-term intranasal treatment with Hexarelin does not interfere with the biological effects of this growth hormone-releasing peptide in short children. Europ. J. Endocrinol. 134, 716–719.
90. Chapman, I.M., Pescovitz, O.H., Murphy, G. et al. (1997) Oral administration of the growth hormone (GH) releasing peptide-mimetic MK-677 stimulates the GH/insulin-like growth factor-I axis in selected GH-deficient adults. J. Clin. Endocrinol. Metab. 82, 3455–3463.
91. Rahim, A., O'Neill, P.A., Shalet, S.M. (1998) Growth hormone status during long-term hexarelin therapy. J. Clin. Endocrinol. Metab. 83, 1644–1649.
92. Murphy, M.G., Plunkett, L.M., Gertz, B.J. et al. (1998) MK-677, an orally active growth hormone secretagogue, reverses diet-induced catabolism. J. Clin. Endocrinol. Metab. 83, 320–325.
93. Pihoker, C., Badger, T.M., Reynolds, G.A., Bowers, C.Y. (1997) Treatment effects of intranasal growth hormone releasing peptide-2 in children with short stature. J. Endocrinol. 155, 79–86.
94. Laron, Z., Frenkel, J., Deghenghi, R., Anin, S., Klinger, B., Silbergeld, A. (1995) Intranasal administration of the GHRP Hexarelin accelerates growth in short children. Clin. Endocrinol. 43, 631–635.
95. Van den Berghe, G., de Zegher, F., Veldhuis, D.J. et al. (1997) The somatotropic axis in critical illness: effect of continuous growth hormone-releasing hormone and GH-releasing peptide-2 infusion. J. Clin. Endocrinol. Metab. 82, 590–599.
96. De Keyzer, Y., Lenne, F., Bertagna, X. (1997) Widespread transcription of the growth hormone-releasing peptide receptor gene in neuroendocrine human tumors. Europ. J. Endocrinol. 137, 715–718.
97. Adams, E.F., Huang, B., Buchfelder, M. et al. (1998) Presence of growth hormone secretagogue receptor messenger ribonucleic acid in human pituitary tumors and rat GH3 cells. J. Clin. Endocrinol. Metab. 83, 638–642.
98. Massoud, A.F., Hindmrsh, P.C., Brook, C.G.D. (1996) Hexarelin-induced growth hormone, cortisol and prolactin release: a dose–response study. J. Clin. Endocrinol. 81, 4338–4341.
99. Grottoli, S., Arvat, E., Gauna, C. et al. (1998) Effects of alprazolam, a benzodiazepine, on ACTH, cortisol, GH and PRL responses to Hexarelin, a peptidyl GH-secretagogue (GHS), in normal subjects and in patients with simple obesity or Cushing's disease. Proc. 80th Meeting of the Endocrine Society, New Orleans, 1998, p. 346.
100. Orth, D.N. (1992) Corticotropin-releasing hormone in humans. Endocr. Rev. 13, 164–191.
101. Korbonits, M., Little, J.A., Forsling, M., Trainer, P.J., Besser, G.M., Grossman, A.B. (1997) The effect of growth hormone secretagogues on GHRH and arginine vasopressin release from the rat hypothalamus *in vitro*. Proc. 79th Meeting of the Endocrine Society, Minneapolis, 1997, p. 153.
102. Elias, K.A., Ingle, G.S., Burnier, J.P. et al. (1995) *In vitro* characterization of four novel classes of growth hormone-releasing peptide. Endocrinology 136, 5694–5699.

103. Hickey, G.J., Drisko, J., Faidley, T. et al. (1996) Mediation by the central nervous system is critical to the *in vivo* activity of the GH secretagogue L-692,585. J. Endocrinol. 148, 371–380.
104. Dickson, S.L., Luckman, S.M. (1997) Induction of c-fos messenger ribonucleic acid in neuropeptide Y and growth hormone (GH)-releasing factor neurons in the rat arcuate nucleus following systemic injection of the GH segretagogue, GH-releasing peptide-6. Endocrinology 138, 771–777.
105. Ramunni, J., Arvat, E., Maccagno, B. et al. (1998) The ACTH-releasing effect of Hexarelin, a synthetic GH-releasing hexapeptide, is sensitive to the negative feed-back of glucocorticoids. Abstract Book: IV Europ. Congr. Endocrinol. Sevilla, 1998, P1–265.
106. Arvat, E., Giordano, R., Ramunni, J. et al. (1999) Effects of the combined administration of hexarelin, a synthetic peptidyl GH secretague, and hCRH on ACTH, cortisol and GH secretion in patients with Cushing's disease. J. Endocrinol. Invest. 22, 23–28.
107. Arvat, E., Giordano, R., Ramunni, J. et al. (1998) Adrenocorticotropin and cortisol hyperresponsiveness to Hexarelin in patients with Cushing's disease bearing a pituitary microadenoma, but not in those with macroadenoma. J. Clin. Endocrinol. Metab. 83, 4207–4211.
108. Jansson, J.O., Svensson, J., Bengtsson, B.A. et al. (1998) Acromegaly and Cushing's syndrome due to ectopic production of GHRH and ACTH by a thymic carcinoid tumour: *in vitro* responses to GHRH and GHRP-6. Clin. Endocrinol. 48, 243–250.

Growth Hormone Secretagogues
Edited by E. Ghigo, M. Boghen, F.F. Casanueva and C. Dieguez
1999 Elsevier Science B.V.

Chapter 13

Effectiveness of Growth Hormone Secretagogues in the Diagnosis and Treatment of GH Secretory Deficiency

BARRY B. BERCU[1] and RICHARD F. WALKER[2]

[1]*Professor of Pediatrics, Pharmacology and Therapeutics, All Children's Hospital, University of South Florida College of Medicine*
[2]*Director of Pharmaceutical Studies and Research Compliance, University of South Florida, St. Petersburg, FL, U.S.A.*

INTRODUCTION

Reduced growth hormone (GH) secretion is a cause of delayed physical development in children as well as a consequence of normal aging (1–8). Deficits in GH secretion can be attributed to pituitary and/or hypothalamic etiologies, with the most common cause being insufficient production and/or secretion of neuroregulatory hormones (9). Intrinsic pituitary defects can also reduce GH production and/or secretion. These principles apply to both GH deficient children and adults. Although the etiologies are disparate, GH deficiency states are almost universally treated with recombinant GH. Only very recently has GH releasing factor been approved for treatment of GH deficiency in children. However, its use as a therapeutic agent has been limited because of variability and unpredictability of growth response.

The progressive decrease in GH secretion during aging could result from several factors. Two of the most obvious are inadequate stimulation of the pituitary gland by hypophysiotropic factors, or degeneration of pituitary-based mechanisms for GH production or secretion. Of the two possibilities, it is most likely that decremental changes in the relationship of the pituitary with GH regulatory hormones of hypothalamic origin are the primary etiologic factors in age-related decline in GH secretion. This view seems inconsistent with many reports that state that the pituitary becomes refractory to stimulation during aging. There is general consensus GH secretion in response to GHRH stimulation *in vivo* declines with advancing age (1–8). Assuming that GHRH is the only stimulatory agent controlling GH secretion, this progressive decrement could result from several factors. One could be that prolonged reduction in pituitary stimulation by GHRH causes desensitization to the

GH secretagogue because hormones often induce their own receptors. Support for reduced stimulation of the pituitary by GHRH comes from evidence that available or appropriate GH secretagogue decline during aging (10–12). GHRH responsivity in aging rats, however, was immediately restored by co-administration with GHRP (13), suggesting that age-related, reduced efficacy of GHRH may be due to the absence or insufficient concentrations of other, as yet undefined, endogenous co-secretagogue (for a detailed overview of this subject, see Ref.14).

About two decades ago, a family of xenobiotic GH secretagogue was discovered which released GH by a mechanism different from GH releasing hormone (GHRH) (15). The prototypic compound for this xenobiotic GH secretagogue family was His-D-Trp-Ala-Trp-D-Phe-Lys-NH_2 or GH releasing hexapeptide (GHRP). GHRP potentiated the action of GHRH when both compounds were administered together; and the potency of GHRH was reduced by passive immunization (16). The reciprocal test was not possible because an endogenous ligand for GHRP has not been identified. The recent report of the identification of the receptor should aid in this effort (17). Because functional dependence of GHRP can be demonstrated by passive immunization against GHRH, the reverse may be true (GHRH activity may be influenced by an endogenous GHRP analog). This possibility that GH secretion may be controlled by more factors than GHRH and somatostatin has been suggested previously (18).

We hypothesize that GHRH and an endogenous analog to GHRP are complementary GH secretagogues that together provide appropriate stimulation of the pituitary gland to sustain normal GH production and release. Using this logic, GHRH and GHRP could be given alone, and in combination, to diagnose pituitary-based causes of inadequate GH

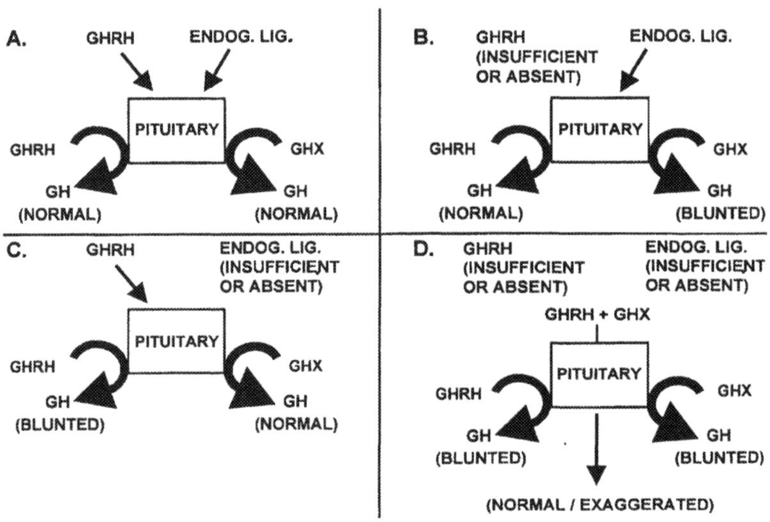

Figure 1. Schematic representation of the hypothetical model in support of the complementary relationship of GHRH and the endogenous ligand for GHRP or GHRP nonpeptide mimic. See text for discussion of the hypothesis. GHX refers to GHRP or GHRP nonpeptide mimic. From Reference (21).

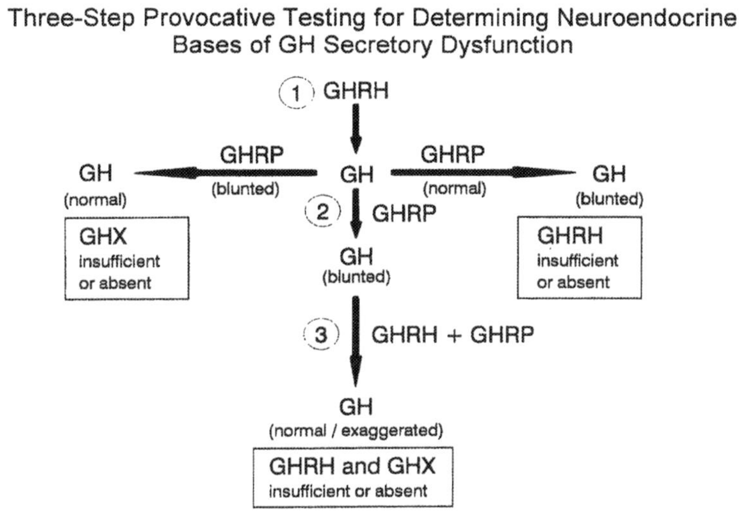

Figure 2. A three-step provocative pituitary function test using responses to sequentially administered and co-administered GH secretagogues as diagnostic parameters for determining the neuroendocrine basis of GH secretory dysfunction. From Ref. (19).

secretion (19). The model (Figures 1 and 2) assumes that a "normal" response to GHRP (or GHRP-mimic) stimulation or GHRH requires the presence of its endogenous analog (i.e. GHRH or GHRP, respectively). A blunted response to either exogenous GH secretagogue is interpreted as indicating a deficiency of its endogenous complement (Figures 1B and 1C). Blunted responses to both exogenous GH secretagogues administered sequentially implies deficiencies of both endogenous complements (Figure 1D). This situation can be differentiated from inherent pituitary abnormalities, such as receptor or second-messenger mediated deficits, by a "normal" response to GHRH and GHRP co-administration. A blunted response following co-administration of both secretagogues would suggest inherent pituitary dysfunction, rather than inadequate endogenous stimuli.

To test our hypothesis, we summarize in this report the following: (1) use of the diagnostic test in children with altered growth patterns, (2) use of the diagnostic test in healthy adults and aging subjects, (3) effectiveness of priming with GHRP in aging men to address potential therapy of GH secretagogue(s) in "normal" aging.

DESCRIPTION OF STUDIES

The GHRP-2 was synthesized by Kaken Pharmaceutical Co., Ltd (Japan). GHRP-2 was provided by Wyeth Ayerst (Radnor, PA), GRF 1-29 by Serono Laboratories, Inc. (Norwell, MA), and GHRH (1-44) NH_2 by ICN Pharmaceuticals (Costa Mesa, CA). GHRH (1-44) NH_2 is equivalent to GRF 1-29 in potency; it was substituted for GRF 1-29 when GRF 1-29 was not commercially available. Recombinant GHRH (1-44) for Study II was supplied by BioNebraska, Inc. (Lincoln, NE).

Study I: Diagnostic testing in children with various growth patterns, in healthy adults and aging subjects

Subjects were recruited from the pediatric endocrine clinics of All Children's Hospital, St. Petersburg, FL, and the University of South Florida College of Medicine, Tampa, FL. Adult subjects were recruited from the local community. The protocol was IRB approved and informed consent was obtained. Children were 3 to 12 years old, including 17 boys and 9 girls. Adult age range was from 29 to 68; 5 men and 2 women. Each subject had a medical history and physical examination. All subjects were fasted from 2200 h until completion of the study, but water was allowed *ad libitum*. Blood pressure and pulse were monitored. There were no adverse effects except flushing with GHRH. The serum samples were frozen and stored at –20°C until assayed.

Briefly, each subject began testing at approximately 0900 hours, half an hour after placement of an intravenous (iv) catheter. For the sequential study, three blood samples for basal concentrations of serum GH were drawn, and GHRP-2 (1 μg/kg) was administered as a bolus. The xenobiotic peptide was administered first in the series because we previously observed that repeated responses to GHRH in the same individual are more consistent if they are preceded by GHRP. Blood samples were then drawn at 5-minute intervals for 20 minutes, during which the GH secretory response occurred, and at longer intervals (30, 45, 60, 90 and 120 minutes). Ninety minutes later GHRH (1 μg/kg) was administered as a bolus, when basal GH concentrations were again established. Specifically, samples were drawn at the new zero time (210 minutes) and then at 215, 220, 225, 240, 255, 270, 300, and 330 minutes. In order to evaluate the combined GHRP-2 plus GHRH (1.0 μg/kg for each secretagogue), study blood samples were drawn at –30, –15, 0, 5, 10, 15, 30, 45, 60, 90, 120, 180, and 210 minutes. The study population was comprised of children with normal GH secretion, and those with GH secretory dysfunction of various clinical diagnoses. The adults were normal volunteers. The children who were normal volunteers were siblings of children being evaluated for growth disturbances. The purpose of evaluating different etiologies of GH secretory dysfunction was an attempt to validate the hypothesis that complementary endogenous GH secretagogues must be present for optimal expression of GHRP or GHRH stimulatory potentials.

A total of 45 tests were performed on 38 subjects. Three GH-deficient and one control child had both sequential and combined GHRP-2 plus GHRH testing. Two adult volunteers had the GHRP-2 study alone repeated. One non-GH-deficient, slow-growing, short-statured child had sequential GHRP-2 plus GHRH testing while on and off clonidine therapy for attention deficit disorder.

GH concentrations were measured by polyclonal radioimmunoassay as previously described (20). Interassay coefficient of variation was 13%, and intraassay coefficient of variation was 9%. All values are expressed as the mean ± SEM. The area under curve (AUC) was calculated for 90 minutes after GHRP-2 and GHRH administration in the sequential study. Statistical analyses were done using ANOVA and Scheffe tests.

Study II: Effectiveness of priming with GHRP in aging men

In order to determine whether the alterations in GH secretory response in aging were irreversible and/or if they were dependent upon a deficiency of GHRP, the second part of the "aging" study was performed. Adolescent and young adult subjects were recruited from the pediatric endocrine clinics. Adult subjects included faculty members from the University of South Florida College of Medicine, as well as local residents from the local community. Male subjects within the young adult group were from 16 to 21 years old. Adult subjects ranged in age from 37 to 68 years. All subjects were fasted from 2200 hours until completion of the study, but water was allowed ad libitum.

During the first part of this study, each subject was administered a standard diagnostic test as noted below. The hypothetical testing model in Figure 1 was previously described under Study I. Figure 2 illustrates the systematic approach for the diagnostic testing. See Study I for the description of the diagnostic testing. In Study I, we had shown that GH secretory responses to GHRP-2 in the younger and older groups of male subjects were comparable. In contrast, the GH secretory responses to the older group of men were significantly lower than that observed in the younger group. In order to determine whether these changes were irreversible and/or if they were dependent upon a deficiency of GHRP, the second part of this study was performed. The object of this second part was to determine whether priming with GHRP-2 would improve the GH secretory responses of the older subjects to challenge with GHRH.

The experimental subjects were selected from the pool of men who participated in the standard diagnostic test. The subjects of interest are those who elicited a blunted response to GHRH while having a normal response to GHRP-2. This change is typical of that which occurs during aging, and was observed in our study population of men over 30 years of age. Those individuals meeting the criteria for this study were then studied according to the following protocol:

1. An initial test was used to compare GH secretory responses to two suboptimal iv doses of GHRP-2 (0.25 μg/kg) administered two hours apart. Since the maximal response to the optimal dosage of GHRP-2 (1 μg/kg) was comparable in young and older men, a suboptimal dosage was used to determine the extent to which the GHRP response is reproducible. This information was required for analysis of data resulting from the subsequent steps presented below.

2. The second test was used to collect quantifiable data prior to manipulation of the GH neuroendocrine axis that could be compared against data from stimulated responses after receptor priming and concentration supplementation had been accomplished. Thus, a diagnostic test was performed on each subject using a submaximal iv dose of GHRP-2 (0.25 μg/kg) alone, followed by co-administration of a bolus iv dose of suboptimal GHRP-2 (0.25 μg/kg) with an optimal dose of GHRH (1 μg/kg) after serum GH concentrations had returned to basal values (120 minutes). The logic of this test is to provide a small enough dosage of GHRP-2 to elicit a measureable GH response so as to determine how much of the total GH release resulting from GHRP-2 and GHRH

could be attributed to GHRP-2. Historical data for GHRH from the same subjects were reviewed to help make this determination.

3. The next procedure was designed to determine whether the GHRH signal transduction process could be enhanced through priming of the GHRP-2 receptor. This procedure involves administering each subject a sc dose of GHRP-2 twice each day (1 mg sc @ 0700 hours and 2100 hours) for 7 to 10 consecutive days. The purpose of GHRP-2 administration was to employ the usual procedure for up-regulating endocrine receptors through the process of "priming". Priming involved the daily stimulation of target tissues for ten consecutive days.

4. The next procedure was to analyze the effect of GHRP-2 priming on the GH secretory response to GHRH. It involved administering the regular diagnostic test to each subject following the priming regimen described above. The purpose of repeating the diagnostic test was to determine if receptor "up-regulation" alone was adequate to restore the age-attenuated response to GHRH. Since the complementary secretagogue to GHRH had been administered one day before administration of the diagnostic test, receptor up-regulation had been accomplished. However, extracellular concentrations of the putative endogenous GHRP analog would be expected to be low. Thus, a continuing poor response to GHRH stimulation after priming with GHRP-2 might be attributed to low local concentrations of the complementary secretagogue. Therefore a final study was performed.

5. The last test involved administering a single iv bolus containing suboptimal concentrations of GHRP-2 (0.25 μg/kg) and optimal concentrations of GHRH (1.0 μg/kg) on the day following the standard diagnostic test. The purpose of this procedure was to obtain supplementary information to that gathered by the procedure described in section 3 above. In this case, co-administration of a low dosage of GHRP-2 was given with GHRH so as to increase extracellular concentrations of the xenobiotic secretagogue. The purpose of this step was to co-stimulate the pituitary in the presence of previously up-regulated GHRP receptors. Comparison of the responses to this procedure with that described in section 1 and section 3 above helped to determine whether the age-related decrement in stimulated GH secretion following GHRH administration in men over thirty could be attributed to receptor down-regulation, low extracellular concentrations of GHRP analog, a combination of the two factors, or none of these factors.

For clinical procedures described in section 1, section 2 and section 4 above, the subjects were denied any food after midnight. The next morning at 0800 hours, an iv infusion with normal saline was started. Blood (1.5 ml) was removed for measurement of GH. Blood was then withdrawn at –30, –15, and 0 minutes before administration of a single iv bolus of GHRP-2. Subsequently, blood samples were taken at +5, 10, 15, 30, 45, 60, 90, and 120 minutes. One and one-half hours later, now designated as –30 minutes, blood samples were withdrawn for determination of basal GH concentrations. Additional samples were taken at –15, and 0 minutes. Then a single iv bolus of GHRH was administered. Blood samples were taken at +5, 10, 15, 30, 45, 60, 90, and 120 minutes. The subjects were allowed ad libitum water during the studies.

For clinical procedure 5, blood samples were drawn according to the following schedule: –30, –15, and 0 minutes. At time 0, GHRP-2 (0.25 μg/kg) iv and GHRH (1.0 μg/kg) iv were administered together as a single bolus. Blood samples were then drawn at +5, 10, 15, 30, 45, 60, 90, 120, 150, 180, and 210 minutes. The serum was kept frozen at –20°C until GH concentrations were determined.

In vivo GH secretion in response to GHRP-2, GHRH, or GHRP-2 with GHRH were compared among the different age groups using analysis of variance, and with Duncan's multiple test. Repeated measures analysis was used to test for time and age-group effects: i.e. to determine whether significant changes in GH secretory profiles occur in response to GH secretagogue administration at different ages. Dose–response curves were compared for parallelism as an indication of mechanistic changes subserving altered sensitivity to the GH secretagogue during aging. Results were expressed as mean ± SEM. A p value < 0.05 was used to define statistically significant differences.

GH Secretagogue testing in children and adults

Table 1 summarizes the peak GH concentrations after GHRP-2 and GHRH during sequential testing. The most profound and consistent results were in the adult volunteers. Note the blunted GH response to GHRH in normal volunteer adults vs. adolescent children (Figure 3). Table 1 reviews the data in GH-deficient, slow-growing, non-GH deficient, and control children. Classical GH deficiency is based on two or more blunted GH stimulation tests of ≤ 10 μg/L, delayed bone age, and poor growth velocity. Non-GH deficient children have a growth velocity less than the 25th percentile for age, normal GH provocative tests and delayed bone age. The ratio of peak GH response to GHRP-2 plus GHRH is similar in all three groups — with the highest ratio in the slow-growing, non-GH deficient group (1.3

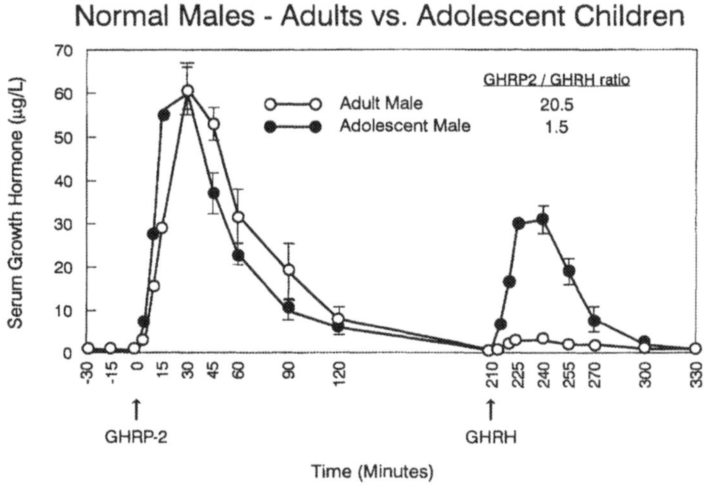

Figure 3. Comparison of changes in serum GH concentrations in adult and adolescent males following iv sequential administration of GHRP-2 and GHRH (N=6/group)(mean ± SEM). From Reference (23).

TABLE 1

PEAK GH CONCENTRATIONS FOLLOWING GHRP-2 AND GHRH IN THE VARIOUS GROUPS

Group	Peak GH after GHRP-2 (μg/L)	Peak GH after GHRH (μg/L)	Peak GHRP-2/ GHRH ratio	Area under curve (μg/L · 90 min) GHRP-2	Area under curve (μg/L · 90 min) GHRH
GH-deficient children	20.1 ± 5.5	19.6 ± 5.1	1.3 ± 0.4	995 ± 371	924 ± 232
(n = 15)	(<0.5–>80)	(0.3 – 70.2)	(0.3 – 5.6)		
Control children	42.2 ± 4.3	39.8 ± 7.8	1.4 ± 0.3	1598 ± 274	2201 ± 437
(n = 8)	(31.7 – 62.0)	(19.2 – 88.9)	(0.4 – 2.9)		
Slowly-growing, non-GH-deficient children (n = 8)	63.6 ± 24.9 (4.9–190.3)	31.4 ± 8.4 (1.5–78.8)	2.9 ± 1.3 (0.4–3.3)	2460 ± 953	1544 ± 449
Adult volunteers (n = 7)	52.0 ± 15.1 (14.6–123.2)	6.8 ± 2.4 (2.5–16.3)	15.0 ± 4.1 (0.9–36.2)	2785 ± 692	285 ± 93
Men (n = 5)	59.3	3.1	20.5		
Women (n = 2)	19.2	16.1	1.2		
Control vs. GH-deficient children	$p < 0.02$	$p < 0.05$			$p < 0.01$
Control children vs. adult volunteers		$p < 0.01$	$p < 0.01$		$p < 0.02$
GH-deficient children vs. adult volunteers	$p < 0.05$		$p < 0.001$	$p < 0.02$	
Slow-growing, non-GH-deficient children vs. adult volunteers		$p < 0.02$	$p < 0.02$		$p < 0.02$
GH-deficient vs. slow-growing, non-GH-deficient children	$p < 0.05$				

Adult peak GH: GHRP-2 vs. GHRH, $p < 0.02$.
Adult area under curve: GHRP-2 vs. GHRH, $p < 0.02$.
From Ref. 21.

and 1.4 for GH deficient and control groups, respectively; 2.9 in the slow-growing, non-GH deficient group; p = NS). As a group, peak GH concentrations after GHRP-2 were lowest in the GH-deficient group, compared with the control children and slow-growing, non-GH deficient group (20.1, 42.2 and 63.6 μg/L, respectively; control vs. GH deficient, $p < 0.02$; GH deficient vs. slow-growing non-GH deficient children, $p < 0.05$).

As a group, peak GH after GHRH was lowest in GH deficient patients. On the other hand, there was marked within-group variation, demonstrating individual differences. The child with the most exaggerated GH response to GHRH had central nervous system damage, causing precocious puberty. Children with direct damage to the pituitary gland had

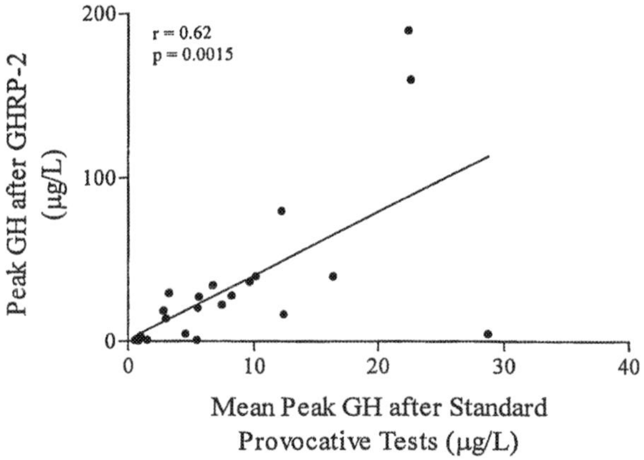

Figure 4. Correlation of peak GH after GHRP-2 vs mean peak GH after standard provocative stimuli in GH-deficient and slow-growing, non-GH-deficient children. From Reference (22).

markedly decreased responses to the combined stimuli. In general, the patients with multiple hormone deficiency had the lowest GH responses to the secretagogues (21).

There was a positive correlation between peak GH after GHRP-2 and mean peak GH after standard provocative testing (Figure 4) ($r = 0.62, p < 0.01$) (22). On the other hand, in the same sequential study, there was no correlation between peak GH after GHRH and the mean peak GH after standard provocative testing (Figure 5) ($r = 0.26, p = 0.24$) (21).

Four children had sequential and combined GHRP-2 plus GHRH studies. Two patients with hypopituitarism (multiple hormone deficiency) had an increased response to the combined stimulus, albeit very low but clearly measurable. This was in comparison to two normal children with exuberant GH responses. In both the normal and hypopituitary children, this "synergistic" relationship was preserved. In an additional child, the sequential GHRP-2 plus GHRH study was done on a regimen on and off clonidine (treatment for

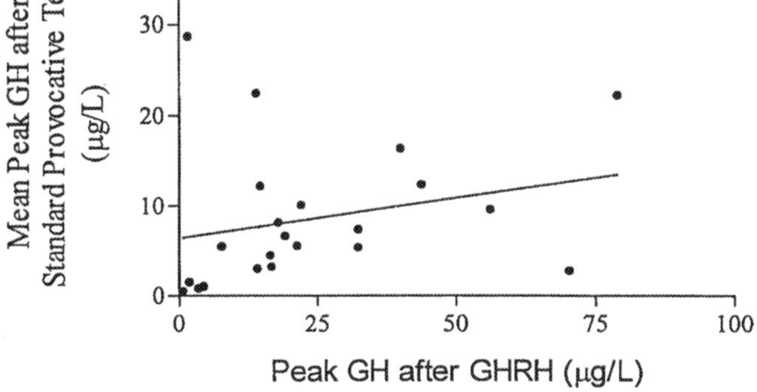

Figure 5. Correlation of peak GH after GHRH vs mean peak GH after standard provocative stimuli in

attention deficit disorder). While taking clonidine, peak GH to GHRP-2 and GHRH was high; whereas, after not taking clonidine for two weeks, the GH responses were relatively reduced (see Ref. (21) for details).

Effectiveness of priming with GHRP in aging men

The data in Figure 3 present and compare the mean GH secretory responses in younger and older subjects following sequential administration of GHRP-2 and GHRH within the framework of the standard diagnostic test. In all subjects, peak serum GH concentrations were significantly greater in response to GHRP-2 administration than GHRH administration. Both groups had exuberant GH responses to GHRP-2. However, profound and consistent differences between the two groups were seen as blunted GH response to GHRH in the older males compared with their younger, adolescent counterparts. In the present study, mean peak serum GH values after GHRP-2 and GHRH for the older and younger groups were 55.0 ± 8.0 and 3.2 ± 0.9 µg/L, vs 68 ± 9 µg/L and 29 ± 6 µg/L, respectively.

Prior to examining the "priming" effects of GHRP-2, submaximal dosages of the xenobiotic peptide were administered sequentially to insure the reliability and reproducibility of its dose response. There were no significant differences in peak stimulated concentrations (23 ± 6 µg/L vs 31 ± 8 µg/L) or areas under the curve (2210 ± 379 µg/L } 90 minutes vs 2351 ± 488 µg/L} 90 minutes) for serum GH between the first and second sequential administrations of GHRP-2 (0.25 µg/kg) during a 4-hour time interval. On the other hand, serum GH concentrations were significantly lower following administration of the 0.25 µg/kg GHRP-2 than for 1.0 µg/kg GHRP-2 (23 ± 6 µg/L vs 58 ± 9 µg/L) demonstrating a dose response for the peptide. Furthermore, co-administration of GHRH with the submaximal dose of GHRP-2 increased the GHRP-2 effect more than the additive effects of GHRP-2 and GHRH alone. As previously reported, mean peak serum GH values for GHRP-2 and GHRH administered alone were 23 ± 6 µg/L and 3.2 ± 0.9 µg/L, respectively, whereas the response to GHRP-2 administered with GHRH increased to 47 ± 9 µg/L. This increase was not additive and the results are consistent with the known synergism between the natural and xenobiotic secretagogues.

The data in Figure 6 show the responses in the older group of subjects to individually administered GHRH following ten days of priming with GHRP-2. As seen in the first part of the figure, priming with GHRP-2 increased the response to GHRP-2, consistent with prior reports from other laboratories. However, the second part of this figure shows that priming with GHRP-2 also significantly increased the response to GHRH when it was administered alone. Thus, priming with GHRP-2 "up-regulated" the mechanism by which GHRH releases GH from the pituitary gland.

Finally, GHRP-2 and GHRH were co-administered as a single bolus to men who had been "primed" with GHRP-2 for ten (10) consecutive days. The peak response to co-administration of the two peptides after priming with GHRP-2 was significantly greater than that before priming (47 ± 9 µg/L vs 69 ± 7 µg/L).

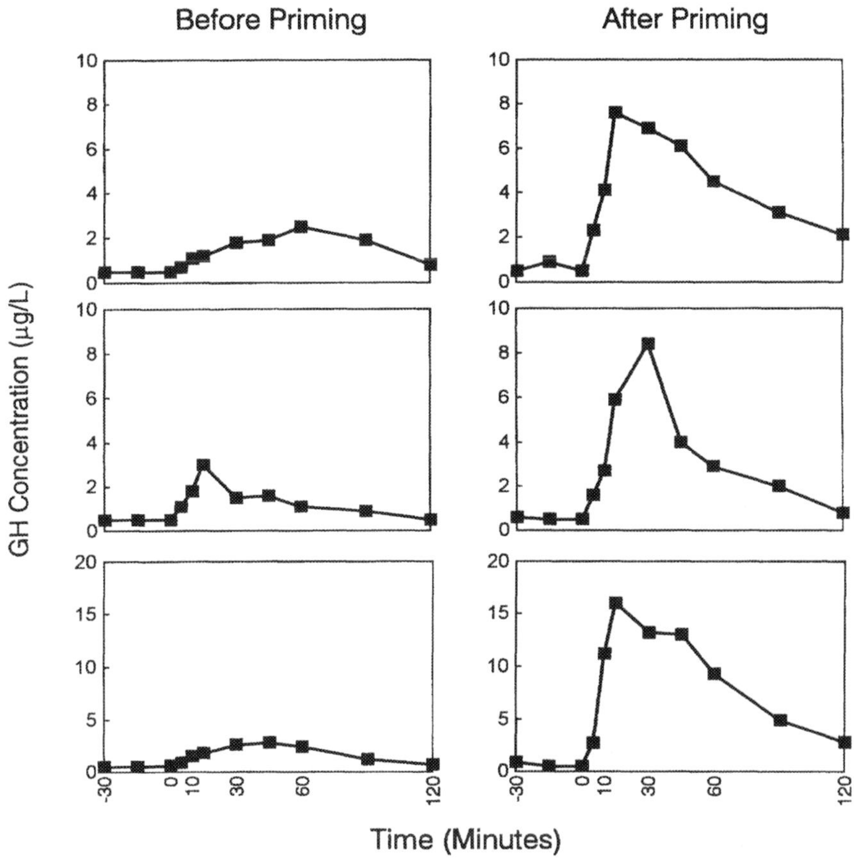

Figure 6. Effects of priming (10 consecutive days of sc administration of GHRP-2) on stimulated GH release in three adult subjects ranging in ages between 37 and 62 years. From Reference (23).

INTERPRETATION OF STUDIES

Here we provide preliminary data to support our previously reported hypothesis and rationale for a three-step provocative testing analysis to determine the neuroendocrine bases of GH secretory function (Figure 2). The results of our study support the hypothesis that GHRH and an endogenous ligand or analog of GHRP (and nonpeptidyl mimics) are complementary GH secretagogues that together provide appropriate stimulation of the pituitary gland to sustain normal GH production and release. By this logic, GHRH and GHRP could be used alone and in combination to diagnose pituitary-based inadequate GH secretion in slow-growing children and "aging" adults. We devised this model to directly test pituitary GH secretory capability (see Figure 1). The model assumes that a "normal" response to exogenous GHRP or GHRH requires the presence of its endogenous complement. A blunted response to either exogenous GH secretagogue is interpreted as indicating a deficiency of its endogenous complement. Blunted responses to both exogenous GH secretagogues administered sequentially, implies deficiencies of both endogenous complements. This condition can be differentiated from inherent pituitary problems, such as those involving receptor or second-messenger deficits by a "normal" response to GHRH and GHRP co-administration. A blunted response following co-administration of both GH secretagogues would indicate inherent pituitary dysfunction, rather than inadequate endogenous stimuli. More "definitive" support for this type of provocative testing would be provided by long-term therapy demonstrating improved growth in children and "metabolic" improvement in adults associated with the prolonged recombinant GH therapy.

Other investigators have compared GHRP alone and in combination with GHRH in various clinical settings (13, 24–27). The observation of synergy was quite consistent. Pombo et al. have also demonstrated an absence of GH secretion to either one or both stimuli in children with neonatal stalk section (28); Popovic et al. have described this in patients with hypothalamo-pituitary disconnection (29). On the other hand, our studies have shown preservation of the relative synergy in a qualitative sense, even when there is significant damage to the pituitary.

The most profound and consistent results were seen in adult volunteers. In all the older adults (more than 30 years old), the peak GH value with GHRP-2 administration was significantly greater than that following GHRH. As previously reported by others, GH release following GHRH provocation decreases with aging, and to a lesser degree, following GHRP provocation. The relative GH output, as expressed by the peak GH ratio (GHRP/GHRH), was most exaggerated in a 33-year-old man. For the 29-year-old woman, the ratio was reversed. These preliminary observations suggest a dichotomous reversal of GH secretory response to these secretagogues at a younger adult age than might be expected. According to our hypothesis, these observations would imply a dimunition in endogenous ligand for GHRP in aging. We have previously reported experimental evidence from animals and relevant clinical studies in children (14, 30–33).

The use of GHRH to evaluate the extent to which GH can be released is invalid, because, after many years of laboratory and clinical studies, it is now recognized that responses to provocative challenges are not dose-dependent (14,34). In fact, maximal GHRH-mediated GH secretion can be amplified by an ancillary factor, such as prostaglandins, arginine, or other such GHRP (19,35). Even when GHRH is administered to humans and animals with normally functioning GH neuroendocrine axes, it stimulated different degrees of GH secretion (14,34). This was attributed to interaction with endogenous GH-regulatory peptides, including somatostatin, or even GHRH itself (28). Blunted responses to GHRH were attributed to interaction of the GHRH stimulus with somatostatin. Alternatively, when GHRH was given during "troughs" of episodic secretion, or during the early-rising stages of each episode, the stimulated response was greater than it was when GHRH was administered at the peak of the endogenous GH secretory episode or during its descending phase (36). Given this normal variation in the response to GHRH provocative stimuli, the diagnostic value of this peptide became limited.

Interpretation of the results must be in the context of the timing of the exogenous bolus secretagogue relative to the phase of the endogenous GH secretory pulse provocative testing reproducibility, and the known difficulties in the diagnosis of growth disorders. The preliminary observations described here may partly explain some of the confusion related to diagnostic testing of children with GH secretory abnormalities. With these considerations in mind, we tested children with various causes of GH secretory dysfunction, ranging from hypopituitarism to that resulting from radiation therapy.

The GH-deficient children had less GH secretion following GHRP-2, but there was overlap with normal and slow-growing, non-GH-deficient children. The most profoundly affected were children with multiple hypothalamic–pituitary deficiency; suggesting that, for most children, the deficiency in GHRH is relative. The same is true for the GHRH provocative responses, but to a smaller degree, suggesting a lesser deficiency in endogenous ligand for GHRP. It is of interest that the relationship of synergy between GHRP-2 and GHRH was preserved in the children who had profound GH deficiency, and even in the child who had hypophysitis, with very little residual pituitary tissue (no pituitary discernible by magnetic resonance imaging). Interestingly, although there was variability in GH secretory patterns of two slow-growing, non-GH-deficient children, the ratio of peak GH to GHRP-2 plus GHRH remained intact. This suggests an important functional qualitative component that these two children might have in common. In our study, there was no correlation between peak GH concentration after GHRH administration with mean peak GH levels to standard GH provocative testing; whereas, there was a positive correlation between peak GH values after GHRP-2 administration and mean peak GH response to standard provocative stimuli.

Interpretation of the GH response must be in the clinical context. For example, a child with short stature might be thought to release GH well within the normal range with a provocative dose of GHRH until it was recognized that the child was receiving clonidine for attention deficit disorder. As noted in our study, after clonidine was removed (and hypothetically, its influence on endogenous GHRH secretion), the GH response was diminished to both GHRP and GHRH in a sequential provocative study.

Radiation of the head for treatment of cancer in children retards growth, presumably by reducing the secretion of endogenous GHRH from hypothalamic neurons. We have suggested a hierarchy of dysfunction (extrahypothalamic neurotransmitters > hypothalamic GHRH or somatostatin secretion > pituitary GH secretion) (37). This hypothesis should also include damage and dysfunction of the neurons producing the endogenous ligand for GHRP. In our study, we demonstrate a variable response consistent with the complexity described in the foregoing, as well as the other potential reasons for variability discussed earlier.

Priming study

During this study, a diagnostic test of pituitary function using comparative responses to GHRH and GHRP was employed to investigate the etiology of age-related changes in human GH secretory dynamics. This test has been successfully used in conjunction with more standard provocative challenges to evaluate pathological GH deficiency in children

(19) and seems to be of value for diagnosing and treating the natural and progressive GH secretory deficits that occur as a consequence of aging.

The first objective of this study was to demonstrate the reliability of responses to the GH secretagogue, especially GHRP-2, which is of pivotal importance in performing the diagnostic test. However, in the present study, it was important to show that the responses to submaximal dosages were reproducible because they were the basis for investigating the etiology of age-related decline in GH secretion. The data corroborated the findings of others that the response to GHRP is dose related (38), and further show that they are reproducible within the same individual when administered hours apart.

It is also important to note that GHRP was always administered before GHRH. The sequence was intentionally used because it stabilized the response to GHRH which has been shown by many laboratories to have significant inter-individual, intra-individual, and temporal variation, presumably because its interaction with the inhibitory effect of endogenous somatostatin. Since GHRP is a functional antagonist of somatostatin, it removed the influence of this negative factor from GHRH allowing a reliable and reproducible response in all subjects.

The first major finding of this study is consistent with previous reports that aging has a significant negative effect upon the efficacy of GHRH in men. Administration of GHRH failed to increase serum concentrations of GH > 4 μg/L in all older subjects participating in this study. In children, responses < 10 μg/L are considered abnormal. By these criteria, all men in the older group were unable to release normal quanta of GH in response to challenge with the naturally occurring, trophic neuropeptide.

Strikingly, this age-related decrement in GHRH efficacy occurred even in a subject that was only 37 years old, demonstrating the remarkable early onset of GH neuroendocrine decline in the human lifespan. In contrast, the "younger" subjects in this study demonstrated robust responses to GHRH, reaching peak GH serum concentrations as high as 40 μg/L in response to a single injection.

It has been previously shown that repeated stimulation or "priming" with GHRH will increase the response to a challenge dose of GHRH. The second major finding, and perhaps the most important discovery of this study, was that priming with GHRP-2 improved the response of GHRH in the older group of subjects. This finding has important implications because it may provide a clue for at least one cause of the age-related decrease in GHRH sensitivity. Synergy between GHRP and GHRH is well known, but heretofore, has only been treated as a scientific curiosity. The findings of this study, taken with our previously published reports showing that passive immunization against GHRH or blockade of GHRH receptors attenuated GHRP efficacy (16), suggest that the complementary activities of the two GH secretagogues may have real functional significance. In other words, GHRH may actually require the naturally occurring ligand for GHRP for full expression of its stimulatory potential. If that is true, and because priming with GHRP partially restored the efficacy of GHRH in "older" subjects, the data suggest that loss or reduction of endogenous GHRP ligand is a major contributory factor to decline in the GH neuroendocrine axis during aging. Bowers and Granda-Ayala have also explored a variation of this by administering chronic daily injections (7–30 days) in normal younger and older men and women (39).

The data from this study also indicate that endogenous GHRP contributes to GH secretory capabilities, not only as a stimulatory agent that must be present in optimal concentrations, but that it maintains the integrity of the signal transduction systems for GHRH including perhaps its receptors and/or second messengers. These conclusions are supported by the fact that one day after older subjects had received their last priming dosage of GHRP, their responses to GHRH were significantly improved. Since GHRP is rapidly metabolized within minutes of its administration (15), none should have been remaining when the challenge dose of GHRH was administered. Since the GHRH response was improved, the data suggest that the GHRH signal transduction mechanism had been enhanced by GHRP priming, and that the improved response was not an expression of additional stimulation. This conclusion is further supported by the final observation that priming improved the response to co-administered GHRH and GHRP-2 in the older subjects. This finding demonstrates that the integrity and activation of the mechanism of GHRH-mediated GH secretion is GHRP dependent; and also that reactivation of the quiescent, aged GH neuroendocrine axis may be possible by supplementation or replacement of endogenous deficits with these compounds that are now being developed as orally active, highly bioavailable, therapeutic agents.

SUMMARY

The data in Study I support the use of a novel diagnostic test for evaluating pituitary function in slowly growing children and aging adults. The objective of the test is to determine whether it is feasible to use GHRP and GHRH as diagnostic tools to investigate the etiology of GH deficiency. Additional diagnostic studies are necessary to corroborate these preliminary observations. We hope that the data resulting from the further application of the principles on which the diagnostic test is based will allow appropriate selection of therapeutic entities, ranging from GHRH or GHRP given separately or in combination, or alternatively recombinant GH, the latter for subjects lacking a pituitary mechanism for GH production or secretion.

The data from Study II indicate that endogenous GHRP contributes to GH secretory capabilities not only as a stimulatory agent that must be present in optimal concentrations, but that it maintains the integrity of the signal transduction systems for GHRH, including perhaps its receptors and/or second messengers. These conclusions are supported by the fact that one day after older subjects had received their last priming dosage of GHRP, their responses to GHRH were significantly improved. Since GHRP is rapidly metabolized within minutes of its administration, none should have been remaining when the challenge dose of GHRH was administered. Since the GHRH response was improved, the data suggest that the GHRH signal transduction mechanism had been enhanced by GHRP priming, and that the improved response was not an expression of additional stimulation. This conclusion is further supported by the final observation that priming improved the response to co-administered GHRH and GHRP-2 in the older subjects. This finding demonstrates that the integrity and activation of the mechanism for GHRH-medicated GH secretion is GHRP dependent, and also that reactivation of the quiescent, aging GH neuroendocrine axis may be possible by supplementation or replacement of endogenous deficits with these compounds that are now being developed as orally active, highly bioavailable, therapeutic agents.

REFERENCES

1. Ceda, G.P., Valenti, G., Butturini, U. and Hoffman, A.R. (1986) Diminished pituitary responsiveness to growth hormone-releasing factor in aging male rats. Endocrinology 118, 2109–2114.
2. Ghigo, E., Goffi, S., Arvat, E., Nicolosi, M., Procopio, M., Bellone, J., Imperiale, E., Mazza, E., Baracchi, G. and Camanni, F. (1990) Pyridostigmine partially restores the GH responsiveness to GHRH in normal aging. Acta Endocrinologica 123, 169–174.
3. Iovino, M., Monteleone, P. and Steardo, L. (1989) Repetitive growth hormone-releasing hormone administration restores the attenuated growth hormone (GH) response to GH-releasing hormone testing in normal aging. J. Clin. Endocrinol. Metab. 69, 910–913.
4. Lang, I., Schernthaner, G., Pietschmann, P., Kurz, R., Stephenson, J.M. and Templ, H. (1987) Effects of sex and age on growth hormone-releasing hormone in healthy individuals. J. Clin. Endocrinol. Metab. 65, 535–540.
5. Shibasaki, T., Shizume, K., Nakahara, M., Masuda, A., Jibiki, K., Demura, H., Wakabayashi, I. and Ling, N. (1984) Age-related changes in plasma growth hormone response to growth hormone-releasing factor in man. J. Clin. Endocrinol. Metab. 58, 212–214.
6. Sonntag, W.E. and Gough, M.A. (1988) Growth hormone releasing hormone induced release of growth hormone in aging male rats: dependence on pharmacological manipulation of engenous somastostatin release. Neuroendocrinology 47, 482–488.
7. Sonntag, W.E., Steger, R.W., Forman, L.J. and Meites, J. (1980) Decreased pulsatile release of growth hormone in old male rats. Endocrinology 107, 1975–1879.
8. Sonntag, W.E., Hylka, V.W., and Meites, J. (1983) Impaired ability of old male rates to secrete growth hormone *in vivo* but not *in vitro* in response to hpGRF(1-44). Endocrinology 113, 2305–2307.
9. Spiliotis, B.E., August, G.P., Hung, W., Sonis, W., Mendelson, W. and Bercu, B.B. (1984) Growth hormone neurosecretory dysfunction: A treatable cause of short stature. JAMA 251, 2223–2230.
10. De Gennaro Colonna, V., Zoli, M., Cocchi, D., Maggi, A., Marrama, P., Agnati, L.F. and Muller, E.E. (1989) Reduced growth hormone releasing factor (GHRF)-like immunoreactivity and GHRF gene expression in the hypothalamus of aged rats. Peptides 10, 705–708.
11. Morimoto, N., Kawakami, F., Makin, S., Chihara, K., Hasegawa, M. and Ibata, Y. (1988) Age-related changes in growth hormone releasing factor and somatostatin in the rat hypothalamus. Neuroendocrinology 47, 459–464.
12. Ono, M., Miki, N. and Shizume, K. (1986) Release of immunoreactive growth hormone-releasing factor (GRF) and somatostatin from incubated hypothalamus in young and old male rats (abst). Neuroendocrinology 43, 111.
13. Walker, R.F., Yang. S.-W. and Bercu, B.B. (1991) Robust growth hormone (GH) secretion in aged female rats co-administered GH-releasing hexapeptide (GHRP-6) or GH releasing hormone (GHRH). Life Sci. 49, 1499–1504.

14. Bercu, B.B. and Walker, R.F. A diagnostic test employing growth hormone secretagogue for evaluating pituitary function in the elderly. In: Growth Hormone Secretagogues. B.B. Bercu and R.F. Walker (eds). Springer-Verlag, New York, pp. 289–305.
15. Bowers, C.Y., Sartor, A.O., Reynolds, G.A. and Badger, T.M. (1991) On the actions of growth hormone-releasing hexapeptide, GHRP. Endocrinology 128, 2027–2035.
16. Bercu, B.B., Yang, S.-W., Masuda, R. and Walker, R.F. (1992) Role of selected endogenous peptides in growth hormone releasing hexapeptide (GHRP-6) activity: analysis of GHRH, TRH, and GnRH. Endocrinology 130, 2579–2586.
17. Howard, A.D., Feighner, S.D., Cully, D.F., Arena, J.P., Liberator, P.A., Rosenblum, C.I. et al. A receptor in pituitary and hypothalamus that functions in growth hormone release. Science 273, 974–977.
18. Goth, M.I., Lyons, C.E., Canny, B.J. and Thorner, M.O. (1992) Pituitary adenylate cyclase activating polypeptide, growth hormone (GH)-releasing peptide and GH-releasing hormone stimulate GH release through distinct pituitary receptors. Endocrinology 130, 939–944.
19. Bercu, B.B. and Walker, R.F. (1996) Evaluation of pituitary function in children using growth hormone secretagogues. J. Pediatr. Endocrinol. Metab. 9, 325–332.
20. Root, A.W., Rosenfeld, R.L., Bongiovanni, G.M. and Eberlein, W.R. (1967) GH assay: The plasma growth response to insulin-induced hypoglycemia in children with retardation of growth. Pediatrics 39, 844–852.
21. Bercu, B.B. and Walker, R.F. (1998) Evaluation of pituitary function using growth hormone secretagogues. In: Growth Hormone Secretagogues in Clinical Practice. B.B. Bercu and R.F. Walker (eds). Marcel Dekker, New York, pp. 285–303.
22. Bercu, B.B. and Walker, R.F. (1997) Growth hormone secretagogues in children with altered growth. Acta Paediatrica 423, 102–106.
23. Walker, R.F. and Bercu, B.B. (1998) Effectiveness of growth hormone (GH) secretagogues for diagnosing and treating GH secretory deficiency in aging men. J. Anti-Aging Medicine 1, 167–168.
24. Tuilpakov, A.N., Bulatov, A.A., Peterkova, V.A., Elizarova, G.P., Volevodz, N.N. and Bowers, C.Y. (1995) Growth hormone (GH)-releasing effects of synthetic peptide GH-releasing peptide-2 and GH-releasing hormone (1-29NH_2) in children with GH insufficiency and idiopathic short stature. Metabolism 9, 1199–1204.
25. Popovic, V., Micic, D., Damjanovic, S., Djurovic, M., Simic, M., Gligorovic, M., Dieguez, C. and Casanueva, F.F. (1996) Evaluation of pituitary GH reserve with GHRP-6. J. Pediatr. Endocrinol. Metab. 9, 289–298.
26. Micic, D., Popovic, V., Doknic, M., Macut, D., Dieguez, C. and Casanueva, F.F. (1998) Preserved growth hormone (GH) secretion in aged and very old subjects after testing with the combined stimulus GH-releasing hormone plus GH-releasing hexapeptide-6. J. Clin. Endocrinol. Metab. 83, 2569–2572.
27. Bercu, B.B., Yang, S.-W., Mauda, R., Hu, C.-S. and Walker, R.F. (1992) Effects of co-administered growth hormone (GH) releasing hormone and GH-releasing hexapeptide on maladaptive aspects of obesity in Zucker rats. Endocrinology 131, 2800–2804.
28. Pombo, M., Barreiro, J, Penalva, A., Peino, R., Dieguez, C. and Casanueva, F.F. (1995) Absence of growth hormone (GH) secretion after the administration of either GH-releasing hormone (GHRH), GH-releasing peptide (GHRP-6), or GHRH plus GHRP-6 in children with neonatal pituitary stalk transection. J. Clin. Endocrinol. Metab. 80, 3180–3184.
29. Popovic, V., Damjanovic, S., Micic, D., Djurovic, M., Dieguez, C. and Casanueva, F.F. (1995) Blocked growth hormone-releasing peptide (GHRP-6)-induced GH secretion and absence of the synergistic action of GHRP-6 plus GH-releasing hormone in patients with hypothalamo-pituitary disconnection: evidence that GHRP-6 main action is exerted at the hypothalamic level. J. Clin. Endocrinol. Metab. 80, 942–947.
30. Walker, R.F, Yang, S.-W., Masuda, R. and Bercu, B.B. (1994) Effects of GH-releasing peptides on stimulated GH secretion in old rats. In: Basic and Clinical Aspects of Growth Hormone II. B.B. Bercu and R.F. Walker (eds). Springer-Verlag, New York, pp. 167–192.

31. Walker, R.F., Ness, G.C., Zhao, A. and Bercu, B.B. (1994) Effects of stimulated GH secretion on age-related changes in plasma cholesterol and hepatic low density lipoprotein messenger RNA concentrations. Mech. Aging Dev. 75, 215–226.
32. Walker, R.F., Engleman, R., Pross, S. and Bercu, B.B. (1994) Effects of growth hormone secretagogues on age-related changes in the rat immune system. Endocrine 2, 857–862.
33. Walker, R.F. and Bercu, B.B. (1998) Effects of a growth hormone releasing peptide-like nonpeptidyl growth hormone secretagogues in physiology and function in aged rats. In: Growth Hormone Secretagogues in Clinical Practice. B.B. Bercu and R.F. Walker. Marcel Dekker, New York, pp. 187–207.
34. Walker, R.F. and Bercu, B.B. (1996) An animal model for evaluating growth hormone secretagogues. In: Growth Hormone Secretagogues. B.B. Bercu and R.F. Walker (eds). Springer-Verlag, New York, pp. 253–287.
35. Bercu, B.B. and Walker, R.F. (1997) Novel growth hormone secretagogues: Clinical applications. The Endocrinologist 7, 51–64.
36. Cho, K.H., Yang, S.W., Hu, C.-S. and Bercu, B.B. (1992) Growth hormone (GH) response to growth hormone-releasing hormone (GHRH) varies with intrinsic growth hormone secretory rhythm in children: Reduced variability using somatostatin pretreatment. J. Pediatr. Endocrinol. Metab. 5, 155–165.
37. Jorgensen, E.V., Schwartz, I.D., Hvizdala, E., Barbosa, J., Phuphanich, S., Shulman, D.I., Root, A.W., Estrada, J., Hu, C.-S. and Bercu, B.B. (1993) Neurotransmitter control of GH secretion in children after cranial radiation therapy. J. Pediatr. Endocrinol. 6, 131–142.
38. Ilson, B.E., Jorkasky, D.K., Curnow, R.T. and Stote, R.M. (1989) Effect of a new synthetic hexapeptide to selectively stimulate growth hormone release in healthy human subjects. J. Clin. Endocrinol. Metab. 69, 212–214.
39. Bowers, C.Y. and Granda-Ayala, R. (1996) GHRP-2, GHRH and SRIF interrelationships during chronic administration of GHRP-2 to humans. J. Ped. Endocrinol. Metab. 9, 261–27.

Growth Hormone Secretagogues
Edited by E. Ghigo, M. Boghen, F.F. Casanueva and C. Dieguez

Chapter 14

Does Desensitisation to Growth Hormone Secretagogues Occur?

ASAD RAHIM and STEPHEN M. SHALET

Department of Endocrinology, Christie Hospital, Withington, Manchester, U.K.

INTRODUCTION

The administration of recombinant human growth hormone (GH) is far from physiological with a bolus dose of GH being administered at night. This does not reflect the natural pulsatile fashion in which GH is released in normal subjects. The use of GH releasing agents with the potential for inducing pulsatile GH secretion has thus been of great interest for some time. The therapeutic use of growth hormone releasing hormone (GHRH) has been limited by several factors including poor bioavailability (1,2), mode of administration and de-sensitisation (3–6). The discovery of growth hormone secretagogues (GHS) (7,8) with reasonable bioavailability, even after oral administration, and their ability to stimulate release of GH in a pulsatile manner has led to both short- and long-term studies to assess the therapeutic potential of this class of drugs. *In vitro* and *in vivo* studies suggest that both short- and long-term administration of GHRH result in de-sensitisation (4–6) of the GH response to GHRH; the use of GHS may similarly be limited if significant de-sensitisation to GHS occurs. Several studies have addressed the issue of de-sensitisation after either continuous infusions or long-term administration.

SHORT-TERM STUDIES

Roh et al. (1997) (9) investigated desensitisation caused by growth hormone releasing peptide-2 (GHRP-2) *in vivo* using calves. GHRP-2 (12.5 μg/kgBW/h) or GH-releasing factor (GRF; 0.125 μg/kgBW/h) were infused for 180 minutes, and 60 minutes after the infusion, a bolus of either GHRP-2 (12.5 μg/kg BW) or GRF (0.125 μg/kg BW) was administered. Infusion of GHRP-2 did not attenuate the GH response to GRF. In contrast, the GH response to a GHRP-2 bolus was attenuated following the GHRP-2 infusion thus demonstrating de-sensitisation to GHRP-2. On completion of the GHRP-2 infusion, two repetitive injections of GHRP-2 (12.5 μg/kg BW) were administered at hourly intervals for up to four hours to assess the duration of post-infusion attenuation. Attenuation of the GH response caused by GHRP-2 was maintained for the full four hours.

Several studies have assessed the effect of continuous GHRP infusion in human subjects. DeBell and colleagues (10) administered a continuous 5.5 hour IV infusion of saline and GHRP-6 at three doses (0.1, 0.3 and 1.0 μg/kg/hr) on different occasions. The GH response to a single IV bolus of GHRP-6 (1.0 μg/kg) was assessed at the end of each infusion. During the saline infusion spontaneous GH peaks occurred at variable times. On initiation of the GHRP-6 infusion at the two lower doses, a single burst of GH release was observed. Similarly, infusion of the highest dose (1.0 μg/kg) of GHRP-6 led to a burst of GH release which was followed by sporadic secretory GH bursts of lesser magnitude during the remainder of the infusion.

Partial attenuation of the GH response to an IV bolus of GHRP-6 was observed after each GHRP-6 infusion. The GH response to the bolus was inversely related to the dose of the preceding infusion. Total GH released, i.e. that released during the infusion and after the bolus, was not different between the three doses.

Huhn and colleagues (11) administered GHRP-6 for 24 hours with each subject receiving four infusions, two saline and two GHRP-6. Using deconvolution analysis, GH secretion during the GHRP-6 infusion was reported to be 8-fold higher compared with saline. Cluster analysis revealed an increase in basal GH levels, the number, height and duration of GH pulses.

Following each infusion a bolus dose of either GHRH or GHRP-6 was administered. Peak GH concentrations and GH secretion rates to a bolus of GHRP-6 were significantly lower after GHRP-6 infusion compared with the saline infusion. The response to GHRH after the GHRP-6 infusion, however was significantly increased. Huhn and colleagues (11) concluded that a 24 hour infusion of GHRP-6 augmented pulsatile GH release but also resulted in attenuation of the subsequent GH response to GHRP-6. Furthermore plasma IGF-I was shown to increase after each GHRP-6 infusion.

Similarly, Jaffe and colleagues (12) administered an IV infusion of GHRP-6 or saline for 34 hours ($n = 9$). Following a loading dose of GHRP-6 (1 μg/kg), subjects were infused with GHRP-6 (1 μg/kg/h). Bolus doses of GHRH (1 μg/kg) and GHRP-6 (1 μg/kg) were then given after the infusion. Integrated GH concentration (IGHC) and parameters of pulsatile GH concentration were calculated for a duration of 18 hours and IGHC was calculated for 2 h after each bolus of GHRP-6 or GHRH. During GHRP-6 infusion, IGHC, maximum pulse amplitude, and mean pulse amplitude all increased significantly. Plasma IGF-I also increased compared with baseline values. No change in interpulse GH concentration or GH pulse frequency was observed. GH responsiveness to GHRH was increased whilst that to GHRP-6 was significantly reduced. Thus in summary prolonged exposure to GHRP-6 led to an increase in spontaneous GH secretion, an increased response to GHRH and de-sensitisation to GHRP-6.

Attenuation of the GH response to a GHS after infusion of the same GHS occurs in both man and animals. Furthermore, the GH response to GHRH is increased suggesting homologous de-sensitisation.

LONG-TERM STUDIES

Several studies have administered repeated bolus doses of a GHS to normal subjects (13–15) and patients with different pathological states including GH deficiency (16), obesity (17) and short stature (18,19). The agent used, route and duration of administration and the dose administered, have differed.

The GHRPs, hexarelin, GHRP-6 and GHRP-2, have been administered to animals and man. Human studies have included children, young adults and elderly subjects.

Long-term administration of hexarelin to elderly subjects has produced conflicting results. Ghigo et al. (14) administered either oral (20 mg = 300 μg/kg body weight tds for 15 days) or intranasal (1.25 mg = 18 μg/kg body weight for eight days) hexarelin to elderly subjects. The GH response to hexarelin was assessed after administration of the first and last dose of hexarelin.

Following intra-nasal administration of hexarelin, an increasing trend in the GH response to hexarelin was reported. The GH-releasing effect of oral hexarelin was maintained after 15 days. A small, but significant, rise in serum IGF-I and IGFBP-3 levels in those treated with the oral preparation, was observed. In contrast, 16 weeks of twice daily administration of subcutaneous hexarelin (1.5 μg/kg body weight) resulted in a significant attenuation of the GH response to hexarelin. In 12 elderly subjects Rahim et al. (15) observed a reduction in the peak GH response and also AUC_{GH} after one week of hexarelin therapy (Figure 1). This decrease, however, was not significant. Compared with baseline, the reductions in AUC_{GH} at the end of week 4 and week 16 were significant and the decrease in peak GH response at the end of week 16 was significant. On an individual basis, only one subject was unable to produce a reasonable response to hexarelin at week 16 with a peak GH response of 3.2 mU/L. Four weeks after completion of hexarelin therapy, both AUC_{GH} and peak GH response to hexarelin increased significantly compared with week 16

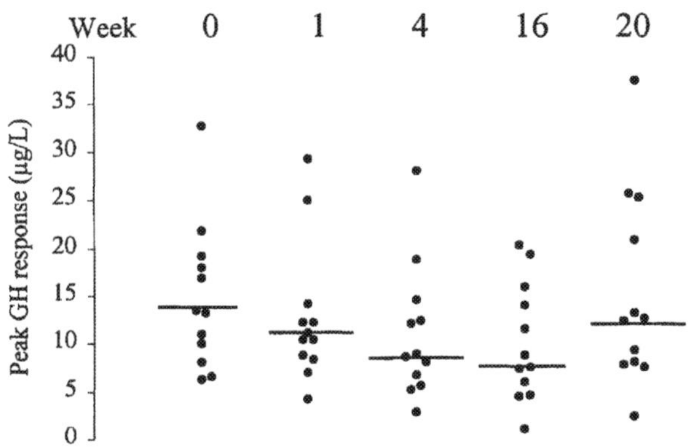

Figure 1. Peak GH response following a subcutaneous injection of hexarelin at baseline (0), weeks 1, 4, 16 and 20. Note the gradual decrease in peak GH response with time whilst receiving therapy, then the increase back to baseline values four weeks after completion of therapy (week 20).

values, and were not significantly different from baseline. Serum IGF-I and IGFBP-3 did not change significantly over the 16 week therapy period.

Thus Rahim et al. (15) reported a partial and reversible attenuation of the GH response with long-term hexarelin therapy. A similar partial and reversible attenuated GH response to hexarelin has been demonstrated by Klinger and colleagues (20). In seven prepubertal, constitutionally short children, Klinger and colleagues (20) administered thrice daily intranasal hexarelin (60 μg/kg body weight) for six to ten months. After one week of therapy, the peak GH response to an intranasal bolus of hexarelin (20 μg/kg body weight) had decreased by 50% and thereafter remained constant throughout the six month period of continuous therapy. At completion of hexarelin therapy ($n = 5$), the mean peak GH response to IV hexarelin had decreased by 75% compared with baseline. Three months after completion of hexarelin therapy, mean peak GH response to intravenous hexarelin had increased to 50% of pretreatment levels ($n = 4$). Despite the attenuated GH response to hexarelin, serum IGF-I increased during the study period and growth rate was greater during hexarelin therapy, compared with that before therapy.

In six old beagle dogs, Cella and colleagues (21) administered subcutaneous hexarelin (500 μg/kg body weight) twice daily for seven weeks, four weeks and one week. Each treatment period was separated by a two-week-off-treatment period and the GH response to subcutaneous hexarelin was assessed at weekly intervals. Hexarelin-stimulated GH release decreased after four weeks of twice daily hexarelin therapy. Two weeks after completion of the seven-week study period, the GH response to hexarelin had increased to pre-treatment levels. During the four-week treatment period, the GH response to hexarelin had decreased within the first two weeks of therapy.

In a recent study, Svensson et al. (17) administered the non-peptidyl GHS MK-677 at a dose of 25 mg once daily to 24 obese males aged 18–50 years for eight weeks. Blood samples were taken over a four hour period at baseline, week 2 and week 8. Peak GH response and serum AUC_{GH} were significantly reduced at week 2 and week 8 compared with baseline. At week 8, both peak GH response and AUC_{GH} were lower than that at week 2 but this change was not significant. Despite this attenuation of the GH response to MK-677, there was an increase in serum IGF-I over the eight-week treatment period.

POSSIBLE MECHANISMS FOR DE-SENSITISATION

The mechanism of decreased GH release in response to chronic therapy with a GHS remains unclear. Several potential mechanisms exist. These include depletion of the pituitary stores of GH, negative feedback by IGF-I, negative feedback by GH and receptor/post-receptor mechanisms.

It is unlikely that depletion of pituitary GH stores plays a significant, if any, role as only homologous de-sensitisation occurs (9,11,12). Indeed the GH response to GHRH has been shown to be higher after an infusion of GHRP whereas the GH response to an acute bolus of the GHRP is attenuated (11,12). If pituitary stores were significantly depleted after the

GHRP infusion then the GH response to GHRH would also be expected to be reduced. Current data clearly demonstrate that this is not the case.

Circulating IGF-I acts directly at the pituitary and hypothalamic levels. At the pituitary, IGF-I inhibits both basal and GHRH-induced GH secretion and suppresses GH gene expression (3,22–24). IGF-I also acts directly at the hypothalamus where it increases somatostatin secretion (25). Several studies have reported an increase in IGF-I levels with chronic GHS therapy and suggested that the increase in IGF-I could explain the reduction in GH release after long-term GHS therapy. This has most recently been suggested by Svensson and colleagues (17). It is likely that IGF-I plays some role in decreased GH release but several studies have also reported de-sensitisation without a concomitant rise in IGF-I. Rahim et al. (15) clearly demonstrated de-sensitisation to hexarelin in normal elderly subjects without an associated increase in IGF-I.

Cella and colleagues (21) reported an increase in GH pulse frequency and amplitude with chronic GHS administration suggesting an overall increase in GH levels. GH appears to have a direct effect at the hypothalamic–pituitary level resulting in a reduction in GH release. The short-term administration of GH results in negative feedback on GH release before IGF-I levels have increased (26) suggesting that the increase in GH is responsible for the observed reduction in GH secretion. Furthermore, somatostatin neurones possess GH receptor mRNA (27) and *in vitro* studies have reported that GH stimulates somatostatin secretion (28). GH may also influence GHRH expression either directly or via IGF-I (29). Recent animal data have also suggested that GH influences hypothalamic GHS-receptor (GHS-R) expression (30). GHS-R expression was found to be increased in GH-deficient rat dwarves and normalised with GH treatment.

Stimulated GH release is similarly affected by negative feedback from GH. Pre-treatment with GH dampens the GH response to provocative stimuli including insulin-induced hypoglycaemia, clonidine and GHRH (31,32), and administration of exogenous GH (33,34) has been shown to attenuate the GH response to hexarelin suggesting that hexarelin-stimulated GH release is subject to partial feedback inhibition by the action of GH on somatostatin and/or GHRH.

Prolonged stimulation with GRF (4) results in a reduction in the number of GRF-binding sites. Bilezikjian et al. (4) demonstrated a reduction in GRF-binding capacity with a decrease of 48% of binding sites in rat anterior pituitary cells after two hours of pre-treatment with rat GRF. Furthermore, the reduction in GRF-binding sites and decreased sensitivity to GRF were reversed 24 hours after washing the cells and allowing them to recover. Thus the attenuated GH response (9–12,15) seen with prolonged exposure to GHS may result from down-regulation of receptor numbers although post-receptor mechanisms may also play a role.

The attenuated GH response to GHS may occur via several mechanisms which include the negative feedback from GH, IGF-I but more plausibly the down-regulation of receptor/post-receptor mechanisms.

De-sensitisation to GHS is a critical issue and occurs after both long- and short-term administration. The therapeutic potential of these agents is likely to be limited by this observation. However, de-sensitisation is usually partial and reversible and may be related to the dose, frequency and duration of GHS administration.

REFERENCES

1. Corpas, E., Harman, S.M. and Blackman, M.R. (1993) Human growth hormone and human aging. Endocr. Rev. 14, 20–39.
2. Thorner, M.O. (1993) On the discovery of growth hormone-releasing hormone. Acta Paediatr. Suppl. 388, 2–7; discussion 8.
3. Ceda, G.P. and Hoffman, A.R. (1985) Growth hormone-releasing factor desensitization in rat anterior pituitary cells *in vitro*. Endocrinology 116, 1334–40.
4. Bilezikjian, L.M., Seifert, H. and Vale, W. (1986) Desensitization to growth hormone-releasing factor (GRF) is associated with down-regulation of GRF-binding sites. Endocrinology 118, 2045–52.
5. Arsenijevic, Y., Rivest, R.W., Eshkol, A., Sizonenko, P.C. and Aubert, M.L. (1987) Plasma growth hormone (GH) response to intravenous GH-releasing factor (GRF) in adult rats: evidence for transient pituitary desensitization after GRF stimulation. Endocrinology 121, 1487–96.
6. Kirk, J.M., Trainer, P.J., Majrowski, W.H., Murphy, J., Savage, M.O. and Besser, G.M. (1994) Treatment with GHRH(1-29)NH2 in children with idiopathic short stature induces a sustained increase in growth velocity. Clin. Endocrinol. Oxf. 41, 487–93.
7. Bowers, C.Y., Momany, F., Reynolds, G.A., Chang, D., Hong, A. and Chang, K. (1980) Structure–activity relationships of a synthetic pentapeptide that specifically releases growth hormone *in vitro*. Endocrinology 106, 663–7.
8. Bowers, C.Y., Momany, F.A., Reynolds, G.A. and Hong, A. (1984) On the *in vitro* and *in vivo* activity of a new synthetic hexapeptide that acts on the pituitary to specifically release growth hormone. Endocrinology 114, 1537–45.
9. Roh, S.G., He, M.L., Matsunaga, N., Hidaka, S. and Hidari, H. (1997) No desensitization of the growth hormone (GH) response between GH-releasing peptide-2 and GH-releasing factor in calves. J. Anim. Sci. 75, 2749–53.
10. DeBell, W.K., Pezzoli, S.S. and Thorner, M.O. (1991) Growth hormone (GH) secretion during continuous infusion of GH-releasing peptide: partial response attenuation. J. Clin. Endocrinol. Metab. 72, 1312–6.
11. Huhn, W.C., Hartman, M.L., Pezzoli, S.S. and Thorner, M.O. (1993) Twenty-four-hour growth hormone (GH)-releasing peptide (GHRP) infusion enhances pulsatile GH secretion and specifically attenuates the response to a subsequent GHRP bolus. J. Clin. Endocrinol. Metab. 76, 1202–08.
12. Jaffe, C.A., Ho, P.J., Demott Friberg, R., Bowers, C.Y. and Barkan, A.L. (1993) Effects of a prolonged growth hormone (GH)-releasing peptide infusion on pulsatile GH secretion in normal men. J. Clin. Endocrinol. Metab. 77, 1641–47.
13. Ghigo, E., Arvat, E., Rizzi, G., et al. (1994) Growth hormone-releasing activity of growth hormone-releasing peptide-6 is maintained after short-term oral pretreatment with the hexapeptide in normal aging. Eur. J. Endocrinol. 131, 499–503.
14. Ghigo, E., Arvat, E., Gianotti, L., et al. (1996) Short-term administration of intranasal or oral Hexarelin, a synthetic hexapeptide, does not desensitize the growth hormone responsiveness in human aging. Eur. J. Endocrinol. 135, 407–12.
15. Rahim, A., O'Neill, P.A. and Shalet, S.M. (1998) Growth hormone status during long-term Hexarelin therapy. J. Clin. Endocrinol. Metab. 83, 1644–1649.

16. Chapman, I.M., Pescovitz, O.H., Murphy, G. et al. (1997) Oral administration of growth hormone (GH) releasing peptide-mimetic MK-677 stimulates the GH/insulin-like growth factor-I axis in selected GH-deficient adults. J. Clin. Endocrinol. Metab. 82, 3455–63.
17. Svensson, J., Lönn, L., Jansson, J.-O. et al. (1997) Two-month treatment of obese subjects with the oral growth hormone (GH) secretagogue MK-677 increases GH secretion, fat-free mass, and energy expenditure. J. Clin. Endocrinol. Metab. 83, 362–369.
18. Laron, Z., Frenkel, J., Deghenghi, R., Anin, S., Klinger, B. and Silbergeld, A. (1995) Intranasal administration of the GHRP hexarelin accelerates growth in short children. Clin. Endocrinol. Oxf. 43, 631–5.
19. Pihoker, C., Badger, T.M., Reynolds, G.A. and Bowers, C.Y. (1997) Treatment effects of intranasal growth hormone releasing peptide-2 in children with short stature. J. Endocrinol. 155, 79–86.
20. Klinger, B., Silbergeld, A., Deghenghi, R., Frenkel, J. and Laron, Z. (1996) Desensitization from long-term intranasal treatment with hexarelin does not interfere with the biological effects of this growth hormone-releasing peptide in short children. Eur. J. Endocrinol. 134, 716–9.
21. Cella, S.G., Cerri, C.G., Daniel, S. et al. (1996) Sixteen weeks of hexarelin therapy in aged dogs: effects on the somatotropic axis, muscle morphology, and bone metabolism. J. Gerontol. A Biol. Sci. Med. Sci. 51, B439–47.
22. Ceda, G.P., Davis, R.G., Rosenfeld, R.G. and Hoffman, A.R. (1987) The growth hormone (GH)-releasing hormone (GHRH)-GH-somatomedin axis: evidence for rapid inhibition of GHRH-elicited GH release by insulin-like growth factors I and II. Endocrinology 120, 1658–62.
23. Yamashita, S. and Melmed, S. (1987) Insulin-like growth factor I regulation of growth hormone gene transcription in primary rat pituitary cells. J. Clin. Invest. 79, 449–52.
24. Yamashita, S., Weiss, M. and Melmed, S. (1986) Insulin-like growth factor I regulates growth hormone secretion and messenger ribonucleic acid levels in human pituitary tumor cells. J. Clin. Endocrinol. Metab. 63, 730–35.
25. Berelowitz, M., Szabo, M., Frohman, L.A., Firestone, S., Chu, L. and Hintz, R.L. (1981) Somatomedin-C mediates growth hormone negative feedback by effects on both the hypothalamus and the pituitary. Science 212, 1279–81
26. Lanzi, R. and Tannenbaum, G.S. (1992) Time-dependent reduction and potentiation of growth hormone (GH) responsiveness to GH-releasing factor induced by exogenous GH: role for somatostatin. Endocrinology 130, 1822–28.
27. Burton, K.A., Kabigting, E.B., Clifton, D.K. and Steiner, R.A. (1992) Growth hormone receptor messenger ribonucleic acid distribution in the adult male rat brain and its colocalization in hypothalamic somatostatin neurons. Endocrinology 131, 958–63.
28. Sheppard, M.C., Kronheim, S. and Pimstone, B.L. (1978) Stimulation by growth hormone of somatostatin release from the rat hypothalamus *in vitro*. Clin. Endocrinol. 9, 583–86.
29. Hurley, D.L. and Phelps, C.J. (1993) Altered growth hormone-releasing hormone mRNA expression in transgenic mice with excess or deficient endogenous growth hormone. Mol. Cell Neurosci. 4, 237–244.
30. Bennett, P.A., Thomas, G.B., Howard, A.D., Feighner, S.D., van der Ploeg, L.H., Smith, R.G. and Robinson, I.C. (1997) Hypothalamic growth hormone secretagogue-receptor (GHS-R) expression is regulated by growth hormone in the rat [see comments]. Endocrinol. 138, 4552–57.
31. Abrams, R.L., Grumbach, M.M. and Kaplan, S.L. (1971) The effect of administration of human growth hormone on the plasma growth hormone, cortisol, glucose, and free fatty acid response to insulin: evidence for growth hormone autoregulation in man. J. Clin. Invest. 50, 940–50.
32. Nakamot, J.M., Gertner, J.M., Press, C.M., Hintz, R.L., Rosenfeld, R.G. and Genel, M. (1986) Suppression of the growth hormone (GH) response to clonidine and GH-releasing hormone by exogenous GH. J. Clin. Endocrinol. Metab. 62, 822–26
33. Massoud, A.F., Hindmarsh, P.C. and Brook, C.G. (1995) Hexarelin induced growth hormone release is influenced by exogenous growth hormone. Clin. Endocrinol. Oxf. 43, 617–21.
34. Arvat, E., Di Vito, L., Gianotti, L. et al. (1997) Mechanisms underlying the negative growth hormone (GH) autofeedback on the GH-releasing effect of hexarelin in man. Metabolism 46, 83–8.

Growth Hormone Secretagogues
Edited by E. Ghigo, M. Boghen, F.F. Casanueva and C. Dieguez

Chapter 15

GHRP-s in Human Obesity

JOHAN SVENSSON, JOHN-OLOV JANSSON and BENGT-ÅKE BENGTSSON

Research Centre for Endocrinology and Metabolism, Sahlgrenska University Hospital, Göteborg, Sweden

INTRODUCTION

The benefit of growth hormone (GH) treatment of GH deficient (GHD) adults is now established. GH replacement has been found to decrease total body fat, with a more marked decrease in visceral fat than in other fat deposits (1). Basal metabolic rate (BMR) has been increased (2). GH treatment has also been found to upregulate the LDL-receptor in the human liver (3) and in most long-term studies of GHD adults, LDL-cholesterol has been decreased by GH treatment (4). Therefore, GH replacement of GHD adults has greatly improved several cardiovascular risk factors in adult GHD, although an increase has been observed in the atherogenic (5–8) lipoprotein(a) (9,10), the importance of this effect being unknown.

Obviously, a reduction of total body fat as well as an improvement of cardiovascular risk factors would be beneficial in obesity. The development of the new class of GH secretagogues, including GH-releasing peptides (GHRP-s) (11) as well as non-peptidyl, orally active GH secretagogues (11), has provided a possibility of increasing GH levels in obesity by oral administration. Furthermore, the members of the GHRP family may be associated with less side-effects than conventional GH treatment since they enhance the pre-existing pulsatile GH secretion (12). This increase in pulsatile GH release is possibly more physiological than the rise in GH levels after subcutaneous injections of GH (13).

GH SECRETION IN HUMAN OBESITY

Obesity, and especially abdominal/visceral obesity, is characterised by low serum levels of GH and IGF-I (insulin-like growth factor-I). With increasing obesity, GH secretion is diminished with a decrease in the mass of GH secreted per burst without any major impact on GH secretory burst frequency (14). Furthermore, the metabolic clearance rate of GH is accelerated (15). By the use of computed tomography (CT), the visceral adiposity has been found to be a major determinant of stimulated (arginine and clonidine) GH secretion in non-obese healthy adults (16). In both young and old males and females, the integrated 24-hour spontaneous GH secretion is negatively related to the visceral fat mass (17).

As found for GH secretion, serum IGF-I has been found to be inversely associated to the percentage of body fat (14). Low serum IGF-I levels were found in a study of males with a predominant visceral obesity, where the low IGF-I concentrations were mainly related to the amount of visceral fat and not to the subcutaneous fat deposits (18).

It seems reasonable to assume that the low GH levels in obesity contribute to the maintenance of the obesity state. To what extent low GH levels also are a cause of developing obesity is unknown. In one study of obese subjects, massive weight loss nearly normalised spontaneous GH secretion and serum IGF-I (19). However, in other studies GH response to provocative testing was not normalised after weight loss (20,21).

In addition to low GH levels, abdominal/visceral obesity is also accompanied by low sex steroid levels as well as an increased activity of the hypothalamic–pituitary–adrenal axis (22–26). The importance of these endocrine aberrations, as well as a possible general modulation of hypothalamic/pituitary hormonal axes by corticotrophin releasing hormone (24,27), is beyond the topic of this review.

CARDIOVASCULAR RISK FACTORS IN VISCERAL OBESITY

The turnover of visceral fat has been found to be higher than in other fat deposits (28–30). With increasing accumulation of visceral fat, the liver via the portal vein is exposed to increased levels of free fatty acids (FFA). Increased levels of FFA decrease hepatic clearance of insulin from the pancreas, and increase gluconeogenesis and the secretion of very low density lipoproteins (VLDL-s) from the liver (31–34). Therefore, visceral fat accumulation may cause increased peripheral levels of insulin, glucose, and VLDL-s, and risk of developing "Syndrome X" (35) (also denominated "The Metabolic syndrome" (36) and "The Insulin Resistance syndrome" (37,38)). "Syndrome X", as well as untreated GH deficiency in adults, is associated with obesity, dyslipoproteinemia, insulin resistance, premature atherosclerosis, and increased cardiovascular morbidity and mortality (4,24,39–41).

GH INTERVENTION

In acromegaly, successful treatment normalises the decreased amount of body fat (42,43). GH treatment of untreated adult GHD massively reduces body fat, the reduction most marked in the visceral region (1). This decrease in body fat has most often not been associated with any major decrease in body weight, since the lean body mass and the extracellular water have concomitantly been increased (4). In obese subjects, the effects of the combination of GH administration and dietary restriction have been investigated in some studies. However, both short term (44) and several weeks (45,46) of combined GH treatment and dietary restrictions were not able to enhance the loss of body fat or body weight compared with saline treatment, although the GH treatment decreased the loss of lean body mass during dietary instruction (44,46).

GH treatment may not enhance weight loss in human obesity, but it may improve cardiovascular risk factors as previously observed during GH treatment of GHD adults (4). In a 9-month GH treatment study of moderately obese males with a predominance of abdominal/visceral obesity (47), GH treatment (without dietary restriction) resulted in a marked decrease of both abdominal subcutaneous and visceral fat. Furthermore, insulin sensitivity improved after 9 months of GH treatment.

GHRP ADMINISTRATION IN HUMAN OBESITY

As mentioned above, stimulated GH secretion is blunted in obesity. GH release is decreased after stimulation with hypoglycaemia, arginine, glucagon, exercise, clonidine, or GH-releasing hormone (GHRH) (16,48–54). The GH responses to GHRP-related substances such as hexarelin (55) and L-692,429 (56) have also been lower in obese subjects than in lean controls. However, in the study by Kirk et al. (56), L-692,429 elicited a higher GH response than GHRH and the response to low dose L-692,429 in fasted obese subjects was similar to that in fed nonobese subjects. In a study of obese subjects by Cordido et al. (57), GHRP-6 also elicited a higher GH response than GHRH, and their combined administration elicited a massive GH response. Furthermore, lowering of FFA by acipimox increased GHRP-6- plus GHRH-mediated GH release in one other study in obesity (58). The findings of the latter two studies suggest that the somatotroph cell is intact in obesity and that increased FFA levels may contribute to the blunted GH secretion in obesity.

The effects of the GHRP-related compounds on body composition in human obesity is previously unknown. In GHD children, hexarelin treatment decreased body fat as measured by skinfold thickness (59). However, in elderly subjects hexarelin treatment did not affect body composition (60). Seven days of MK-677 treatment in healthy male volunteers reversed diet-induced catabolism, suggesting that prolonged MK-677 treatment may diminish the loss of lean body mass seen during catabolic stages (61).

TWO-MONTH TREATMENT OF OBESE SUBJECTS WITH MK-677

In a randomised, double-blind, and placebo controlled study (62), we investigated the effects of treatment with the oral, non-peptidyl GH secretagogue MK-677 on GH secretion and body composition in otherwise healthy obese subjects with abdominal/visceral obesity. Twenty-four obese males aged 19 to 49 years, with a body mass index > 30 kg/m^2 and a waist:hip ratio > 0.95, were treated with MK-677 25 mg or placebo for 8 weeks.

Eight-hour profiles of GH were performed after tablet intake at initiation of treatment and at 2 and 8 weeks of treatment. MK-677 treatment significantly increased serum peak

GH values and serum GH area under curve (AUC) values obtained from these profiles throughout MK-677 treatment, although the initial GH response was of a significantly higher magnitude than the GH responses after MK-677 administration at 2 and 8 weeks of GH treatment. (The initial MK-677 administration induced a 33-fold increase in peak GH and a 20-fold increase GH AUC when compared to the corresponding values after placebo administration. At study end, there remained a 16-fold increase in peak GH and an 8-fold increase in GH AUC). Serum IGF-I was significantly increased compared with placebo by approximately 40 percent throughout MK-677 treatment. Serum IGF-binding protein-3 (IGFBP-3) was also significantly increased throughout MK-677 treatment.

The reason for the dampening of the GH response to MK-677 from the initiation of treatment to 2 weeks of treatment is unknown. A similar, initial dampening of GH levels was found after MK-677 administration to adults with idiopathic GHD of childhood onset (63). The negative feedback on GH secretion that is exerted by increased serum IGF-I levels (64) may be of importance as well as a possible homologous desensitisation (65,66). Anyhow, 8 weeks MK-677 treatment of obese subjects does not completely deplete the pituitary reserve of GH, possibly due to a stimulatory effect on GH synthesis.

After the initial MK-677 administration, serum peak and AUC values of prolactin and cortisol were significantly increased compared with placebo. At 2 and 8 weeks of treatment, only the response in prolactin AUC remained. Furthermore, urine concentrations of free cortisol and 17-OH-deoxycorticosteroids were not changed by MK-677 treatment. These results are in line with previous studies (12,67–70): an initial cortisol response to MK-677 is rapidly downregulated while a minor prolactin response persists.

MK-677 treatment significantly increased body weight by approximately 3 kg. The increase in body weight appears to mainly consist of an increase in fat-free mass, since fat-free mass estimated by dual energy x-ray absorptiometry (DEXA) was also significantly increased by approximately 3 kg (Figure 1a). Similarly, body cell mass was significantly increased as estimated by a four-compartment model based on total body potassium and total body water assessments (Figure 1c). To what extent the increase in the fat-free body mass consisted of body water is unknown. However, total body water estimated by the use of tritiated water was not significantly increased, why it is unlikely that the whole increase in fat-free mass consisted of body water.

Total body fat, both as estimated by DEXA (Figure 1b) and by the four-compartment model (Figure 1d), was unchanged. Also, FFA levels were unchanged, not indicating any increased lipolysis. Compared with the previous experience from GH treatment of GH deficient adults and obese subjects, this result was surprising. However, GH levels remain increased for 12 hours after a subcutaneous injection of GH (71), while MK-677 increases the pre-existing pulsatile pattern of GH release (13). This difference in pattern of serum GH may possibly contribute to the different results in total body fat since in GH deficient adults, a continuous infusion of GH reduces body fat even more effectively than subcutaneous injections of GH (72).

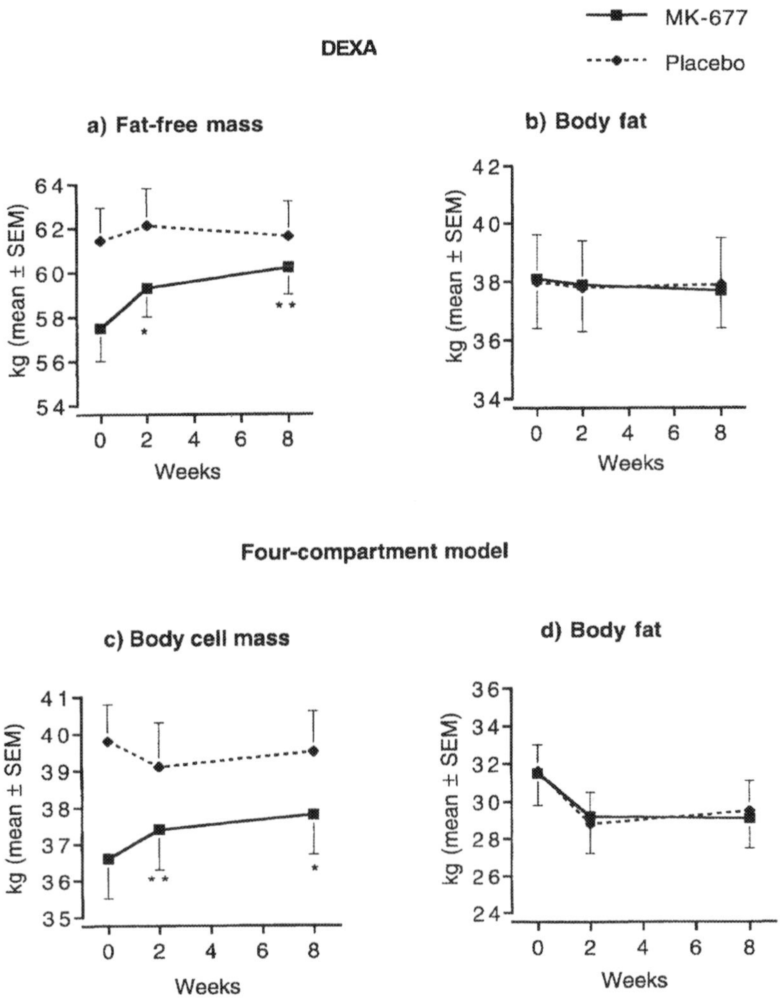

Figure 1. Body composition during two months treatment with MK-677 25 mg or placebo daily in obese subjects. (a) and (b) show fat-free mass and body fat respectively, as assessed by DEXA. (c) and (d) show body cell mass and body fat respectively, as calculated using the four-compartment model. The vertical bars indicate the SE for the mean values shown. P-values are based on a between-group analysis of percent change from baseline. Reproduced, with permission, from Svensson, J., Lönn, L., Jansson, J.-O. et al. (1998) Two-month treatment of obese subjects with the oral growth hormone (GH) secretagogue MK-677 increases GH secretion, fat-free mass, and energy expenditure. J. Clin. Endocrinol. Metab. 83, 362–369, © The Endocrine Society. $^{*}p < 0.05$, $^{**}p < 0.01$.

The total visceral fat volume, as estimated by 5-scan abdominal CT determinations, was also unaffected by MK-677 treatment. However, correlation analysis revealed an inverse baseline correlation between serum IGF-I and visceral fat volume (Figure 2a), confirming the previously discussed negative association between serum IGF-I and visceral fat mass. Furthermore, an inverse correlation was found between the change in serum IGF-I and the change in visceral fat volume at 8 weeks of MK-677 treatment (Figure 2b). This finding indicates that a higher dose of, or a prolonged treatment period of MK-677 may decrease visceral fat mass.

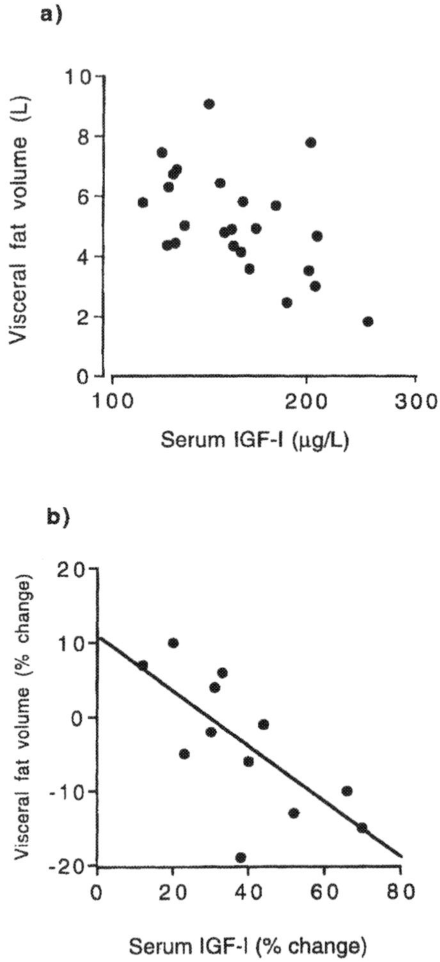

Figure 2. Correlations, (a) in the total study population at baseline, between serum IGF-I and visceral fat volume ($r = -0.5; p < 0.01$) and (b) in the MK-677 treatment group, between the percent change in serum IGF-I and the percent change in visceral fat volume at the end of the study period ($r = -0.7; p < 0.01$). Note the logarithmic scale for the x-axis in Figure 2a. Reproduced, with permission, from Svensson, J., Lönn, L., Jansson, J.-O. et al. (1998) Two-month treatment of obese subjects with the oral growth hormone (GH) secretagogue MK-677 increases GH secretion, fat-free mass, and energy expenditure. J. Clin. Endocrinol. Metab. 83, 362–369, © The Endocrine Society.

BMR was significantly increased by MK-677 treatment at 2 weeks, but the increase was not significant at 8 weeks. In line with this finding, an initial increase in BMR was downregulated during a 9-month GH treatment study of moderately obese subjects (73).

The increase in body weight, and the absence of a fat reduction during MK-677 treatment, may possibly be caused by an increased food intake. In the rat, GHRH has a stimulatory effect on appetite (74), and intracerebroventricular injection of the GHRP KP-102 has been reported to increase food intake (75). Furthermore, systemic injection of GHRP-6 activates cells containing the appetite-stimulating (76) neuropeptide Y (NPY) in the arcuate nucleus of the rat hypothalamus (77). GHRP-6 may also affect human NPY-producing cells (78). It is difficult to understand how MK-677 treatment could increase body weight, concomitant with an increase in BMR, without an increase in food intake. However, the dietary questionnaires used were not able to detect any increase in food intake.

MK-677 treatment did not affect fasting levels of glucose or insulin while the 2-hour values after administration of 75 g glucose indicated an impairment of glucose homeostasis at 2 weeks, and with some attenuation, at 8 weeks. This finding is in line with a previous study of GH deficient adults, where an initial impairment of insulin resistance had disappeared after 6 months of GH treatment (79).

GENERAL CONCLUSION

In abdominal/visceral obese males, GH treatment has previously reduced total and visceral fat and improved cardiovascular risk factors, although body weight has not been greatly affected. In a 2-month treatment study of obese males with the oral GH secretagogue MK-677, we show that MK-677 substantially increases GH and IGF-I levels as well as fat-free body mass. We did not find a decrease in body fat although a strong inverse relation was found between the changes in serum IGF-I and visceral fat. There is need for further studies in human obesity addressing whether a higher dose of or a longer treatment period with MK-677 or other GHRP related substances can reduce body fat.

ACKNOWLEDGEMENTS

This work was supported by grants from the Swedish Medical Research Council (No. 11621 and 9894) and from Merck Research Laboratories, Rahway, NJ, USA. We are indebted to Lena Wirén, Anne Rosén, Ingrid Hansson and Annika Reibring at the Research Centre for Endocrinology and Metabolism for their skilful technical support.

REFERENCES

1. Bengtsson, B.-Å., Edén, S., Lönn, L. et al. (1993) Treatment of adults with growth hormone (GH) deficiency with recombinant human GH. J. Clin. Endocrinol. Metab. 76, 309–317.
2. Stenlöf, K., Johansson, J.-O., Lönn, L., Sjöström, L. and Bengtsson, B.-Å. (1997) Diurnal variations in twenty-four-hour energy expenditure during growth hormone treatment of adults with pituitary deficiency. J. Clin. Endocrinol. Metab. 82, 1255–1260.
3. Rudling, M., Norstedt, G., Olivecrona, H., Reihnér, E., Gustafsson, J.-Å. and Angelin, B. (1992) Importance of growth hormone for the induction of hepatic low density lipoprotein receptors. Proc. Natl. Acad. Sci. USA. 89, 6983–6987.
4. De Boer, H., Blok, G.-J. and Van Der Veen, E. (1995) Clinical Aspects of Growth Hormone Deficiency in Adults. Endocr. Rev. 16, 63–86.
5. Rosengren, A., Wilhelmsen, L., Eriksson, E., Risberg, B. and Wedel, H. (1990) Lipoprotein(a) and coronary heart disease: a prospective case-control study in a general population sample of middle aged men. Br. Med. J. 301, 1248–1251.

6. Uterman, G. (1989) The mysteries of lipoprotein (a). Science 246, 904–910.
7. Scott, J. (1991) Lipoprotein (a). Br. Med. J. 303, 663–664.
8. Scanu, A. (1992) Lipoprotein (a). A genetic risk factor for premature coronary heart disease. JAMA 267, 3326–3329.
9. Edén, S., Wiklund, O., Oscarsson, J., Rosén, T. and Bengtsson, B.-Å. (1993) Growth hormone treatment of growth hormone-deficient adults results in a marked increase in Lp(a) and HDL cholesterol concentrations. Arterioscler. Thromb. 13, 296–301.
10. Johannsson, G., Oscarsson, J., Rosén, T. et al. (1995) Effects of 1 year of growth hormone therapy on serum lipoprotein levels in growth hormone-deficient adults; influence of gender and apo(a) and apoE phenotypes. Arterioscler. Thromb. Vasc. Biol. 15, 2142–2150.
11. Korbonits, M. and Grossman, A. (1995) Growth hormone-releasing peptide and its analogues. Novel stimuli to growth hormone release. Trends Endocrinol. Metab. 6, 43–49.
12. Chapman, I., Bach, M., Van Cauter, E. et al. (1996) Stimulation of the growth hormone (GH)-Insulin-like growth factor-I axis by daily oral administration of a GH secretagogue (MK-677) in healthy elderly subjects. J. Clin. Endocrinol. Metab. 81, 4249–4257.
13. Smith, R., Van Der Ploeg, L., Howard, A. et al. (1998) Peptidomimetic regulation of growth hormone secretion. Endocr. Rev. 18, 621–645.
14. Veldhuis, J., Liem, A., South, S. et al. (1995) Differential impact of age, sex steroid hormones, and obesity on basal versus pulsatile growth hormone secretion in men as assessed in an ultrasensitive chemiluminescence assay. J. Clin. Endocrinol. Metab. 80, 3209–3222.
15. Veldhuis, J., Iranmesh, A.H.K., Waters, M., Johnson, M. and Lizzerade, G. (1991) Dual defects in pulsatile growth hormone secretion and clearance subserve the hyposomatotropism in man. J. Clin. Endocrinol. Metab. 72, 51–59.
16. Vahl, N., Jørgensen, J., Jurik, A. and Christiansen, J. (1996) Abdominal obesity and physical fitness are major determinants of age associated decline in stimulated GH secretion in healthy adults. J. Clin. Endocrinol. Metab. 81, 2209–2215.
17. Clasey, J., Weltman, A., Weltman, J. et al. Abdominal visceral fat is related to 24-h growth hormone release in both young and older men and women. Paper read at the 79th Annual Meeting of the Endocrine Society, June 11–14 1997, Minneapolis, Minnesota, USA.
18. Mårin, P., Kvist, H., Lindstedt, G., Sjöström, L. and Björntorp, P. (1993) Low concentrations of insulin-like growth factor-I in abdominal obesity. Int. J. Obesity 17, 83–89.
19. Rasmussen, M., Hviberg, A., Juul, A. et al. (1995) Massive weight loss restores 24-hour growth hormone release profiles and serum insulin-like growth factor-I levels in obese subjects. J. Clin. Endocrinol. Metab. 80, 1407–1415.
20. Jung, R.T., Campbell, R.G., James, W.P.T. and Callingham, B.A. (1982) Altered hypothalamic and sympathetic response to hypoglycaemia in familial obesity. Lancet 1, 1043–1046.
21. Kopelman, P.G., Pilkington, T.R.E., White, N. and Jeffcoate, S.L. (1980) Evidence for existence of two types of massive obesity. Br. Med. J. 281, 82–83.
22. Mårin, P., Darin, N., Amemeia, T., Andersson, B., Jern, S. and Björntorp P. (1992) Cortisol secretion in relation to body fat distribution in obese premenopausal women. Metabolism. 41, 882–886.
23. Pasquali, R., Cantobelli, S., Casimirri, F. et al. (1993) The hypothalamic–pituitary–adrenal axis in obese women with different patterns of body fat distribution. J. Clin. Endocrinol. Metab. 77, 341–346.
24. Björntorp, P. (1993) Visceral obesity: A "Civilization syndrome". Obes. Res. 1, 206–222.
25. Mårin, P., Holmäng, S. and Gustafsson, C. et al. (1993) Androgen treatment of abdominally obese men. Obes. Res. 1, 245–251.
26. Lapidus, L., Bengtsson, C., Larsson, B., Pennert, K. and Sjöström, L. (1984) Distribution of adipose tissue and risk of cardiovascular disease and death: a 12 year follow-up of participants in the population study of women in Gothenburg, Sweden. Br. Med. J. 289, 1257–1261.
27. Chrousos, G. and Gold, P. (1992) The concept of stress and stress system disorders. JAMA 267, 1244–1252.

28. Mårin, P., Andersson, B. and Ottosson, M. et al. (1992) The morphology and metabolism of intraabdominal adipose tissue in men. Metabolism 41, 1242–1248.
29. Rebuffé-Scrive, M., Andersson, B., Olbe, L. and Björntorp, P. (1989) Metabolism of adipose tissue in intraabdominal depots of nonobese men and women. Metabolism 41, 453–458.
30. Rebuffé-Scrive, M., Andersson, B., Olbe, L. and Björntorp, P. (1990) Metabolism of adipose tissue in intraabdominal depots in severely obese men and women. Metabolism 39, 1021–1025.
31. Williamsson, J., Kreisberg, R. and Felts, P. (1966) Mechanism for the stimulation of gluconeogenesis by fatty acids in perfused rat liver. Proc. Natl. Acad. Sci. USA. 56, 247–254.
32. Svedberg, J., Strömblad, G., Wirth, A., Smith, U. and Björntorp, P. (1991) Fatty acids in the portal vein of the rat regulate hepatic insulin clearance. J. Clin. Invest. 88, 2054–2058.
33. Nurjhan, N., Consoli, A. and Gerich, J. (1992) Increased lipolysis and its consequences on gluconeogenesis in non-insulin-dependent diabetes mellitus. J. Clin. Invest. 89, 169–175.
34. Fukuda, N. and Ontko, J. (1984) Interactions between fatty acid synthesis, oxidation and esterification in the production of triglyceride-rich lipoproteins by the liver. J. Lipid Res. 25, 277–327.
35. Reaven, G. (1988) Role of insulin resistance in human disease. Diabetes 37, 1595–1607.
36. Herberg, L., Bergmann, M., Hennings, U., Major, E. and Gries, F.A. (1972) Influence of diet on the metabolic syndrome of obesity. Isr. J. Med. Sci. 8, 822–823.
37. Knospe, S. and Kohler, E. (1981) Impaired hormonal regulation of adenosine 3′,5′-monophosphate release in adipose from hyperglycemic sand rats *in vitro*. Horm. Metab. Res. 13, 434–437.
38. Nakamura, R., Emmanoel, D. and Katz, A. (1983) Insulin binding sites in various segments of the rabbit nephron. J. Clin. Invest. 72, 388–392.
39. Reaven, G. (1995) Pathophysiology of insulin resistance in human disease. Physiol. Rev. 75, 473–486.
40. Rosén, T. and Bengtsson, B.-Å. (1990) Premature mortality due to cardiovascular diseases in hypopituitarism. Lancet 336, 285–288.
41. Rosén, T., Edén, S., Larsson, G., Wilhelmsen, L. and Bengtsson, B.-Å. (1993) Cardiovascular risk factors in adult patients with growth hormone deficiency. Acta Endocrinol. 129, 195–200.
42. Bengtsson, B.-Å., Brummer, R.-J., Edén, S. and Bosaeus, I. (1989) Body composition in acromegaly. Clin. Endocrinol. 30, 121–130.
43. Bengtsson, B.-Å., Brummer, R.-J., Edén, S., Bosaeus, I. and Lindstedt, G. (1989) Body composition in acromegaly: The effect of treatment. Clin. Endocrinol. 31, 481–490.
44. Clemmons, D., Snyder, D., Williams, R. and Underwood, L. (1987) Growth hormone administration conserves lean body mass during dietary restriction in obese subjects. J. Clin. Endocrinol. Metab. 64, 878–883.
45. Snyder, D., Clemmons, D. and Underwood, L. (1988) Treatment of obese, diet-restricted subjects with growth hormone for 11 weeks: effects on anabolism, lipolysis and body composition. J. Clin. Endocrinol. Metab. 67, 54–61.
46. Drent, M., Wever, L., Adèr, H. and van der Veen, E. (1995) Growth hormone administration in addition to a very low calorie diet and an exercise program in obese subjects. Eur. J. Endocrinol. 132, 565–572.
47. Johannsson, G., Mårin, P., Lönn, L. et al. (1997) GH treatment of abdominally obese men reduces abdominal fat mass, improves glucose and lipoprotein metabolism, and reduces diastolic blood pressure. J. Clin. Endocrinol. Metab. 82, 727–734.
48. Bell, J., Donald, R. and Espiner, E. (1970) Pituitary response to insulin-hypoglycemia in obese subjects before and after fasting. J. Clin. Endocrinol. Metab. 31, 546–551.
49. Sims, E.A.H., Danforth, E., Horton, E.S. et al. (1973) Endocrine and metabolic effects of experimental obesity in man. Recent Prog. Horm. Res. 29, 457–496.
50. Copinschi, G., Wegienka, L., Hane, S. and Forsham, P.H. (1967) Effect of arginine on serum levels of insulin and growth hormone in obese subjects. Metabolism 16, 485–491.
51. Glass, A., Burman, K., Dahms, W. and Bohem T. (1981) Endocrine function in human obesity. Metabolism 30, 89–104.

52. Finer, H., Price, P., Grossman, A. and Besser, G. (1987) The effects of enkephalin analogue on pituitary hormone release in human obesity. Horm. Metab. Res. 19, 68–70.
53. Cordido, F., Dieguez, C. and Casanueva, F. (1990) Effect of central cholinergic neurotransmission enhanced by pyridostgmine on the growth hormone secretion elicited by clonidine, arginine, or hypoglycemia in normal and obese subjects. J. Clin. Endocrinol. Metab. 70, 1361–1370.
54. Williams, T., Berelowitz, M., Joffe, S. et al. (1984) Impaired growth hormone responses to growth hormone-releasing factor in obesity. A pituitary defect reversed with weight reduction. N. Engl. J. Med. 311, 1403–1407.
55. Grottoli, S., Maccario, M., Procopio, M. et al. (1996) Somatotrope responsiveness to hexarelin, a synthetic hexapeptide, is refractory to the inhibitory effect of glucose in obesity. Eur. J. Endocrinol. 135, 678–682.
56. Kirk, S., Gertz, B., Schneider, S. et al. (1997) Effect of obesity and feeding on the growth hormone (GH) response to the GH secretagogue L-692,429 in young men. J. Clin. Endocrinol. Metab. 82, 1154–1159.
57. Cordido, F., Penalva, A., Dieguez, C. and Casanueva, F. (1993) Massive growth hormone (GH) discharge in obese subjects after the combined administration of GH-releasing hormone and GHRP-6: Evidence for a marked somatotroph secretory capability in obesity. J. Clin. Endocrinol. Metab. 76, 819–823.
58. Cordido, F., Peino, R., Penalva, A., Alvarez, C., Casanueva, F. and Dieguez, C. (1996) Impaired growth hormone secretion in obese subjects is partially reversed by acipimox-mediated plasma free fatty acid depression. J. Clin. Endocrinol. Metab. 81, 914–918.
59. Laron, Z., Frenkel, J., Deghenghi, R., Anin, S., Klinger, B. and Sibergeld, A. (1995) Intranasal administration of the GHRP hexarelin accelerates growth in short children. Clin. Endocrinol. 43, 631–635.
60. Rahim, A., O'Neill, P. and Shalet, S. (1998) Growth hormone status during long-term hexarelin therapy. J. Clin. Endocrinol. Metab. 83, 1644–1649.
61. Murphy, M.G., Plunkett, L.M., Gertz, B.J. et al. (1988) MK-677, an orally active growth hormone secretagogue, reverses diet-induced catabolism. J. Clin. Endocrinol. Metab. 83, 320–325.
62. Svensson, J., Lönn, L., Jansson, J.-O. et al. (1998) Two-month treatment of obese subjects with the oral growth hormone (GH) secretagogue MK-677 increases GH secretion, fat-free mass, and energy expenditure. J. Clin. Endocrinol. Metab. 83, 362–369.
63. Chapman, I., Pescovitz, O., Murphy, G. et al. (1997) Oral administration of growth hormone (GH) releasing peptide-mimetic MK-677 stimulates the GH/insulin-like growth factor-I axis in selected GH-deficient adults. J. Clin. Endocrinol. Metab. 82, 3455–3463.
64. Hartman, M., Clayton, P., Johnston, M. et al. (1993) A low dose euglycemic infusion of recombinant insulin-like growth factor I rapidly suppresses fasting-enhanced pulsatile growth hormone secretion in humans. J. Clin. Invest. 91, 2453–2462.
65. Huhn, W., Hartmann, M., Pezzoli, S. and Thorner, M. (1993) Twenty-four hour growth hormone (GH)-releasing peptide (GHRP) infusion enhances pulsatile GH secretion, and specifically attenuates the response to a subsequent GHRP bolus. J. Clin. Endocrinol. Metab. 76, 1202–1208.
66. Jaffe, C., Ho, P., Demott-Friberg, R., Bowers, C. and Barkan, A. (1993) Effects of a prolonged growth hormone (GH)-releasing peptide infusion on pulsatile GH secretion in normal men. J. Clin. Endocrinol. Metab. 77, 1641–1647.
67. Ghigo, E., Arvat, E., Gianotti, L. et al. (1994) Growth hormone-releasing activity of hexarelin, a new synthetic hexapeptide, after intravenous, subcutaneous, intranasal, and oral administration in man. J. Clin. Endocrinol. Metab. 78, 693–698.
68. Gertz, B., Barrett, J., Eisenhandler, R. et al. (1993) Growth hormone response in man to L-692,429, a novel nonpeptide mimic of growth hormone-releasing peptide-6. J. Clin. Endocrinol. Metab. 77, 1393–1397.

69. Bowers, C., Reynolds, G., Durham, D., Barrera, C., Pezzoli, S. and Thorner, M. (1990) Growth hormone (GH) releasing peptide stimulates GH release in normal men and acts synergistically with GH releasing hormone. J. Clin. Endocrinol. Metab. 70, 975–982.
70. Copinschi, G., Van Onderbergen, A., L'Hermite-Balériaux, M. et al. (1996) Effects of a 7-day treatment with a novel, orally active growth hormone (GH) secretagogue, MK-677, on 24-hour GH profiles, insulin-like growth factor-I, and adrenocortical function in normal young men. J. Clin. Endocrinol. Metab. 81, 2776–2782.
71. Oscarsson, J., Johannsson, G., Johansson, J.-O., Lundberg, P.-A., Lindstedt, G. and Bengtsson, B.-Å. (1997) Diurnal variation in serum insulin-like growth factor (IGF)-I and IGF binding protein-3 concentrations during daily subcutaneous injections of recombinant human growth hormone in GH-deficient adults. Clin. Endocrinol. 46, 63–68.
72. Johansson, J.-O., Oscarsson, J., Bjarnason, R. and Bengtsson, B.-Å. (1996) Two weeks of daily injections and continuous infusion of recombinant human growth hormone (GH) in GH-deficient adults: I. Effects on insulin-like growth factor-I (IGF-I), GH and IGF-binding proteins, and glucose homeostasis. Metabolism 45, 362–369.
73. Karlsson, C., Stenlöf, K., Johannsson, G. et al. (1998) Effects of growth hormone treatment on the leptin system and on energy expenditure in abdominally obese men. Eur. J. Endocrinol. 138, 408–414.
74. Vaccarino, F., Bloom, F., Rivier, J., Vale, W. and Koob, G. (1985) Stimulation of food intake in rats by centrally administered hypothalamic growth hormone-releasing factor. Nature 314, 167–168.
75. Okada, K., Ishii, S., Minami, S., Sugihara, H., Shibasaki, T. and Wakabayashi, I. (1996) Intraventricular administration of the growth hormone releasing peptide KP-102 increases food intake in free-feeding rats. Endocrinology 137, 5155–5158.
76. Clark, J., Kalra, P., Crowley, W. and Kalra, S. (1984) Neuropeptide Y and human pancreatic polypeptide stimulate feeding in rats. Endocrinology 115, 427–429.
77. Dickson, S., Luckman, S. (1997) Induction of c-fos mRNA in NPY and GRF neurones in the rat arcleus nucleus following systemic injection of the growth hormone secretagogue, GHRP-6. Endocrinology 138, 771–777.
78. Jansson, J.-O., Svensson, J., Bengtsson, B.-Å. et al. (1998) Acromegaly and Cushing's syndrome due to ectopic production of GHRH and ACTH by a thymic carcinoid tumour: *in vitro* responses to GHRH and GHRP-6 [Case report]. Clin. Endocrinol. 48, 243–250.
79. Fowelin, J., Attvall, S., Lager, I. and Bengtsson, B.-Å (1993) Effects of treatment with recombinant human growth hormone on insulin sensitivity and glucose metabolism in adults with growth hormone deficiency. Metabolism 42, 1443–1447.

Growth Hormone Secretagogues
Edited by E. Ghigo, M. Boghen, F.F. Casanueva and C. Dieguez

Chapter 16

Effects of Growth Hormone Secretagogues on in vivo Substrate Metabolism in Humans

NIELS MØLLER, JENS OTTO JORGENSEN and JENS SANDAHL CHRISTIANSEN

Institute of Experimental Clinical Research, University of Aarhus and Medical Dep M (Endocrinology & Diabetes), Aarhus Kommunehospital, Aarhus, Denmark

INTRODUCTION

Being controlled primarily by the two hypothalamic peptides somatostatin (inhibitory) and GH-releasing hormone (GHRH —stimulatory), GH secretion may be amplified by the use of either somatostatin antagonists or GHRH agonists (1). Though the natural ligand(s) remain(s) elusive, a new class of GH releasing peptides, secretagogues or peptidomimetica has recently been identified (2). These compounds, which bear resemblance to benzodiazepines, at the same time potentiate the effects of GHRH and act as functional somatostatin antagonists by interference with specific receptors. Given the facts that amplification of the natural pulsatile pattern of GH release is induced and that biological activity is maintained after oral administration, it is possible that GH secretagogues in the future will be used clinically in catabolic states where the metabolic effects of GH are desirable.

The overall biological response to administration of GH secretagogues will be determined by two factors, namely: (i) any intrinsic, GH-independent effects of the agent administered and (ii) the quality and quantity of GH secretion imposed.

Little information is available as to whether GH secretagogues have any direct effects on intermediary metabolism in humans. Studies in animals have suggested that GH secretagogues may have widespread effects in the brain and interfere with secretion of neuropeptide Y and dopamine, effects which may lead to increased secretion of ACTH, cortisol and prolactin and increased appetite (2). On the other hand studies in which peptidomimetica have been given to humans have only reported minute elevations of circulating levels of cortisol and prolactin (3–6). These alterations in all likelihood will have little impact on intermediary metabolism. Thus the principal metabolic effect of GH secretagogues will be that of GH. This view is supported by a substantial number of studies reporting pure GH effects after administration in terms of increased fasting blood glucose concentrations, hyperinsulinaemia and protein retention (4–6).

The present review will therefore focus on the metabolic effects of GH. It should evidently be considered that the overall biological effect imposed will be dependent on the subjects recruited for treatment and on the doses employed. In some cases physiological conditions may be re-established and in others acromegaloid states inflicted.

BACKGROUND – GH AND METABOLISM

Early studies in the 1920s and 1930s by Bernardo Houssay established that extracts from the pituitary gland had profound effects on glucose metabolism. These studies showed that removal of the pituitary gland increased the sensitivity to insulin in normal animals and diminished the severity of diabetes in depancreatized animals and that administration of pituitary extracts led to insulin resistance and in some cases frank diabetes (7,8). Concomitantly it was observed that anterior lobe extracts are ketogenic and growth promoting (8,9) and recognized that these actions were caused by distinct hormones. The notion that the diabetogenic, ketogenic and growth promoting effects of secretion from the pituitary were caused by a single hormone was first advanced by Shipley and Long (9). After the (partial) purification of human GH a number of important studies showed that exposure to large amounts of pituitary extracts of GH in normal, GH-deficient and diabetic human volunteers stimulated lipolysis and led to hyperglycemia (10–13) and it was also reported that GH, when perfused locally through the brachial artery, consistently caused acute inhibition of muscle glucose uptake in the forearm of normal subjects (14–16). The next major break-through was the identification of insulin-like growth factors (IGFs) and the subsequent moulding of the concept that GH regulated IGF-I synthesis accounts for a large proportion of the anabolic impact of GH (17).

Over the years, understanding of the metabolic role of the GH/IGF-I axis has been limited by a number of factors. Supplies of both hormones have been scarce and at times impure and the mode of administration has often produced unphysiologic conditions. Additional confusion has arisen from studies — in particular *in vitro* studies — reporting both "insulin-like" and "insulin antagonistic" effects of GH on glucose and lipid metabolism (18). In general the "insulin-like" effects (i.e. inhibition of lipolysis, stimulation of glucose uptake and augmentation of lipogenesis) are observed transiently and early, are easily exhaustible and are most readily seen in GH deprived tissues, whereas they — if at all present —seem extremely volatile in humans exposed to GH levels within the physiological range. Induction of "insulin-like" activity requires GH concentrations 30-fold above "insulin-antagonistic" activity (18). It is possible that "insulin-like" actions are generated by small molecular fragments of GH (19) or by local IGF release, but any biological significance of this disconcerting phenomenon still remains uncertain. Furthermore the precise biological consequences of the presence of binding proteins and of molecular heterogeneity and isoforms of GH in the circulation are presently unclear.

NORMAL PHYSIOLOGY – GH AND METABOLISM

It has been estimated that in normal young humans GH pulses are discharged roughly every second or third hour and that an average of 45 mg GH is released with each secretory episode thus adding up to total 24-hour GH secretion close to 0.5 mg (20). This secretory pattern is amplified during fasting and stress conditions, whereas meals in general inhibit GH release (21,22), implying that the main impact of GH either lies in the postabsorptive, fasting states and during stress or in the transition phase from these states to the fed periprandial state.

Some earlier studies have administered very high doses of GH. Pulsatile and continuous administration of more moderate amounts of GH between 70 and 400 mg to healthy postabsorptive humans reveals a clear dose-dependent stimulation of lipolysis, circulating levels of free fatty acids (FFA) and glycerol and increased lipid oxidation rates, as assessed by indirect calorimetry (23–25). The most spectacular impact of a physiologic GH pulse is a peak increase in FFA concentrations in the magnitude of 100% after 2–3 hours, suggesting that a prime target for GH is stimulation of lipolysis in adipose tissue. There is some evidence that the lipolytic sensitivity to GH is increased during fasting (26). Interestingly an investigation of young healthy subjects showed that the nocturnal mean peak of GH preceded that of free fatty acids by 2 h (27), a time lag very close to the one found after GH bolus administration, thus supporting that GH acts as an important regulator of diurnal fluctuations in release and oxidation of lipids. This concept is further supported by studies showing that lack of nocturnal GH compromises the expected overnight surge of lipid fuels (28,29) and studies suggesting a temporal and dimensional correlation between nocturnal GH and concentrations of lipid intermediates (30,31).

The immediate effects of GH on postabsorptive glucose metabolism are more subtle. Though muscle utilization of glucose is already low a further suppression of glucose uptake is typically seen after acute GH exposure (14–16, 23–25). The increase in lipid oxidation is offset by a decrease in glucose oxidation, total glucose turnover remains unaffected and — in consequence — non-oxidative glucose turnover increases. To what extent these phenomena are secondary to increased lipid availability and subsequent "Randle" substrate competition (32) is not known, though it has been shown that co-infusion of GH with nicotinic acid (an antilipolytic agent) abolishes the effects of GH on glucose tolerance (33). The coexistence of decreased glucose oxidation and suppressed muscle glucose uptake in the presence of unchanged glucose turnover implies that GH promotes non-oxidative glucose utilization in some non-muscle compartment of the body. Neither the tissue, nor the biochemical pathways responsible for this flux of glucose are known. Stimulated lipogenesis in adipose tissue or liver seems implausible, since ongoing lipogenesis as opposed to the observed decrease in respiratory exchange ratio would increase this parameter. More likely GH may increase gluconeogenesis and glucose cycling in e.g. splanchnic tissues/liver, adipose tissue or skin. Large doses of GH have been reported to decrease postabsorptive, net splanchnic glucose output acutely, compatible with increased glucose uptake (34) and *in vitro* experiments have shown increased gluconeogenesis from either alanine or lactate in canine kidney cortex incubated with GH (35). In addition studies in acromegalic patients have revealed a 50% increase in glucose/glucose-6-phosphate cycling (36), which could explain the major part of the increased glucose turnover recorded in these patients. Besides, it has been described that overnight exposure to high levels of GH in normal humans stimulated gluconeogenesis, as judged by the incorporation of labelled carbon dioxide into glucose (37). Finally dogs treated with high GH doses (1 mg/kg/day) for 4–6 days showed more than a doubling of liver glycogen content — from 5 to 11g/100g of liver (38).

Albeit circumstantial, current evidence therefore suggests that the explicit stimulation of lipolysis by GH is accompanied by a proportional decrease in glucose oxidation and an increase in non-oxidative glucose disposal, conceivably in the form of gluconeogenesis and glucose storage.

The direct impact of GH on protein metabolism in humans is not well described. The protein sparing effects of prolonged GH exposure are unquestionable, but a majority of investigations in this area have employed high dose GH administration for several days, thus inducing "short-term acromegaly" (10, 39–43). This invariably leads to stimulation of lipolysis, hyperinsulinaemia and stimulation of insulin-like growth factor-I (IGF-I) activity. Because all of these compounds have potent protein anabolic properties, distinction between direct and indirect effects becomes perplexing. The studies above do however clearly show that GH causes nitrogen retention as evidenced by decreased urinary excretion rates for urea, creatinine and ammonium. There is additional evidence that massive GH exposure may preferentially stimulate protein synthesis; in contrast insulin is believed to restrict breakdown, whereas IGF-I may be capable of affecting both processes (44,45). This theory has received some support from acute perfusion studies (46,47), but could not be confirmed in a controlled study (48).

The effects of GH on hepatic nitrogen metabolism are also poorly elucidated. Experiments in hypophysectomized rats have indicated that GH may act on the liver to decrease urea synthesis and in parallel increase glutamate release, thereby diminishing hepato–renal clearance of the circulating nitrogen pool (49). These findings have now been reproduced in humans after prolonged GH exposure (50). In this connection it should again be emphasized that many of the effects of GH could be secondary to activation of lipolysis; the protein sparing actions of lipid intermediates are well documented (51–53).

GROWTH HORMONE DEFICIENCY

Assessment of fuel metabolism in patients with growth hormone deficiency (GHD) has been entangled by factors such as the absence of clear-cut diagnostic criteria and the inclusion of heterogenous populations, varying as regards age, development, additional pituitary insufficiency/replacement therapy and body composition. In particular the influence of pituitary replacement therapy and the coexisting obesity often present in GHD may have distorted the picture.

It is well described that subjects with GHD are prone to fasting hypoglycaemia and nitrogen wasting (54–61). In contrast postabsorptive blood glucose concentrations and glucose turnover are normal (61–63), but merely a short period of fasting may induce significant hypoglycaemia due to a mismatch between glucose production and utilization (57). Some of the above studies have reported decreased levels of lipid fuel intermediates, whereas others observe an increased circulating pool of these substrates. The reason for this inconsistency may be, that at times fasting and subsequent discrete hypoglycemia may prompt secretion of counterregulatory stress hormones, such as epinephrine, glucagon and cortisol, and lead to an overall catabolic substrate response with lipid mobilization. Furthermore some of the GHD patients studied have been overweight and may had access to an increased mobilizable lipid mass (64,65).

Many, but not all (66) studies report that subjects with GH deficiency are hypersensitive to the actions of insulin. The hypersensitivity resides in muscle, liver and adipose tissue (61) and is also apparent after an oral or intravenous glucose challenge (62,63). Interestingly there is evidence that GH treatment may cause a biphasic response, i.e. impaired insulin sensitivity after some weeks, followed by restoration of the initial sensitivity to insulin (67). On the whole it remains dubious whether obese patients with GHD are insulin resistant in excess of their obesity. In this context it should be underlined that replacement therapy with GH does not in any way inappropriately increase the risk of impaired glucose tolerance or frank diabetes mellitus in GHD patients, but simply restores native physiologic conditions.

A specific clinical problem pertaining to treatment of growth hormone deficient subjects is the observation that "insulin-like" effects of GH may prevail. Press and coworkers have reported that GH administration on alternate days increases the risk for hypoglycemia (68). They showed that when large amounts of GH (50–60 mg thrice weekly) were given to three children below 5 yr of age, fasting hypoglycemia (plasma glucose concentrations below 2.5 m mol/l) could on some occasions be recorded 30 to 60 h after the GH injections. It therefore appears advisable to initiate GH treatment or treatment with secretagogues in these patients with low doses administered frequently — this approach should also counteract the risk of side effects related to fluid retention.

INSULIN SENSITIVITY AND DIABETES — GH

Since the demonstration of elevated circulating concentrations of GH in type 1 diabetic patients (69) the role of GH in metabolic regulation in diabetes has attracted much interest. Insulin dependent diabetic subjects are highly susceptible to the insulin antagonistic effects of GH, since they are deprived of residual beta cell function and the capability of generating compensatory hyperinsulinaemia. There is little doubt that type 1 diabetic patients in general are exposed to excessive amounts of circulating GH (70,71) and a recent survey has estimated that GH concentrations during poor control are increased 2–3 fold (72), which extrapolates to a diurnal secretion rate between 0.5 and 1 mg in more strictly controlled patients (20,72). A number of studies have administered amounts considerably above.

Administration of a bolus of 210 mg GH, intended to mimic a pulsatile episode, to well controlled diabetic subjects led to a marked transient elevation of circulating lipid intermediates together with more subtle changes in glucose metabolism, in a manner very similar to observations in normal man (73). This suggests that in well insulinized diabetic subjects modest GH bursts may serve as beneficial metabolic regulators, preserving carbohydrate and protein at the cost of promoted lipid consumption.

Hypoglycemia is presently an inevitable consequence of insulin therapy in type 1 diabetes and and there is now mounting evidence that intact GH secretion is important in combatting prolonged hypoglycemia, in particular in patients with impaired secretion of glucagon and catecholamines (74,75). It is however also possible that GH may accelerate late post-hypoglycemic hyperglycemia (76).

Over the past decades much attention has been paid to the deleterious metabolic actions of GH and it has consistently been shown that sustained exposure to high levels of GH results in poor glycemic control and increased insulin requirements. The studies by Press et al. have clearly defined the capacity of GH to deteriorate metabolic control in type 1 diabetes (77,78). These experiments showed that administration of hourly 100 mg GH pulses after a latency of several hours induced dramatic 100% increases in circulating glucose values together with marked increments in circulating lipid fuels.

The effects of GH on insulin sensitivity have been thoroughly assessed and it has repeatedly been demonstrated that continuous infusion of large amounts (1.5 mg) of GH impaired both hepatic and peripheral insulin sensitivity of normal man after 12 h (79,80). A later study employing more moderate amounts of both GH and insulin showed that GH impaired hepatic and peripheral insulin sensitivity after approximately 2 h, that the impairment of peripheral insulin sensitivity largely resided in muscle and that GH despite light hyperinsulinaemia promoted lipolysis (81). There is also evidence that GH acts to diminish both insulin and glucose dependent glucose disposal (82). Presently it is unclear, whether modification of the glucose transporters and key glucoregulatory enzymes are involved in GH induced impairment of insulin action; it has been shown that short-term GH exposure blunts the activity of glycogen synthase in striated muscle (83).

Information on the effects of GH on insulin sensitivity in type 1 diabetic subjects, is surprisingly sparse. There is evidence that GH worsens peroral glucose tolerance (78,84) and Periello et al. recently showed impairment of both hepatic and peripheral insulin sensitivity after nocturnal exposure to more than 800 mg GH (85); it therefore seems fair to extrapolate from data obtained in normal subjects. In the course of diabetic ketoacidosis circulating concentrations of GH are in general high (86,87). This may evidently contribute to the pronounced insulin resistance and to the life-threatening ketosis (88).

It has been suggested that nocturnal surges of GH could be responsible for the so-called "dawn phenomenon", i.e. an increase in the insulin requirements in the early morning hours (85,89,90), though opinions on the topic are not unanimous (91,92). Again it should be considered that a majority of studies have employed rather bulky doses (0.6–0.8 mg) of GH, thereby perhaps overestimating the role of GH. As suggested it is possible that increased early morning insulin requirements may be caused by transient sleep correlated decrements in glucose turnover and insulin demands, and a subsequent normalization of these parameters at arousal — waning of insulin action from precedent meals may also be involved (91,93).

On the whole it is still beyond doubt that GH contributes significantly to the overall insulin resistance of type 1 diabetes and also acts as an initiator of the vicious circles leading to acute metabolic derangement (94).

ACROMEGALY AND PHARMACOLOGICAL USE OF GH

Active acromegaly unmistakably unveils the diabetogenic potential of GH. It is remarkable that these actions of GH prevail in spite of substantial compensatory hyperinsulinaemia; virtually all studies describe 2–3 fold elevations of basal concentrations of insulin in acromegalic patients (95–98).

Under these hormonal circumstances small increments in circulating glucose concentrations and elevated glucose turnover are characteristic (95,97,98). Little information is available regarding lipid metabolism in patients with acromegaly. There are however explicit suggestions that the disease is characterized by increased levels of circulating lipid intermediates, increased muscle uptake of these intermediates and an increased rate of lipid oxidation in a magnitude of 40–50% (98). These abnormalities are accompanied by increased rates of total energy expenditure and suppressed rates of glucose oxidation.

Increased energy expenditure in acromegaly has been recognized for many years (99) and may relate to substrate cycling, to increased levels of IGF-I or perhaps to increased thyroid activity (100). Despite the increased metabolic rate nitrogen excretion is apparently still normal in acromegalic subjects (98).

When hyperinsulinaemic glucose clamps are performed to assess insulin sensitivity in acromegalic subjects, it becomes evident that the actions of insulin on both glucose and lipid metabolism are blunted (95,98). It is also clear that the restrictive effects on insulin action are due to defects in both hepatic and extrahepatic glucose metabolism. The peripheral insulin resistance is largely due to insulin resistance in striated muscle (98). As mentioned, these aberrations may in part be caused by stimulation of lipolysis leading to peripheral substrate competition, together with a possible augmentation of gluconeogenesis.

It is striking that the abnormalities of fuel metabolism are completely resolved a few months after successful surgery in acromegalic patients and — conversely — that the very same abnormalities may be imposed after only 2 weeks GH treatment in normal humans (98,101). Nevertheless, it should be underlined that prescription of large amounts of GH for e.g. therapeutic purposes leads to substantial hyperinsulinaemia and insulin resistance; such alterations may after many years cause premature atherosclerosis and hypertension (102), as reported in patients with acromegaly (103–105). Little is known about the potential effects of relatively short periods of GH induced hyperinsulinaemia and insulin resistance on long term morbidity and mortality from cardiovascular events.

REFERENCES

1. Thorner, M.O., Vance, M.L., Horwath, E. and Kovaccs, K. (1992) The anterior pituitary. In: Williams textbook of endocrinology, 8th ed. J.D. Wilson and D.W. Foster (eds). Saunders, Philadelphia, pp. 221–310.
2. Smith, R.G., Van der Ploeg, L.H.T., Howard, A.D., Feighner, S.D., Cheng, K., Hickey, G.J., Wyratt, M.M.J., Fisher, M.H., Nargund, R.P. and Patchett, A.A. (1997) Peptidomimetic regulation of growth hormone secretion. Endocrine Reviews 18, 621–45.
3. Copinschi, G., Van Onderbergen, A., L'Hermite-Baleriaux, M., Mendel, C.M., Caufriez, A., Leproult, R., Bolognese, J.A., De Smet, M., Thorner, M.O. and Van Cauter, E. (1996) Effects of a 7-day treatment with a novel orally active growth hormone (GH) secretagogue MK-677 on 24-hour GH profiles, insulin-like growth factor I, and adrenocortical function in normal young men. J. Clin. Endocrinol. Metab. 81, 2776–2782.
4. Chapman, I.M., Bach, M.A., Van Cauter, E., Farmer, M., Krupa, D., Taylor, A.M., Schilling, L.M., Cole, K.Y., Skiles, E.H., Pezzoli, S.S., Hartman, M.L., Veldhuis, J.D., Gormley, G.J. and Thorner, M.O.. (1996) Stimulation of growth hormone (GH)-insulin-like growth factor I axis by daily oral administration of a GH secretagogue (MK-677) in healthy elderly subjects. J. Clin. Endocrinol. Metab. 81, 4249–4257.
5. Chapman, I.M., Pescovitz, O.H., Murphy, G., Treep, T., Cerchio, K.A., Krupa, D., Gertz, B., Polvino, W.J., Skiles, E.H., Pezzoli, S.S. and Thorner, M.O. (1997) Oral administration of growth hormone (GH) releasing peptide mimeting MK-677 stimulates the GH/insulin-like growth factor I axis in selected GH-deficient adults. J. Clin. Endocrinol. Metab. 82, 3455–3463.
6. Murphy, M.G., Plunkett, L.M., Gertz, B.J., He, W., Wittreich, J., Polvino, W.M. and Clemmons, D.R. (1998) MK-677, an orally active growth hormone secretagogue, reverses diet-induced catabolism. J. Clin. Endocrinol. Metab. 83, 320–325.
7. Houssay, B.A. (1936) The hypophysis and metabolism. N. Eng. J. Med. 214, 961–86.
8. Young, F.G. (1940) The pituitary gland and carbohydrate metabolism. J. Clin. Endocrinol. 26, 345–351.
9. Shipley, R.A. and Long, C.H.N. (1938) Studies on ketogenic activity of anterior pituitary; relation of ketonaemia to ketonuria in rat; method for assay of ketogenic activity; nature of ketogenic principle. Biochem. J. 32, 2242–2256.
10. Beck, J.C., McGarry, E.E., Dyrenfurth, I. and Venning, E.H. (1957) Metabolic effects of human and monkey growth hormone in man. Science 125, 884–85.
11. Ikkos, D., Luft, R., Gemzell, G.A. and Olivecrona, H. (1958) Effect of human growth hormone in hypophysectomised diabetic subjects. Lancet i, 721–22.
12. Raben, M.S. and Hollenberg, C.H. (1959) Effect of growth hormone on plasma fatty acids. J. Clin. Invest. 39, 484–88.
13. Henneman, D.H. and Henneman, P.H. (1960) Effects of human growth hormone on levels of blood and urinary carbohydrate and fat metabolites in man. J. Clin. Invest. 40, 1239–45.
14. Zierler, K. and Rabinowitz, R. (1963) Roles of insulins and growth hormones based on studies of forearm metabolism in man. Medicine 42, 385–95.
15. Rabinowitz, D., Klassen, D.A. and Zierler, K.L. (1965) Effect of human growth hormone on muscle and adipose tissue metabolism in the forearm of man. J. Clin. Invest. 44, 51–61.
16. Fineberg, S.E. and Merimee, T.J. (1974) Acute metabolic effects of human growth hormone. Diabetes 23, 499–504.
17. Froesch, E.R., Hussain, M.A., Schmid, C. and Zapf, J. (1996) Insulin-like growth factor I: Physiology, metabolic effects and clinical uses. Diabetes/Metabolism Reviews 12, 195–215.
18. Chipkin, S.R., Scecowka, J., Tai, L.-R., Kostyo, J.L. and Goodman, H.M. (1989) Different growth hormone-receptor interactions mediate insulin-like and lipolytic responses of rat adipose tissue. Endocrinology 125, 450–458.
19. Lim, N., Ng, F.M., Wu, Z.M., Ede, N. and Hearn, T.W. (1992) Hypoglycemic action of a novel constrained analog of human growth hormone (6-13). Endocrinology 131, 835–40.

20. Hartman, M.L., Faria, A.C.S., Vance, M.L., Johnson, M.L., Thorner, M.O. and Veldhuis, J.D. (1991) Temporal structure of *in vivo* growth hormone secretory events in humans. Am. J. Physiol. 260, E101–10.
21. Ho, K.Y., Veldhuis, J.D., Johnson, M.L., Furianetto, R., Evans, W.S., Alberti, K.G.M.M. and Thorner, M.O. (1988) Fasting enhances growth hormone secretion and amplifies the complex rhythms of growth hormone secretion in man. J. Clin. Invest. 81, 968–75.
22. Johnston, D.G., Davis, R.R. and Prescott, R.W.G. (1985) Regulation of growth hormone secretion in man: a rewiev. J. Roy. Soc. Med. 78, 319–27.
23. Møller, N., Jørgensen, A.O.L., Alberti, K.G.M.M., Flyvbjerg, A. and Schmitz, O. (1990) Short-term effects of growth hormone on fuel oxidation and regional substrate metabolism in normal man. J. Clin. Endocrinol. Metab. 70, 1179–86.
24. Møller, N., Jørgensen, J.O.L., Schmitz, O., Møller, J., Christiansen, J.S., Alberti, F.G.M.M. and Ørskov, H. (1990) Effects of a growth hormone pulse on total and forearm substrate fluxes in humans. Am. J. Physiol. 258, E86–91.
25. Møller, N., Schmitz, O., Møller, J., Pørksen N. and Jørgensen, J.O.L. (1991) Dose–response studies on the metabolic effects of a growth hormone pulse in humans. Metabolism 41, 172–175.
26. Møller, N., Pørksen, N., Ovesen, P. and Alberti, K.G.M.M. (1993) Evidence for increased sensitivity of fuel mobilization to growth hormone during short-term fasting in humans. Hormone & Metabolic Research 25, 175–79.
27. Rosenthal, M.J. and Woodside, W.F. (1998) Nocturnal regulation of free fatty acids in healthy young and elderly men. Metabolism 37, 645–48.
28. Boyle, P.J., Avogaro, A., Smith, L., Bier, D.M., Pappu, A.S., Illingworth, D.R. and Cryer, P.E. (1992) Role of GH in regulating nocturnal rates of lipolysis and plasma mevalonate levels in normal and diabetic humans. Am. J. Physiol. 263, E168–172.
29. Cersosimo, E., Danou, F., Persson, M. and Miles, J.M. (1996) Effects of pulsatile delivery of basal growth hormone on lipolysis in humans. Am. J. Physiol. 271, E123–26.
30. Edge, J.A., Pal, B.R., Harris, D.A., Matthews, D.R., Phillips, P.E. and Dunger, D.B. (1993) Evidence for a role for insulin and growth hormone in overnight regulation of 3-hydroxybutyrate in normal and diabetic adolescents. Diabetes Care 16, 1011–18.
31. Hagstrom-Toft, E., Bolinder, J., Ungerstedt, U. and Arner, P. (1997) A circadian rhytm in lipid mobilization which is altered in IDDM. Diabetologia 40, 1070–78.
32. Randle, P.J., Garland, P.B., Hales, C.N. and Newsholme, E.A. (1963) The glucose–fatty acid cycle: its role in insulin sensitivity and the metabolic disturbances of diabetes mellitus. Lancet i, 785–89.
33. Davidson, M.B. and Bernstein, J.M. (1975) The effect of nicotinic acid on growth hormone-induced lipolysis and glucose intolerance. J. Lab. Clin. Med. 81, 568–76.
34. Adamson, U., Wahren, J. and Cerasi, E. (1977) Influence of growth hormone on splanchnic glucose production in man. Acta Endocrinol. 86, 803–12.
35. Rogers, S.A., Karl, I.E. and Hammerman, M.R. (1989) Growth hormone directly stimulates gluconeogenesis in canine proximal tubule. Am. J. Physiol. 257, E751–56.
36. Karlander, S., Vranic, M. and Efendic, S. (1986) Increased glucose turnover and glucose cycling in acromegalic patients with normal glucose tolerance. Diabetologia 29, 778–83.
37. Butler, P., Kryshak, E. and Rizza, R. (1991) Mechanism of growth hormone induced postprandial carbohydrate intolerance in humans. Am. J. Physiol. 260, E513–20.
38. Altszuler, N., Rathgeb, I., Winkler, B., DeBodo, R.C. and Steele, R. (1968) The effects of growth hormone on carbohydrate and lipid metabolism in the dog. Ann. N.Y. Acad. Sci. 148, 441–58.
39. Ikkos, D., Luft, R. and Gemzell, C.A. (1958) The effect of human growth hormone in man. Lancet I, 720–21.
40. Henneman, P.H., Forbes, A.P., Moldawer, M., Dempsey, E.F. and Carroll, E.L. (1960) Effects of human growth hormone in man. J. Clin. Invest. 39, 1223–38.
41. Bray, G.A., Raben, M.S., Londono, J. and Gallagher, T.F. (1971) Effects of triiodothyronine, growth hormone and anabolic steroids on nitrogen excretion and oxygen consumption of obese patients. J. Clin. Endocrinol. 33, 293–300.

42. Manson, J.McK. and Wilmore, D.W. (1986) Positive nitrogen balance with human growth hormone and hypocaloric intravenous feeding. Surgery 100, 188–196.
43. Horber, F.F. and Haymond, M.W. (1990) Human growth hormone prevents the protein catabolic side effects of prednisone treatment. J. Clin. Invest. 86, 265–72.
44. Tessari, P.R., Trevisan, R., Inchiostro, S., Biolo, G., Nosadini, G., DeCreutzenberg, S.V., Duner, E., Tiengo, A. and Crepaldi, G. (1986) Dose–response curves of effects of insulin on leucine kinetics in humans. Am. J. Physiol. 251, E334–42.
45. Jacob, R., Barret, E., Piewe, G., Fagin, K.D. and Sherwin, R.S. (1989) Acute effects of insulin-like growth factor I on glucose and amino acid metabolism in the awake fasted rat. J. Clin. Invest. 83, 1717–23.
46. Fryburg, D.A., Gelfand, R.A. and Barrett, E.J. (1991) Growth hormone acutely stimulates forearm muscle protein synthesis in normal humans. Am. J. Physiol. 260, E499–504.
47. Fryburg, D.A., Louard, R.J., Gerow, K.E., Gelfand, R.A. and Barrett, E.J. (1992) Growth hormone stimulates skeletal muscle protein synthesis and antagonizes insulins antiproteolytic action in humans. Diabetes 41, 424–29.
48. Copeland, K.C. and Nair, K.S. (1994) Acute growth hormone effects on amino acid and lipid metabolism. J. Clin. Endocrinol. Metab. 78, 1040–47.
49. Welbourne, T., Joshi, S. and McVie, R. (1989) Growth hormone effects on hepatic glutamate handling *in vivo*. Am. J. Physiol. 257, E959–62.
50. Wolthers, T., Grøfte, T., Møller, N., Vilstrup, H. and Jørgensen, J.O.L. (1996) Effects of long-term growth hormone (GH) and triiodothyronine (T3) administration on functional hepatic nitrogen clearance in normal man. J. Hepatol. 24, 313–19.
51. Sherwin, R.S., Hendler, R.G. and Felig, P. (1975) Effect of ketone infusions on amino acid and nitrogen metabolism in man. J. Clin. Invest. 55, 1382–90.
52. Tessari, P., Nissen, S.L., Miles, J. and Haymond, M.W. (1986) Inverse relationship of leucine flux and oxidation to free fatty acid availability *in vivo*. J. Clin. Invest. 77, 575–81.
53. Nair, K.S., Welle, S.L., Halliday, D. and Campell, R.G. (1988) Effect of 3-hydroxybutyrate on whole-body leucine kinetics and fractional mixed skeletal muscle protein synthesis in humans. J. Clin. Invest. 82, 198–205.
54. Goodman, H.G., Grumbach, M.M. and Kaplan, S.L. (1968) Growth and growth hormone II. A comparison of isolated hormone deficiency and multiple pituitary deficiencies in 35 patients with idiopathic hypopituitary dwarfism. N. Eng. J. Med. 278, 57–68.
55. Merimee, T.J., Felig, P., Marliss, E., Fineberg, E. and Cahill, G.C. (1971) Glucose and lipid homeostasis in the absence of human growth hormone. J. Clin. Invest. 50, 574–82.
56. Haymond, M.W., Karl, I., Weldon, V.V. and Pagliara, A.S. (1976) The role of growth hormone and cortisol on glucose and gluconeogenic substrate regulation in fasted hypopituitary children. J. Clin. Endocrinol. Metab. 42, 846–56.
57. Bougneres, P.-F., Artavia-Loria, E., Ferre, P., Chaussain, J.-L. and Job, J.-C. (1985) Effects of hypopituitarism and growth hormone replacement therapy on the production and utilization of glucose in childhood. J. Clin. Endocrinol. Metab. 61, 1152–57.
58. Dahms, W.T., Owens, R.P., Calhan, S.C., Kerr, D.S. and Danish, R.K. (1989) Urea synthesis, nitrogen balance and glucose turnover in growth-hormone-deficient children before and after growth hormone administration. Metabolism 38, 197–203.
59. Wolsdorf, J.I., Sagdeghi-Nejad, A. and Senior, B. (1983) Hypoketonaemia and age related fasting hypoglycemian growth hormone deficiency. Metabolism 32, 457–62.
60. Tamborlane, W.V., Genel, M., Gianfredi, S. and Gertner, J.M. (1984) The effect of small but sustained elevations in circulating growth hormone on fuel metabolism in growth hormone deficiency. Ped. Res. 18, 212–15.
61. Jørgensen, J.O.L., Møller, J., Alberti, K.G.M.M., Schmitz, O., Christiansen, J.S. and Møller, N. (1993) Effects of a low basal growth hormone (GH) elevation on day-to-day fuel metabolism: Studies in GH deficient patients and healthy untreated subjects. J. Clin. Endocrinol. Metab. 77, 1589–96.

62. Costin, G., Kogut, M.D. and Frasier S.D. (1980) Effect of low-dose human growth hormone on carbohydrate metabolism in children with hypopituitarism. J. Pediatr. 80, 796–803.
63. Lippe, B.M., Kaplan, S.A., Golden, M.P., Hendricks, S.A. and Scott, M.L. (1981) Carbohydrate tolerance and insulin receptor binding in children with hypopituitarism: Responses after acute and chronic growth hormone administration. J. Clin. Endocrinol. Metab. 53, 507–13.
64. Jørgensen, J.O.L., Pedersen, S.A., Thuesen, L., Hansen, T.I., Skakkebæk, N. and Christiansen, J.S. (1989) Beneficial effects of growth hormone treatment in GH-deficiency. Lancet 1, 1221–25.
65. Salomon, F., Cuneo, R.C., Hesp, R. and Sonksen, P.H. (1989) The effects of treatment with recombinant human growth hormone on body composition and metabolism in adults with growth hormone deficiency. N. Eng. J. Med. 321, 1797–1803.
66. Hew, F.L., Koschman, M., Christopher, M., Rantzau, C., Vaag, A., Ward, G., Nielsen, H.B. and Alford, F. (1996) Insulin resistance in growth hormone deficient adults: defects in glucose utilization and glycogen synthase activity. J. Clin. Endocrinol. Metab. 81, 555–564.
67. Fowelin, J., Attwall, S., Lager, I. and Bengtsson, B.-Å. (1993) Effects of treatment with recombinant human growth hormone on insulin sensitivity and glucose metabolism in adults with growth hormone deficiency. Metabolism 42, 1443–47.
68. Press, M., Notarfrancesco, A. and Gelel, M. (1987) Risk of hypoglycemia with alternate-day growth hormone injections. Lancet i, 1002–04.
69. Hansen, A.P. and Johansen, K. (1970) Diurnal patterns of blood glucose, serum free fatty acids, insulin, glucagon and growth hormone in normals and juvenile diabetics. Diabetologia 6, 27–38.
70. Schaper, N.C. (1990) Growth hormone secretion in type 1 diabetes: A review. Acta Endocrinol. 122, 7–12.
71. Asplin, C.M., Faria, A.C.S., Carlsen, E.C., Vacaro, V.A., Barr, R.E., Iranmanesh, A., Lee, M.M., Veldhuis, J.D. and Evans, W.S. (1989) Alterations in the pulsatile mode of growth hormone release in men and women with insulin-dependent diabetes mellitus. J. Clin. Endocrinol. Metab. 69, 239–45.
72. Møller, N., Schmitz, O., Møller, J. and Butler, P.C. (1992) Effects of a growth hormone pulse on substrate metabolism in insulin dependent (type 1) subjects. J. Clin. Endocrinol. Metab. 75, 432–36.
73. Kollind, M., Adamson, U., Lins, P.-E. and Curstedt, T. (1998) Importance of growth hormone for blood glucose regulation following insulin-induced nocturnal hypoglycemia in insulin-dependent diabetes mellitus. Acta Med. Scand. 223, 159–64.
74. De Feo, P., Perriello, G., Torlone, E., Ventura, M.M., Santeusanio, F., Brunetti, P., Gerich, J.E. and Bolli, G.B. (1989) Demonstration of a role for growth hormone in glucose counter-regulation. Am. J. Physiol. 256, E835–43.
75. Boyle, P.I. and Cryer. P.E. (1991) Growth hormone, cortisol, or both are involved in defense against, but are not critical to recovery from hypoglycemia. Am. J. Physiol. 260, E395–402.
76. Fowelin, J., Attwall, S., von Schenk, H., Smith, U. and Lager, I. (1989) Combined effect of cortisol and growth hormone on late post-hypoglycemic insulin resistance in humans. Diabetes 38, 1357–64.
77. Press, M., Tamborlane, W.V. and Sherwin, R.S. (1984) Importance of raised growth hormone levels in mediating the metabolic derangements of diabetes. N. Eng. J. Med. 310, 810–15.
78. Press, M., Tamborlane, W.V. and Sherwin, R.S. (1986) Effect of insulin on growth hormone-induced metabolic derangements in diabetes. Metabolism 35, 956–59.
79. Rizza, R.A., Mandarino, L.J. and Gerich, J.E. (1982) Effects of growth hormone on insulin action in man. Diabetes 31, 663–69.
80. Bratusch-Marrain, P.R., Smith, D. and DeFronzo, R.A. (1982) The effect of growth hormone on glucose metabolism and insulin secretion in man. J. Clin. Endocrinol. Metab. 55, 973–82.
81. Møller, N., Butler, P.C., Antsiferov, M. and Alberti, K.G.M.M. (1989) Effects of growth hormone on insulin sensitivity and forearm metabolism in normal man. Diabetologia 32, 105–10.
82. Ørskov, L., Schmitz, O., Jørgensen, J.O.L., Arnfred, J., Abildgaard, N., Christiansen, J.S., Alberti, K.G.M.M. and Ørskov, H. (1989) Influence of growth hormone on glucose induced

glucose uptake in normal men as assessed by the hyperglycemic clamp technique. J. Clin. Endocrinol. Metab. 68, 276–82.
83. Bak, J.F., Møller, N. and Schmitz, O. (1991) Effects of growth hormone on fuel utilization and muscle glycogen synthase activity in normal humans. Am. J. Physiol. 260, E736–42.
84. Bratusch-Marrain, P., Waldhausl, W., Grubeck-Lobenstein, B., Korn, A., Vierhapper, H. and Nowotny, P. (1981) The role of "diabetogenic" hormones on carbohydrate and lipid metabolism following oral glucose loading in insulin dependent diabetics: effects of acute hormone administration. Diabetologia 21, 387–93.
85. Periello, G., De Feo, P., Torlone, E., Fanelli, C., Santeusanio, F., Brunetti, P. and Bolli, G.B. (1990) Nocturnal spikes of growth hormone secretion cause the dawn phenomenon in type 1 (insulin-dependent) diabetes mellitus by decreasing hepatic (and extrahepatic) sensitivity to insulin in the abscence of insulin waning. Diabetologia 33, 52–59.
86. Unger, R.H. (1965) High growth hormone levels in diabetic ketoacidosis. J. Am. Med. Ass. 191, 945–47.
87. Cryer, P.E. and Daughaday, W.H. (1970) Diabetic ketosis. Serial plasma growth hormone concentrations during therapy. Diabetes 19, 519–523.
88. Schade, D.S., Eaton, P. and Peake, G.T. (1978) The regulation of plasma ketone body concentration by counter-regulatory hormones in man. II Effects of growth hormone in diabetic man. Diabetes 27, 916–24.
89. Campell, P.J., Bolli, G.B., Cryer, P.E. and Gerich, J.E. (1985) Pathogenesis of the dawn phenomenon in patients with insulin dependent diabetes mellitus. N. Eng. J. Med. 312, 1473–79.
90. Beaufrere, B., Beylot, M., Metz, C., Ruitton A., Francois, R., Riou, J.P. and Mornex, R. (1988) Dawn phenomenon in type 1 (insulin dependent) diabetic adolescents: influence of nocturnal growth hormone secretion. Diabetologia 31, 607–11.
91. Blackard, W.G., Barlascini, C.O., Clore, J.N. and Nestler, J.E. (1989) Morning insulin requirements. Critique of dawn and meal phenomena. Diabetes 38, 273–77.
92. Skor, D.A., White, N.H., Thomas, L. and Santiago, J.V. (1985) Influence of growth hormone on overnight insulin requirements in insulin-dependent diabetes. Diabetes 34, 135–39.
93. Clore, J.N., Nestler, J.E. and Blackard, W.G. (1989) Sleep associated fall in glucose disposal and hepatic glucose output in normal humans. Putative signaling mechanism linking peripheral and hepatic events. Diabetes 38, 285–90.
94. Ørskov, H. (1985) Growth hormone hyperproduction inducing some of the vicious circles in diabetes mellitus. Acta Med. Scand. 217, 343–46.
95. Hansen, I., Tsalikian, E., Beaufrere, B., Gerich, J., Haymond, M. and Rizza, R. (1986) Insulin resistance in acromegaly; defects in both hepatic and extrahepatic insulin action. Am. J. Physiol. 250, E269–73.
96. Bolinder, J., Ostman, J., Werner, S. and Arner, P. (1986) Insulin action in human adipose tissue in acromegaly. J. Clin. Invest. 77, 1201–06.
97. Karlander, S., Vranic, M. and Efendic, S. (1986) Increased glucose turnover and glucose cycling in acromegalic patients with normal glucose tolerance. Diabetologia 29, 778–83.
98. Møller, N., Schmitz, O., Jørgensen, J.O.L., Astrup, J., Bak, J.F., Christensen, S.E., Alberti, K.G.M.M. and Weeke, J. (1992) Basal and insulin stimulated substrate metabolism in patients with active acromegaly before and after adenomectomy. J. Clin. Endocrinol. Metab. 74, 1012–19.
99. Ikkos, D., Ljunggren, H. and Luft, R. (1956) Basal metabolic rate in relation to body size and cell mass in acromegaly. Acta Endocrinol. 21, 237–44.
100. Møller, J., Jørgensen, J.O.L., Møller, N., Christiansen, J.S. and Weeke, J. (1992) Effects of growth hormone administration on fuel oxidation and thyroid function in normal man. Metabolism 41, 728–31.
101. Møller, N., Møller, J., Jørgensen, J.O.L., Ovesen, P., Schmitz, O., Alberti, K.G.M.M. and Christiansen, J.S. (1993) Impact of 2 weeks high dose growth hormone treatment on basal and insulin stimulated substrate metabolism in humans. Clinical Endocrinology 39, 577–81.
102. Reaven, G.M.(1988) Banting Lecture 1988. Role of insulin resistance in human disease. Diabetes 37, 1595–1607.

103. Wright, A.D., Hill, D.M., Lowy, C. and Fraser, T.R. (1970) Mortality in acromegaly. Quart. J. Med. 153, 1–16.
104. Alexander, L., Appleton, D., Hall, R., Ross, W.M. and Wilkinson, R. (1980) Epidemiology of acromegaly in the Newcastle region. Clin. Endocrinol. 12, 71–79.
105. Bengtson, B., Eden, S., Ernest, I., Oden, A. and Sjogren, B. (1988) Epidemiology and long-term survival in acromegaly. Acta Med. Scand. 223, 327–35.

Growth Hormone Secretagogues
Edited by E. Ghigo, M. Boghen, F.F. Casanueva and C. Dieguez

Chapter 17

Growth Hormone Secretagogues. Physiological Role and Clinical Implications

CARLOS DIEGUEZ[2], VERA POPOVIC[1], DRAGAN MICIC[1], ALFONSO LEAL-CERRO[3], ANGELA PEÑALVA[2], RICARDO V. GARCIA-MAYOR[4], MANUEL POMBO[2] and FELIPE F. CASANUEVA[2]

[1]*Institute of Endocrinology, University Clinical Center, Belgrade, Yugoslavia*
[2]*Departments of Physiology, Medicine and Pediatrics, University of Santiago de Compostela, Spain*
[3]*Endocrinology Unit, Hospital Virgen del Rocio, Sevilla, Spain*
[4]*Internal Medicine, Hospital Xeral Cies, Vigo, Spain*

INTRODUCTION

In addition to GHRH and somatostatin, several other neuropeptides and neurotransmitter pathways, as well as a variety of peripheral feedback signals regulate GH secretion, either by acting directly at the anterior pituitary level, or by modulating GHRH and somatostatin release from the hypothalamus (1–4).

In recent years, considerable attention has been focused on a synthetic hexapeptide, so-called GHRP-6, which was developed by a combination of conformational energy calculation, synthesis and biological activity testing (5–9). More recently, a second generation of GHRP-6 analogues has been developed, both peptidic and non-peptidic (10–12). They all share some common features, such as being potent releasers of GH in all species tested so far, after administration through different routes, such as i.v., ip, intranasally as well as by the oral route (13–20). All the compounds developed so far seem to exhibit a high binding affinity to the recently cloned GHS-receptor (21). Taking into account that this receptor is mainly expressed in the pituitary and the CNS, it provides the biochemical basis regarding the mechanism and loci of action of these compounds on GH release (21–24). Nevertheless, there is some strong evidence suggesting the existence of additional receptor subtypes which may exhibit different affinities for these compounds. Thus, in some tissues, the specific binding of iodinated hexarelin can be displaced by peptidyl analogues of GHRP-6 but not by MK-0677 (25). Furthermore, although all peptidyl and non-peptidyl GHS are potent releasers of GH, some of them are also able to stimulate the pituitary–adrenal axis, leading to increased plasma ACTH and cortisol levels, while others do not (26). Therefore, it is likely that further subtypes of GHS are waiting to be cloned. There is no doubt that all GHS act directly on the pituitary. It would be surprising if they did not, considering that their development was guided by their *in vitro* GH-releasing capability. After activating their specific receptor, GHS elicit GH secretion through a different signalling system from GHRH (27–29). Furthermore these two peptides also seem to act on different somatotroph sub-populations. With the use of the reverse haemolytic plaque assay, GHRP was shown to increase the number of somatotrophs releasing GH, without altering the amount of hormone released by each individual cell (30). On the other hand, GHRH stimulates both the number and amount of GH secreted per cell (31). A direct pituitary effect was also supported by data obtained *in vivo*. Thus GHRP-6 has been found to be able to increase plasma GH levels in two different animal models, where the anterior pituitary was not under hypothalamic influence, namely rats with hypothalamic ablation, and in hypophysectomized-transplanted rats bearing two hypophysis under the renal capsule (14). Similar findings were also reported in hypothalamo–pituitary disconnected sheep, indicating a pituitary site of action of these peptides (32). However, as these peptides exhibit a much greater potency in terms of GH release *in vivo* than *in vitro*, it soon became quite clear that they may also be acting at hypothalamic level. Furthermore, the *in vitro* potency of these peptides was much greater in hypothalamic–pituitary incubate than in monolayer cultures of rat anterior pituitary cells.

Although the structure of a natural GHRP-like ligand still remains to be known, considerable insight has been gained in recent years regarding the role and mechanism of action of these compounds in the regulation of GH secretion. In this chapter, we will review data obtained in humans that have allowed a greater insight into the mechanisms of action of GHS in the regulation of GH secretion. Furthermore, we will summarise data regarding GH responses to these secretagogues in different disease states, highlighting their pathophysiological implication and their potential from a diagnostic point of view in the clinical setting.

GHS AND GH SECRETION IN PATIENTS WITH HYPOTHALAMO–PITUITARY DISCONNECTION

Data obtained in humans also support a hypothalamic action of GHS. In adult subjects with hypothalamo–pituitary disconnection due to hypothalamic lesions of tumoral origin, and therefore disrupted communication between the hypothalamus and the adenohypophysis, GH responses to GHRP-6 have been assessed (33). The fact that GHRH-induced GH secretion in these patients was roughly similar to the control group indicates that, in functional terms, the pituitary tissue remains intact. Interestingly, when the stimulus administered was the GHS, GHRP-6, which in normal controls is more efficacious than GHRH, the patients with hypothalamo–pituitary disconnection showed no response at all. These data indicate that GHRP-6 releases GH secretion in man by acting at hypothalamic level. An obvious criticism of these conclusions could be that the pituitary GHS-receptor could be down-regulated. However, this is unlikely considering that decreased circulating GH levels lead to increased levels of the pituitary GHRH-R as well as the GHS-R (34). Furthermore, in those patients with hypothalamo–pituitary disconnection, the GHRH + GHRP-6 mediated GH release was not different from the action of GHRH alone, with an absence of synergism. Finally, similar findings have been reported in children with neonatal pituitary stalk transection due to perinatal damage, with the ensuing complete GH deficiency and pan-hypopituitarism of variable degrees. In these patients, and in comparison with controls, the GHRH-mediated GH release was minimal (35). This finding is in clear contrast with data from adult patients with hypothalamo–pituitary disconnection, in whom GHRH-stimulated GH secretion was normal. It seems that one hypothalamic factor, probably GHRH, is crucial for the development and maintenance of the normal somatotroph population in the neonatal and early childhood period, but its absence in the adult period does not introduce any change in somatotroph number or responsiveness. This hypothesis is supported by results in the rat, in which pretreatment with anti-GHRH serum in neonatal rats induces permanent damage in GH secretion and growth, but in adults is devoid of long term effects (36). In any event, both the acute GH release of either GHRH or GHRP-6 as well as the synergistic action exerted by the combined administration of the two peptides were severely blocked in patients with perinatal pituitary stalk transection (35). In these patients, GHRP-6 was less effective than GHRH, suggesting that the main action of the hexapeptide is exerted at hypothalamic level, through as yet undetermined mechanisms. Finally, the absence of a GH response to GHRH + GHRP-6 indicates that this simple, cheap and risk-free test may be a diagnostic tool for immediate identification of patients with pituitary stalk transection from patients with other causes of idiopathic GH deficiency.

GH DEFICIENCY

Pituitary GH reserve can be assessed by substances that act directly at the somatotroph, such as GHRH, or by a variety of metabolic and neuropharmacological tests acting at the hypothalamic level such as hypoglycaemia, clonidine or L-dopa (1–4). In the light of the powerful GH releasing capabilities of GHS, the role of these compounds in this clinical setting has been assessed. Due to the different physiopathological mechanisms involved in the genesis of childhood- and adult-onset GH deficiency, we will review these two entities independently.

Childhood-onset GH deficiency

Although the criteria for the diagnosis of GHD are not clear-cut, in general terms, pituitary GH reserve in children with idiopathic GHD is established by a decreased GH response after the administration of hypothalamic stimuli such as clonidine, hypoglycaemia and propanolol-exercise in the presence of adequate auxological parameters, indicative of short stature and delayed growth (37–39). In contrast to the aforementioned stimuli, assessment of spontaneous GH secretion over 24 hours appears to be of poor diagnostic value (40), while GH responses to exogenous administered GHRH are normal (80–90%) in the majority of patients (38,39). The latter data indicate that in the majority of these patients pituitary GH reserve is largely maintained (39).

Studies in normal children have shown that acutely administered i.v. GHRP-6 is a potent GH releasing substance, independently of the age and sex of the subjects (41). Nevertheless,

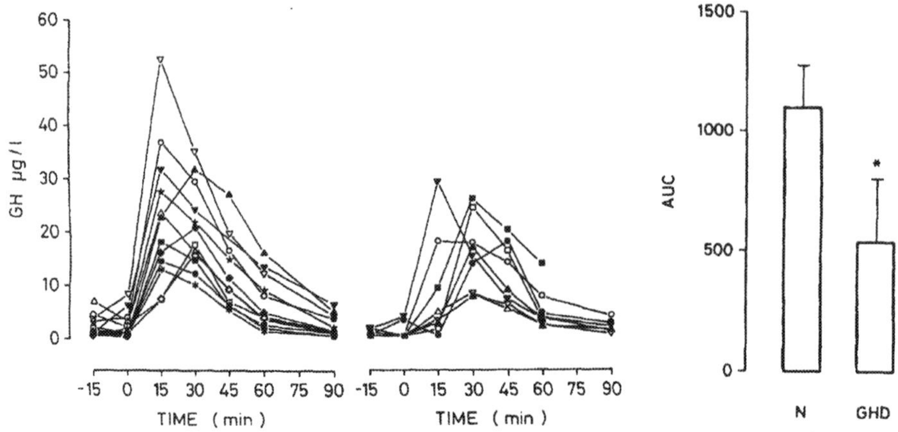

Figure 1. Individual levels and area under the secretory curve (AUC) of GH responses to GHRP-6 in normal (N) (right) and children with growth hormone deficiency (GHD) (left).

the possible pattern of response to this GHS in children with idiopathic GH deficiency was difficult to predict, as the exact mechanism of action of these GHS is largely unknown. Although the main action of these compounds in increasing plasma GH levels appears to be exerted at the hypothalamus, they also act directly at the pituitary.

On a group mean basis, GH response following the administration of GHRP-6 either by the oral or i.v. route has been shown to be lower in patients diagnosed with GHD by auxological parameters and conventional GH-testing than normal controls (42,43). Nevertheless, it was also found that many patients with low GH response to other tests, i.e. hypoglycaemia, still exhibited a marked GH response to exogenously administered GH-secretagogues. In fact, on an individual basis, a considerable degree of overlap was observed between patients with idiopathic GHD and normal controls (Figure 1) (43). Therefore the possible diagnostic application of these compounds in the diagnosis of childhood-onset idiopathic-GHD is quite limited. Nevertheless, the finding that these compounds exert marked GH-releasing capability in many of these patients opens up the possibility of assessing their therapeutic potential during chronic administration (see Chapter 20).

Adult-onset GHD

In recent years the health problems of adults with GHD and the benefits of GH replacement therapy have received considerable attention. Although more studies are still needed, data so far available indicate that GHD in adults is associated with changes in body composition, increased prevalence of cardiovascular morbidity and shortened life expectancy. Thus, it seems reasonable that adults with GHD may benefit from GH replacement therapy (44–52).

However, the diagnosis of GHD in adults is particularly difficult, for several reasons: the advantage of measuring spontaneous rather than stimulated GH secretion remains controversial (40). There are no clear biological markers of GH action at the tissue level, such as linear growth in children (53). IGF-1 levels do not provide a marker of GH secretion in adults, since they are markedly influenced by nutritional and metabolic status, IGF-l and IGFBP-3 being within the normal range in more than 60% of patients with GHD (53,54). For these reasons, the current consensus is that within an appropriate clinical context, the diagnosis of this ailment in adults must be established biochemically, by means of a GH provocative test. Due to its generalised use, the insulin tolerance test (ITT) is at present the recommended test, with severe GH deficiency being defined as the GH response lower than the arbitrary cut-off of 3 μg/L. However, the ITT-test has been challenged, due to its low degree of reproducibility and lack of clear-cut normative limits. It has been recognised that other tests such as the arginine-GHRH test (55) are promising alternatives. In the search for an alternative test to ITT, the possibility of using the combined administration of GHRH + GHRP-6 as the test of choice has been put forward. The combined administration of these two peptides is considered as the most potent GH releaser known to date, with excellent reproducibility and absence of side effects (54,56). Other theoretical advantages of this test included: (1) unaffected by metabolic variables such as FFA or glucose; (2) unaffected by gender or gonadal status; (3) unaffected by thyroid status or diabetes mellitus; (4) not affected by previous treatment with GH; (5) unaffected by age or adiposity; (6) unaffected by the circadian rhythm.

Assessment of the potential diagnostic application of this test has shown that in normal controls, the mean peak GH response to GHRH + GHRP-6 was 61 + 2.8 μg/L, while in patients with adult-onset GH-deficiency, defined according to the Growth Research Society (GRS) consensus guidelines, it was 4.7 + 0.3 μg/L. Of great interest, was the finding that with a cut-off value of 15 μg/L, this test gave a 100% sensitivity and 0% of false positives

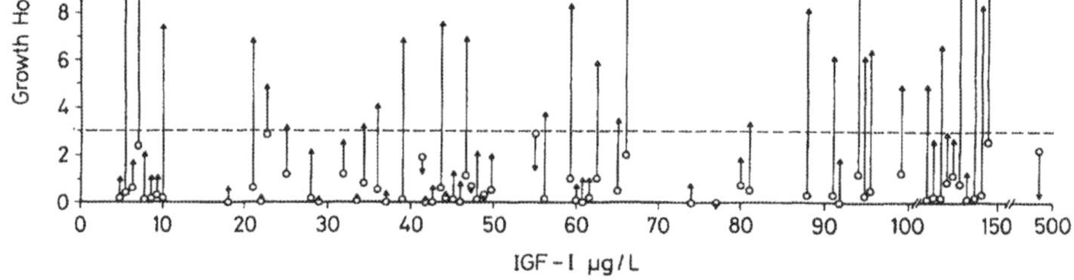

Figure 2. A large group of adult patients with organic hypothalamopituitary disease who were diagnosed as having severe growth hormone (GH) deficiency by an insulin hypoglycaemia test (ITT)-mediated GH peak < 3 μg/l that appear plotted against the respective IGF-1 basal values (open circles). After testing with a combination of GH-releasing hormone (GHRH) and GHRP-6, some of them experienced an increased release of GH (arrows) that was always under the cutoff point of 15 μg/l for this test. Ref. (1).

(100% specificity). Finally, its diagnostic capability was maintained by measuring GH only at two time points (15 and 30 min) after the administration of GHRH + GHRP-6 (Figure 2). Altogether, these data suggest that the GHRH + GHRP-6 is a convenient, safe and reliable test for the diagnosis of adult-onset GH (1).

OBESITY

GH secretion in response to all provocative stimuli is decreased in patients with obesity. The disrupted GH secretion in obesity has relevant clinical implications, considering that the resulting low GH levels may contribute to perpetuating the obese status. Although a partial enhancement of GH clearance cannot be discarded, it is undisputed that the main alteration in obesity is a reduced somatotroph responsiveness. On a theoretical basis, this could be due to decreased hypothalamic GHRH release, increased somatostatinergic tone, greater inhibition by peripheral signals such as FFA or an alteration at the level of the somatotroph (4).

In obese subjects, GHRP-6 induced GH secretion, although lower than in normal subjects was much greater than those after GHRH. The GH response to GHRP-6 in these patients was markedly enhanced by pyridostigmine supporting the view of the existence of increased somatostatinergic tone in obesity (57,58). Even more interesting, was the finding that combined administration of GHRH and GHRP-6 induced a massive discharge of GH. Such a response, rules out the idea that the somatotroph cell is altered, at least in terms of secretory capability, and indicates that the GH characteristic of obesity is a functional and reversible state (59).

Since pyridostigmine increased GH responses to either GHRH or GHRP-6 administered independently, both in normal and in obese subjects, it is possible that the lower GH response to the combined GHRH plus GHRP-6 administration in obese subjects could have been due to increased somatostatinergic tone. However pretreatment with pyridostigmine, in either normal or obese subjects, failed to further increase the GH response to the combined administration of GHRH + GHRP-6 (60,61). This lack of effect of pyridostigmine could be due to the fact that the massive GH discharge induced by this stimulus represents the full secretory capacity of the somatotroph. Alternatively, it is possible that GH responses to the combined administration of GHRH + GHRP-6 are largely independent of somatostatinergic tone. In support of this possibility, it should be noted that, while somatostatin completely prevented *in vitro* GH responses to GHRH and GHRP-6 when administered alone, it was unable to exert a similar effect when both compounds were administered in combination *in vitro* (12). Also, while *in vivo* pre-treatment with atropine, which presumably increases endogenous somatostatinergic tone, abolished GH responses to either GHRH or GHRP-6, when administered alone it only exerted a partial inhibition of GH responses to the combined administration of GHRH + GHRP-6 (61). Therefore, these findings suggested that other hormonal or metabolic alterations could be contributing to the impaired GH response to GHRH + GHRP-6 observed in obese subjects (Figure 3).

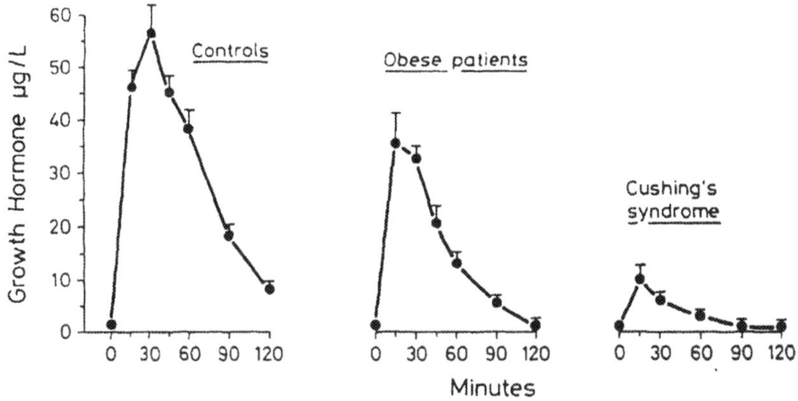

Figure 3. Mean ± SE serum GH levels in controls, obese and patients with Cushing's syndrome after testing with a combination of GH-releasing hormone (GHRH) and GHRP-6.

Data gathered over the last few years have shown that free-fatty acids (FFAs) play a marked inhibitory role on GH secretion, possibly by acting directly on the somatotrophs (62). On the other hand, plasma FFA levels are increased in patients with obesity. The possibility that FFAs might be involved in the GH secretory alterations in obesity was studied by assessing the GH response to different GH secretagogues after FFA reduction by acipimox, a lipid-lowering drug devoid of serious side effects. Interestingly, GH response to either GHRH, GHRP-6 or combined administration of both peptides was enhanced by FFA suppression, thus indicating that abnormally high FFA levels may be a contributing factor for the disrupted GH secretory mechanisms of obesity (63,64).

In conclusion, taken together, these data suggest that although somatotroph secretory capacity is largely preserved in obese subjects, decreased GH secretion can be accounted for by alteration at both central and peripheral level. Data supporting an increase in hypothalamic somatostatinergic tone, as well as increased inhibition of GH secretion by elevated FFA levels in obese subjects, have now come to light. Nevertheless, the possible involvement of other factors involved in the regulation of GH secretion that are also altered in obesity, such as leptin, remains to be explored.

CUSHING'S SYNDROME

Chronic hypercortisolism, as in Cushing's syndrome, is associated with blunted GH secretion in response to all stimuli tested so far (1,2). Although excessively simplistic, glucocorticoid-induced GH blockade may be explained by three different mechanisms, operating either alone or in combination: (a) a deficit in the endogenous release of GHRH, (b) a tonic hypersecretion of somatostatin and (c) a direct inhibitory action upon the somatotrophs. The first possibility seems unlikely, considering that in Cushing's syndrome patients, GH secretion is blunted after the administration of saturating doses of exogenous GHRH, and that repetitive administration (priming) of GHRH rarely increased the GHRH induced GH release (65). A similar lack of experimental support exists in the case of somatostatin, since pre-treatment with pyridostigmine failed to restore GH secretion on chronic hypercortisolism (66). The remaining possibility has been tested, assessing GH responses to GHRP-6. Neither GHRP-6 alone nor GHRH + GHRP-6 were able to increase GH secretion in patients with Cushing's syndrome. This lack of GH response to the combined administration of GHRH plus GHRP-6 is especially relevant when compared with other GH-hyposecretory states like obesity and ageing (Figure 3). Thus while in the later two the GH blockade is a functional and reversible state, the GH impairment of chronic hypercortisolism is not overridden by this potent combined stimulus, indicating that glucocorticoids exert a permanent inhibition upon somatotroph cells, therefore suggesting a direct alteration at the pituitary level and by different mechanisms from those in obesity (67).

ANOREXIA NERVOSA

Anorexia nervosa patients characteristically show elevated basal levels of GH and abnormal GH response to most GH-stimuli. It is unknown at present whether these alterations in GH secretion merely reflect the malnutrition state or if they reflect basic alterations in the CNS pathways involved in the neuroregulation of GH secretion (68–71).

One of the characteristic features of these patients is an enhanced GH response to exogenously administered GHRH (70). However, the fact that a similar increase can be observed in normal subjects after short-term hypocaloric diet suggests that this increase could be related to decreased food-intake in these patients (72). In contrast, GH responses to hexarelin were similar in normal-weight women with normal food-intake, in patients with anorexia nervosa, in women with amenorrhea due to weight loss and the women on a hypocaloric diet. These observations may imply that, in clear contrast to the GHRH stimulus, GH responses to GH secretagogues are poorly dependent on the metabolic state of the subjects (73).

Also of interest was the assessment of GH responses to a sequential test performed using the administration of hexarelin as a first stimulus followed 120 min later by GHRH. In normal subjects the administration of hexarelin elicited a normal GH response, and completely inhibited the response to GHRH administered 120 min later (Figure 4) (74). This heterologous desensitisation was not observed when the sequential test was performed in a large group of 14 patients with anorexia nervosa (73). The absence of hexarelin-induced desensitisation was not due to the low calorie intake of the patients with anorexia nervosa, since normal weight controls after 72 hours of fasting, or women with secondary amenorrhea due to voluntary weight loss showed a similar pattern of response to the one observed in normal women (65). The basic mechanisms underlying this peculiar absence of hexarelin-induced desensitisation are still unclear, as is whether this alteration may play a role in the elevated GH secretion observed in these patients. Further studies are needed in order to assess whether this sequential test may be of use in the diagnosis and follow-up of patients with this disease.

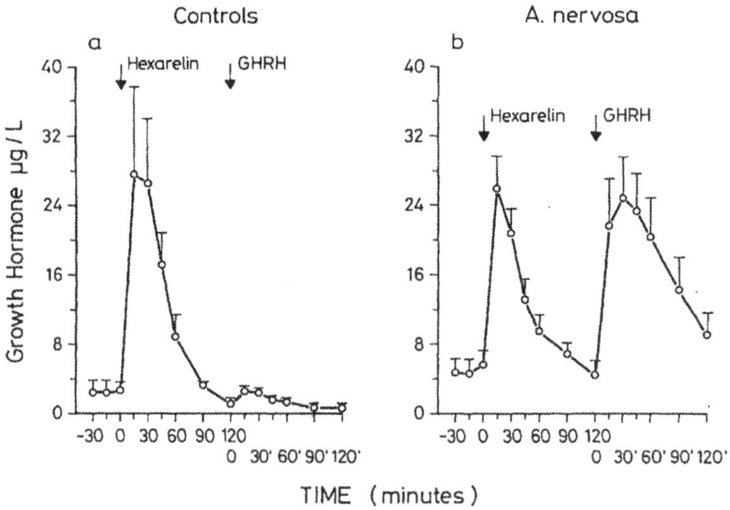

Figure 4. Serum GH levels (mean ± SE) after administration of hexarelin (1 µg/kg i.v.) at 0 min, followed by GH releasing hormone (GHRH) (1 µg/kg i.v.) 120 min later (0) in (a) six normal-weight women acting as control; (b) 14 women with anorexia nervosa. Ref. (73).

ACROMEGALY

Although considerable advances have been made in the understanding of the aetiology and clinical management of GH-secreting pituitary tumours, some pathophysiological aspects remain unknown. For example, the paradoxical responses to a variety of hormonal and metabolic stimuli, such as glucose, TRH, and dopamine agonists and its development in the process of autonomy of the GH-secreting adenoma from its hypothalamic control. Taking into account the relevant role played by GHS in the control of GH secretion in normal subjects, its action in acromegalic patients was assessed. Furthermore, the presence of GHS-receptor mRNA transcripts and *in vitro* actions of GHS in cultured tumour cells obtained after surgery have been reported (see chapter by A. Grossman in this book).

In a similar manner to normal subjects, patients with active acromegaly exhibited a brisk response after challenge with either GHRP-6 alone or combined administration of GHRH + GHRP-6 (75–77). The only noticeable differences were that the GH responses were enhanced compared with those in normal subjects, a fact that is not at all surprising, considering the large number of somatotroph cells, and that in these patients no potentiating effect was observed. The combined stimulus was not different from the arithmetical addition (sum) of each hormone. These data suggest that for the potentiating effect of both compounds to occur, a normal hypothalamo–pituitary connection must be present (75).

Interestingly, on an individual basis, when a plot was performed for each subject with the responses to GHRH and GHRP-6, two different patterns appeared. Some subjects, either acromegalic or controls, showed roughly similar responses to GHRH and GHRP-6, with the values distributed on the bisector sides. Similarly interesting was the fact that a large group of subjects (seven controls and four acromegalic patients) displayed a modest response to GHRH, whereas they showed a greater response after being challenged with GHRP-6. It is not known at present whether the acromegalic patients who presented a poor GH response to GHRH and an important one to GHRP-6 belong to the group of patients with tumour-expressing mutations of GTP-binding proteins (Gsp oncogene), characterised by high sensitivity to somatostatin and poor response to GHRH (78). The scarce GH response after GHRH treatment in these later tumours is not due to a defective GHRH receptor, but, on the contrary, to its permanent activation. This fact is relevant, as in somatotrophs with defective GHRH receptors, GHRP-6 is ineffective. Beside the possible interpretations, it could be interesting to incorporate the response to GHRP-6 in the phenotypic pattern of a subgroup of GH-secreting tumours.

PROLACTINOMAS

GH release after either GHRP-6 alone or GHRH plus GHRP-6 was fully preserved in patients with microprolactinomas, and did not differ before and after treatment with bromocriptine. Thus these data indicate that alterations in PRL secretion do not influence GH responses to GHS. In contrast, patients with macroprolactinomas have blunted responses of GH after either GHRH and GHRP-6. Furthermore the synergistic effect usually observed in normal subjects following combined administration of GHRH and GHRP-6 was severely compromised. GH responsiveness to and synergistic interaction between GHRH and GHRP-6 recovered after shrinkage of macroprolactinomas with bromocriptine, although not to the normal range (Figure 5). These data suggest that the main action of GHS on GH release was exerted at hypothalamic level, and that responsiveness is improved by the re-establishment of the hypothalamo–pituitary connections (79). Nevertheless, it is intriguing that despite long-term monitoring of some patients after treatment, over a period of two years, they still failed to fully recover normal responsiveness. It is possible that this could be due to a permanent damage of the hypothalamo–pituitary axis in these patients, which could result in a state of partial GH deficiency (79,80). Further studies are needed to answer this question.

SUMMARY

When studying GH secretion in different clinical settings, i.e. natural models of deranged GH secretion, there is a clear impression that there is a difference in GH responses to GHRH and GHRPs in the sense that there is a large variability in the stimulatory action of GHRH contrasted with the greater reproducibility of GH response to GHS. In different metabolic states, e.g obesity and anorexia nervosa, the GH response after GHRH is more impaired than after GHRPs stimulation. On the other hand, in different neuroendocrine pathologies the GH response after GHRPs is more impaired than after GHRH. Thus both secretagogues provide separate information on the physiological state of the somatotrophs, and both should be used in clinical practice. Finally, the combined administration of GHRH

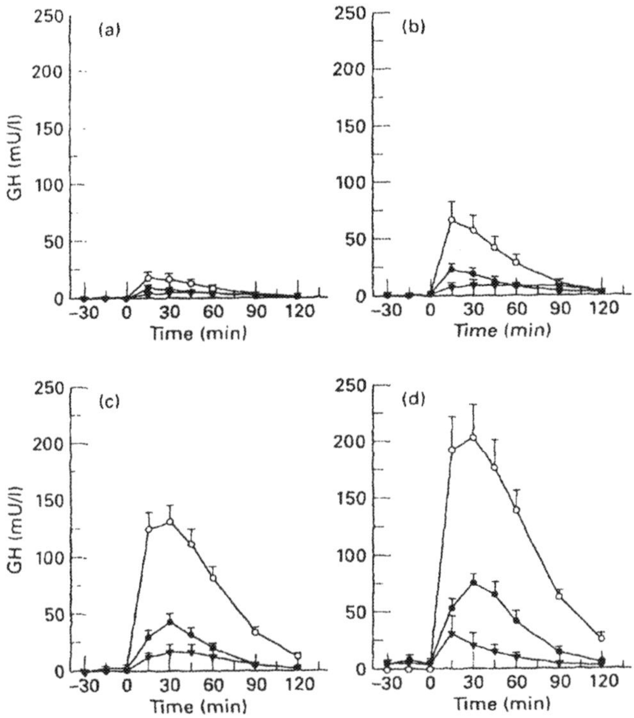

Figure 5. GH response to GH secretagogues in patients with macroprolactinomas (a) before and (b) after bromocriptine treatment, and patients with microprolactinomas (c) before and (d) after bromocriptine. (▼) GHRH, (●) GHRP-6 and (○) GHRH + GHRP-6. Ref. (79).

and GHRP-6, or similar compounds, appears to be the most potent stimulus in terms of GH release in man. Therefore, taking into account that this test is almost completely devoid of side effects, it offers a great reliability in the assessment of pituitary GH reserve. The possibility of using the combined administration of GHRH + GHRP-6 test in the diagnosis of adult-onset GH-deficiency should be considered.

ACKNOWLEDGEMENTS.

This work was supported by grants from the Fondo de Investigación Sanitaria, Spanish Ministry of Health and the Xunta de Galicia.

REFERENCES

1. Casanueva, F.F. and Dieguez, C. (1999) Growth hormone secretagogues: Physiological role and clinical utility. Trends in Endocrinology and Metabolism 10, 30–38.
2. Dieguez, C., Page, M. and Scanlon, M. (1988) Growth hormone neuroregulation and its alterations in disease states. Clin. Endocrinol. 28, 109–143.
3. Casanueva, F.F. (1992) Physiology of growth hormone secretion and action. Endocrinol. Metab. Clin. N. Am. 21, 483–517.

4. Dieguez, C. and Casanueva, F.F. (1995) Influence of metabolic substrates and obesity on growth hormone secretion. Trends in Endocrinology and Metabolism 6, 55–59.
5. Momany, F.A., Bowers, C.Y., Reynolds, G.A., Chang, D., Hong, A. and Newlander, K. (1980) Design, synthesis, and biological activity of peptides which release growth hormone *in vitro*. Endocrinology 108, 31–39.
6. Momany, F.A., Bowers, C.Y., Reynolds, G.A., Hong, A. and Newlander, K. (1984) Conformational energy studies and *in vitro* and *in vivo* activity data on active growth hormone releasing peptide. Endocrinology 114, 1531–1536.
7. Bowers, C.Y. (1993) GH releasing peptides — structure and kinetics. J. Pediatr. Endocrinol. 6, 21–31.
8. Bowers, C.Y., Momany, F.A., Reynolds, G.A. and Hong, A. (1984) On the *in vitro* and *in vivo* activity of a new synthetic hexapeptide that acts on the pituitary to specifically release growth hormone. Endocrinology 114, 1537–1545.
9. Bowers, C.Y., Sartor, A.O., Reynolds, G.A. and Badger, T.M. (1991) On the actions of the growth hormone-releasing hexapeptide, GHRP. Endocrinology 128, 2027–2035.
10. Smith, R.G., Cheng, K., Schoen, W.R., Pong, S.-S., Hickey, G.J., Jacks, T.M., Butler, B.S., Chan, W.W.-S., Chaung, L.-Y.P., Judit, F., Taylor, A.M., Wyvratt, Jr. M.J. and Fisher, M.H. (1993) A nonpeptidyl growth hormone secretagogue. Science 260, 1640–1643.
11. Wyvratt, Jr. M.J. (1996) Non-peptidyl growth hormone secretagogues. In: Growth hormone secretagogues. B.B. Bercu and R.F. Walker (eds). Springer-Verlag, New York, pp. 103–117.
12. Smith, R.G., Van der Ploeg, L.H.T., Howard, A.D., Feigner, S.D., Cheng, K., Hickey, G.J., Wyvratt, M.J., Fisher, M.H., Nargund, R.P. and Patchett, A.A. (1997) Peptidomimetic regulation of growth hormone secretion. Endocrine Rev. 18, 621–645.
13. Peñalva, A., Pombo, M., Mallo, F., Barreiro, J., Carballo, A., Alvarez, C., Coya, R., Casanueva, F.F. and Dieguez, C. (1993) Mechanism of action of the growth hormone releasing hexapeptide (GHRP-6) on *in vivo* growth hormone secretion. In: Two decades of experience on growth. M. Pombo and R.G. Rosenfeld (eds). Raven Press, New York, pp. 253–260.
14. Mallo, F., Alvarez, C.V., Benitez, L., Burguera, B., Coya, R., Casanueva, F.F. and Dieguez, C. (1993) Regulation of His-DTrp-Ala-Trp-DPhe-Lys-NH_2 (GHRP-6)-induced GH secretion in the rat. Neuroendocrinol. 57, 247–256.
15. Walker, R.F., Codd, E.E., Barone, F.C., Nelson, A.H., Goodwin, T. and Campbell, S.A. (1990) Oral activity of the growth hormone releasing peptide His-DTrp-Ala-Trp-DPhe-Lys-NH_2 in rats, dogs and monkeys. Life Sci. 47, 29–36.
16. Muruais, J., Peñalva, A., Dieguez, C. and Casanueva, F.F. (1993) Influence of endogenous cholinergic tone and α-adrenergic pathways on the growth hormone responses to His-DTrp-Ala-Trp-DPhe-Lys-NH_2 in the dog. J. Endocrinol. 138, 211–218.
17. Ilson, B.E., Jorkasky, D.J., Curnow, R.T. and Stole, R.M. (1989) Effect of a new synthetic hexapeptide to selectively stimulate growth hormone release in healthy human subjects. J. Clin. Endocrinol. Metab. 69, 212–214.
18. Thorner, M.O., Vance, M.L., Rogol, A.D., Blizzard, R.M., Veldhuis, J.D., Van Cauter, D., Copinschi, G., Bowers, C.Y. (1990) Growth hormone-releasing hormone and growth hormone-releasing peptide as potential therapeutic modalities. Acta Paediatr. Scand. 367, 29–32.
19. Ghigo, E., Arvat, E., Gianotti, L., Imbimbo, B.P., Lenaerts, V., Deghenghi, R. and Camanni, F. (1994) Growth hormone-releasing activity of hexarelin, a new synthetic hexapeptide, after intravenous, subcutaneous, intranasal, and oral administration in man. J. Clin. Endocrinol. Metab. 78, 693–698.
20. Bowers, C.Y., Reynolds, G.A., Durham, D., Barrera, C.M., Pezzoli, S.S. and Thorner, M.O. (1990) Growth hormone (GH)-releasing peptide stimulates GH release in normal men and acts synergistically with GH-releasing hormone. J. Clin. Endocrinol. Metab. 79, 975–982.
21. Howard, A.D., Feighner, S.C., Cully, D.F., Arena, J.P., Liberator, P.A., Rosemblum, C.I. et al. (1996) A receptor in pituitary and hypothalamus that functions in growth hormone release. Science 273, 974–976.

22. Cood, E.E.C., Shu, A.I.L. and Walker, R.F.C. (1989) Binding of a growth hormone releasing hexapeptide to specific hypothalamic and pituitary binding sites. Neuropharmacology 28, 1339–114.
23. Shimon, I., Yan, X. and Melmed, S. (1998) Human fetal pituitary expresses functional growth hormone-releasing peptide receptors. J. Clin. Endocrinol. Metab. 83, 174–178.
24. Tannenbaum, G.S., Lapointe, M., Beaudet, A. and Howard, A.D. (1998) Expression of GH secretagogue-receptors by GH-releasing hormone neurons in the mediobasal hypothalamus. Endocrinology 139, 4420–4423.
25. Muccioli, G. et al. (1998) Tissue distribution of GHRP receptors in humans. Proceedings of 6th European Congress of Endocrinology, Sevilla, Spain. Abstr. OR 17-7.
26. Rasmussen, M.H., Sogaard, B., Ynddal, L., Groes, L., Helmgaard, L. and Nordholm, L. (1998) Ipamorelin — a very potent novel growth hormone secretagogue. Proceedings of 80th Annual Meeting of the American Endocrine Society. New Orleans, USA. Abstr. P1-185.
27. Chen, C., Wu, D. and Clarke, I.J. (1996) Signal transduction system employed by synthetic GH-releasing peptides in somatotrophs. J. Endocrinol. 148, 381–386.
28. Cheng, K., Chan, W., Barreto, A., Convey, E. and Smith, R. (1989) The synergistic effects of His-D-TrpAla-TrpD-Phe-Lys-NH_2 on growth hormone (GH)-releasing factor stimulated GH release and intracellular adenosine 3,5-monophosphate accumulation in rat pituitary cell culture. Endocrinology 124, 2791–2798.
29. Goth, M.L., Lyons, C.E., Canny, B.J. and Thorner, M.O. (1992) Pituitary adenylate-cyclase activating polipeptide, GH-releasing peptide and GHRH stimulate GH release through distinct pituitary receptors. Endocrinology 130, 939–944.
30. Bowers, C.Y. (1996) Xenobiotic growth hormone secretagogues: growth hormone releasing peptides. In: Growth hormone secretagogues. B.B. Bercu and R.F. Walker (eds). Springer-Verlag, New York, pp. 9–28.
31. Deghenghi, R. (1996) Growth hormone releasing peptides. In: Growth hormone secretagogues. B.B. Bercu and R.F. Walker (eds). Springer-Verlag, New York, pp. 85–102.
32. Fletcher, T.P., Thomas, G.B., Willoughby, J.O. and Clarke, I.J. (1994) Constitutive growth hormone secretion in sheep after hypothalamo–pituitary disconnection and the direct *in vivo* pituitary effect of Growth Hormone Releasing Peptide-6. Neuroendocrinology, 57, 247–256.
33. Popovic, V., Damjanovic, S., Micic, D., Djurovic, M., Dieguez, C. and Casanueva, F.F. (1995) Blocked growth hormone-releasing peptide (GHRP-6)-induced GH secretion and absence of the synergic action of GHRP-6 plus GH-releasing hormone in patients with hypothalamo–pituitary disconnection: evidence that GHRP-6 main action is exerted at the hypothalamic level. J. Clin. Endocrinol. Metab. 80, 942–947.
34. Kamegai, J., Wakabayashi, I., Miyamoto, K., Unterman, T.G., Kineman, R.D. and Frohman, L.A. (1998) Growth hormone-dependent regulation of pituitary GH secretagogue receptor (GHS-R) mRNA levels in the spontaneous dwarf rat. Neuroendocrinology 68, 312–318.
35. Pombo, M., Barreiro, J., Peñalva, A., Peino, R., Dieguez, C. and Casanueva, F.F. (1995) Absence of growth hormone (GH) secretion after the administration of either GH-releasing hormone (GHRH), GH-releasing peptide (GHRP-6) or GHRH plus GHRP-6, in children with neonatal pituitary stalk transection. J. Clin. Endocrinol. Metab. 80, 3180–3184.
36. Cella, S.G., Locatelli, V., Mennini, T. et al. (1990) Deprivation of growth hormone-releasing hormone early in the rat's neonatal life permanently affect somatotropic function. Endocrinology 127, 1625–1634.
37. Shalet, S.M., Toogood, A., Rahim, A. and Brennan, B.M.D. (1998) The diagnosis of growth hormone deficiency in children and adults. Endocrin. Rev. 19, 203–223.
38. Zadik, Z., Chalew, S.A., Raiti, S. and Kowarski, A.A. (1985) Do short children secrete insufficient growth hormone? Pediatrics 76, 355–360.
39. Zadik, Z., Chalew, S.A., Gilula, Z. and Kowarski, A.A. (1990) Reproducibility of growth hormone testing procedures: comparison between 24-hour integrated concentration and pharmacological stimulation. J. Clin. Endocrinol. Metab. 71, 1127–1130.

40. Rose, S.R., Ross, J.L., Uriarte, M., Barnes, K.M., Cassorla, F.G. and Cutler, G.B. (1988) The advantage of measuring stimulated as compared with spontaneous growth hormone levels in the diagnosis of growth hormone deficiency. N. Engl. J. Med. 319, 201–207.
41. Peñalva, A., Pombo, M., Carballo, A., Barreiro, J., Casanueva, F.F. and Dieguez, C. (1993) Influence of sex, age and adrenergic pathways on GH responses to GHRP-6. Clin. Endocrinol. 38, 87–91.
42. Bowers, C.Y., Alster, D.K. and Frentz, J.M. (1992) The growth hormone-releasing activity of a synthetic hexapeptide in normal men and short statured children after oral administration. J. Clin. Endocrinol. Metab. 74, 292–298.
43. Pombo, M., Barreiro, J., Peñalva, A., Mallo, F., Casanueva, F.F. and Dieguez, C. (1995) Plasma growth hormone response to growth hormone-releasing hexapeptide (GH-RP-6) in children with short stature. Acta Paediatr. 84, 904–908.
44. Casanueva, F.F. (1996) Diagnosis of growth hormone deficiency in adulthood. Eur. J. Endocrinol. 135, 168–170.
45. Jorgensen, J.O.L., Pedersen, S.A., Thuesen, I. et al. (1989) Beneficial effects of growth hormone treatment in GH-deficient adults. Lancet ii, 1221–1224.
46. Salomon, F., Cuneo, R.C., Hesp, R., Sonksen, P.H. (1989) The effects of treatment with recombinant human growth hormone on body composition and metabolism in adults with growth hormone deficiency. N. Engl. J. Med. 321, 1797–1803.
47. Rudman, D., Feller, A.G., Nagraj, H.S. et al. (1990) Effects of human growth hormone in men over 60 years old. N. Engl. J. Med. 323, 1–6.
48. Cuneo, R.C., Salomon, F., McGauley, G.A. and Sonksen, P.H. (1992) The growth hormone deficiency syndrome in adults. Clin. Endocrinol. 37, 387–397.
49. Withehead, H.M., Boreham, C., Mcllrath, E.M., Sheridan, B., Kennedy, L., Atkinson, A.B. et al. (1992) Growth hormone treatment of adults with growth hormone deficiency: results of a 13-month placebo controlled cross-over study. Clin. Endocrinol. 36, 45–52.
50. Lamberts, S.W.J., Valk, N.K. and Binnerie, A. (1992) The use of growth hormone in adults: a changing scene. Clin. Endocrinol. 37, 111–115
51. Moller, J., Jorgensen, J.O.L., Lauersen, T. et al. (1993) Growth hormone dose regimens in adult GH deficiency: effect on biochemical markers and metabolic parameters. Clin. Endocrinol. 39, 403–408.
52. Ross, R.J.M. (1993) Growth hormone replacement in adults: what dose? Clin. Endocrinol. 39, 401–402.
53. Hoffman, D.M., O'Sullivan, A.J., Baxter, R.C. and Ho, K.K.Y. (1994) Diagnosis of growth hormone deficiency in adults. Lancet 343, 1064–1068.
54. Leal-Cerro, A., Garcia, E., Astorga, R., Casanueva, F.F. and Dieguez, C. (1995) Growth hormone (GH) responses to the combined administration of GH-releasing hormone plus GH-releasing peptide 6 in adults with GH deficiency. Eur. J. Endocrinol. 132, 712–715.
55. Ghigo, E., Arvat, E., Gianotti, L., Ramunni, J., DeVito, L., Maccagno, B., Grotolli, S., Camanni, F. (1996) Human aging and the GH.IGF-1 Axis. J. Pediatric Endocrinol. Metab. 9, 271–278.
56. Micic, D., Popovic, V., Kendereski, A., Macut, D.J., Casanueva, F. and Dieguez, C. (1995) Growth hormone (GH) secretion after the administration of GHRP-6 or GHRH plus GHRP-6 does not decline in late adulthood. Clin. Endocrinol. 42, 191–194.
57. Cordido, F., Casanueva, F.F. and Dieguez, C. (1989) Cholinergic receptors activation by pyridostigmine restores growth hormone (GH) responsiveness to GH releasing hormone administration in obese subjects. J. Clin. Endocrinol. Metab. 68, 290–293.
58. Cordido, F., Dieguez, C. and Casanueva, F.F. (1990) Effect of central cholinergic neurotransmission enhancement by pyridostigmine on the growth hormone secretion elicited by clonidine, arginine or hypoglycemia in normal and obese subjects. J. Clin. Endocrinol. Metab. 70, 1361–1370.
59. Cordido, F., Peñalva, A., Dieguez, C. and Casanueva, F.F. (1993) Massive discharge in obese subjects after the combined administration of growth hormone releasing hormone and GHRP-6.

Evidence for a marked somatotroph secretory capability in obesity. J. Clin. Endocrinol. Metab. 76, 819–823.
60. Cordido, F., Peñalva, A., Peinó, R., Casanueva, F.F. and Dieguez, C. (1995) Effect of combined administration of growth hormone (GH)-releasing hormone, GH-releasing peptide-6 and pyridostigmine in normal and obese subjects. Metabolism 44, 1–4.
61. Peñalva, A., Carballo, A., Pombo, M., Casanueva, F.F. and Dieguez, C. (1993) Effect of growth hormone (GH)-releasing hormone (GHRH), atropine, pyridostigmine and hypoglycemia on GHRP-6-induced GH secretion in man. J. Clin. Endocrinol. Metab. 76, 168–17.
62. Alvarez, C., Mallo, F., Burguera, B., Cacicedo, L., Dieguez, C. and Casanueva, F.F. (1991) Evidence for a direct pituitary inhibition by free fatty acids of *in vivo* growth hormone responses to growth hormone releasing hormone in the rat. Neuroendocrinology 53, 185–189.
63. Cordido, F., Peinó, R., Peñalva, A., Alvarez, C.V., Casanueva, F.F. and Dieguez, C. (1996) Impaired growth hormone secretion in obese subjects is partially reversed by acipimox-mediated plasma free fatty acid depression. J. Clin. Endocrinol.Metab. 81, 914–918.
64. Peinó, R., Cordido, F., Peñalva, A., Alvarez, C.V., Dieguez, C. and Casanueva, F.F. (1996) Acipimox-mediated plasma free fatty acid depression *per se* stimulates growth hormone (GH) secretion in normal subjects and potentiates the response to other GH-releasing stimuli. J. Clin. Endocrinol. Metab. 81, 909–913.
65. Leal-Cerro, A., Pumar, A., Villamil, F., Astorga, R., Dieguez, C. and Casanueva, F.F. (1993) Growth hormone releasing hormone priming increases growth hormone secretion in patients with Cushing's syndrome. Clin. Endocrinol. 38, 399–403.
66. Leal-Cerro, A., Pereira, J.L., Garcia-Luna, P.P., Astorga, R., Cordido, F., Casanueva, F.F. and Dieguez, C. (1990) Effect of the enhancement of endogenous cholinergic tone with pyridostigmine on growth hormone (GH) responses to GH-releasing hormone in patients with Cushing's syndrome. Clin. Endocrinol. 33, 291–295.
67. Leal-Cerro, A., Pumar, A., Garcia-Garcia, E., Dieguez, C. and Casanueva, F.F. (1994) Inhibition of growth hormone release after the combined administration of GHRH and GHRP-6 in patients with Cushing's syndrome. Clin. Endocrinol. 41, 649–654.
68. Brown, G.M., Garfinkel, P.E., Jeunievic, N., Moldofski, H. and Stancer, H.C. (1977) Endocrine profiles in anorexia nervosa. In: Anorexia Nervosa. R. Vigerski (ed). Raven Press, New York, pp. 123–131.
69. Vigerski, R.M. and Loriaux, D.L. (1977) Anorexia nervosa as a model of hypothalamic dysfunction. In: Anorexia Nervosa. R. Vigerski (ed). Raven Press, New York, pp. 132–134.
70. Casanueva, F.F., Borras, C.G., Burguera, B., Muruais, C., Fernandez, M. and Devesa, J. (1987) Steroids and neuroendocrine function in anorexia nervosa. J. Steroid Biochem. 27, 635–640.
71. Marinis, L., Folli, 0., D'Amico, C., Manicic, A., Samba, P., Tofani, A., Oradei, A. and Barbarino, A. (1988) Differential effects of feeding on the ultradian variation of the growth hormone (GH) response to GH-releasing hormone in normal subjects and patients with obesity and anorexia nervosa. J. Clin. Endocrinol. Metab. 66, 598–604.
72. Kelijman, M., Frohman, L.A. (1988) Enhanced growth hormone (GH) responsiveness to GH-releasing hormone after dietary manipulation in obese and nonobese subjects. J. Clin. Endocrinol. Metab. 66, 489–494.
73. Popovic, V., Micic, D., Djurovic, M., Obradovic, S., Casanueva, F.F. and Dieguez, C. (1997) Absence of desensitization by hexarelin to subsequent GH releasing hormone-mediated GH secretion in patients with anorexia nervosa. Clin. Endocrinol. 46, 539–543.
74. Mici, D., Popovic, V., Kendereski, A., Peino, R., Dieguez, C. and Casanueva, F.F. (1996) The sequential administration of growth hormone (GH)-releasing hormone followed 120 min later by hexarelin, as an effective test to assess the pituitary GH reserve in man. Clin. Endocrinol. 45, 543–551.
75. Popovic, V., Damjanovic, S., Micic, D., Petakov, M., Dieguez, C. and Casanueva, F. (1994) Growth hormone (GH) secretion in active acromegaly after the combined administration of GH-releasing hormone and GH-releasing peptide-6. J. Clin. Endocrinol. 79, 456–460.

76. Hanew, K., Ulsumi, A., Sugawara, A., Shimiyu, Y. and Abe, K. (1994) Enhanced GH responses to combined administration of GHRP and GHRH in patients with acromegaly. J. Clin. Endocrinol. Metab. 78, 509–512.
77. Ciccarelli, E., Grottoli, S., Razzore, P., Arvat, E., Gianotti, L., Deghendi, R., Camanni, F. and Ghigo, E. (1995) Effects of hexarelin, a synthetic GH-releasing peptide, on prolactin secretion in acromegalic and in hyperprolactinemia patients. J. Endocrinol. 144, 319.
78. Spada, A. and Vallar, L. (1993) Gs mutations in pituitary tumors. In: Molecular and clinical advances in pituitary disorders. S. Melmed (ed). Endocrine Res. Ed., Los Angeles, pp. 29–34.
79. Popovic, V., Simic, C., Llic, L.J., Micic, D., Damjanovic, S., Djurovic, M., Obradovic, S., Dieguez, C. and Casanueva, F.F. (1998) Growth hormone secretion elicited by GHRH, GHRP-6 or GHRH plus GHRP-6 in patients with microprolactinoma and macroprolactinoma before and after bromocriptine therapy. Clin. Endocrinol. 48, 103–108.
80. Ciccarelli, E., Grotolli, S., Gianotti, E., Camanni, G. and Ghigo, E. (1996) Hexarelin, a synthetic growth hormone releasing peptide, stimulates prolactin secretion in acromegalic but not in hyperprolactinemic patients. Clin. Endocrinol. 44, 67–71.

Growth Hormone Secretagogues
Edited by E. Ghigo, M. Boghen, F.F. Casanueva and C. Dieguez

Chapter 18

Growth Hormone Secretagogues in Critical Illness

GREET H. VAN DEN BERGHE

Department of Intensive Care Medicine, University Hospital Gasthuisberg, University of Leuven, B-3000 Leuven, Belgium

THE SOMATOTROPIC AXIS IN CRITICAL ILLNESS

The human body responds to life-threatening insults such as shock, sepsis, multiple trauma or extensive burns with activated lipolysis, gluconeogenesis, proteolysis and inflammation (1,2). These adaptive changes are thought to provide host defence and metabolic substrates for survival. Despite the adequacy of these defence mechanisms, the above-mentioned conditions used to be lethal in the pre-ICU era. Although the development of modern intensive care enabled resuscitation of patients suffering from these conditions, it still cannot prevent the development of a chronic, intensive care-dependent phase, which is hallmarked on one hand by the inability of feeding to restore protein content in vital organs and tissues while re-esterification of fatty acids allows fat stores to accrue (3,4) and, on the other hand, by impaired capacity of the immune system to respond appropriately to an additional toxic or infectious challenge (5).

Anterior pituitary function, known to play a crucial role in normal metabolic and immunological homeostasis, changes during the course of critical illness. The alterations documented to occur immediately after the onset of life-threatening disease or trauma appear to be quite different from those observed in the chronic intensive care-dependent phase. This type of biphasic endocrine response to severe and sustained stress has recently been described in detail for the somatotropic axis (6–10).

During the first hours or days after trauma, surgery or the onset of an infectious disease, the circulating growth hormone (GH) levels are elevated (11) (Figure 1). The normal nocturnal pattern of GH serum concentrations, consisting of peaks alternating with virtually undetectable troughs, is altered by acute illness: the amount of GH released from the somatotropes is increased, peak GH levels as well as interpulse concentrations are high (12,13) and the GH pulse frequency is somewhat elevated (Figure 1). It is still unclear which factor ultimately controls this stimulation of GH release from the somatotropes in response

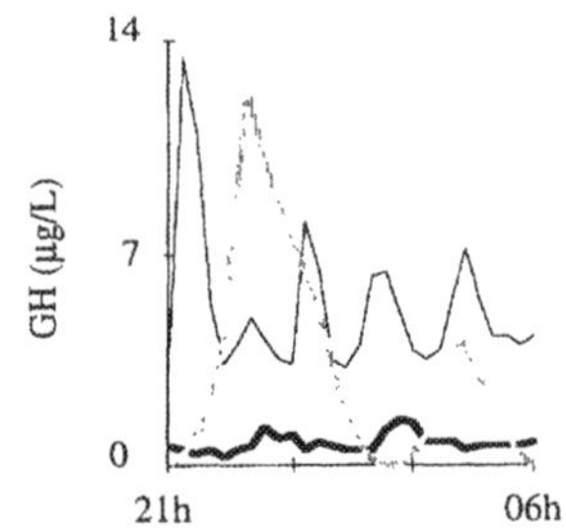

Figure 1. Illustration of a normal (gray line) nocturnal GH concentration profile and the differences between the initial phase of critical illness (first hours to a few days after onset, thin black line) and the prolonged intensive care-dependent phase (thick black line). Reproduced with permission from Ref. (46).

to stress. More frequent withdrawal of the inhibitory somatostatine (SRIH) and/or an increased availability of stimulatory hypothalamic GH-releasing factors could hypothetically be involved. Secondly, serum concentrations of insulin-like growth factor-I (IGF-I) are low (12–15). The concurrence of elevated GH and low IGF-I levels has been interpreted as peripheral resistance to GH, that may be related to decreased GH-receptor expression (15). Thirdly, there are changes in the circulating IGF-binding proteins, which regulate IGF-I plasma half-life and bioavailability (16). The low serum concentrations of IGF-I are associated with low levels of IGFBP-3 and acid-labile subunit (14,15,17); the synthesis of these three polypeptides is normally upregulated by GH and, together, they form a 150 kD ternary complex in the circulation (16). In the acute phase of severe illness, there is increased presence of IGFBP-3-protease activity in plasma thought to result in increased dissociation of IGF-I from the ternary complex and a shortening of IGF-I plasma half-life (14,17). IGFBP-1, which normally binds only a small amount of IGF-I as compared to IGFBP-3, remains in the circulation either in elevated or normal concentrations (18,19).

Inflammatory cytokines may be among the mediators of the aforementioned changes. In addition, nutritional factors may be involved, as most conditions of acute stress are accompanied by starvation or at least a degree of protein malnutrition (15, 20–22). It has been hypothesised — but it remains unproved — that the abundantly released GH during acute stress exerts direct lipolytic, insulin-antagonizing and proinflammatory actions, while its indirect somatotropic effects are attenuated. This constellation of changes, in balance with an activation of the adrenocortical axis, has been interpreted as a mechanism to provide essential substrates for survival, to postpone anabolism and to contribute to host defence (10).

If recovery does not follow within days, critical illness becomes protracted and intensive care support is further needed, often for weeks or months. This chronic intensive care-dependent phase is characterized by a different set of changes in the somatotropic axis (6–10). Firstly, the pattern of GH secretion has been characterized as having a severely reduced pulsatile fraction (Figure 1) whereas the non-pulsatile or tonic fraction is (still) somewhat elevated and the number of pulses is still relatively high (8,9). This pattern results in mean nocturnal GH serum concentrations that are low–normal (8,9). Moreover, GH appears to be released in an erratic fashion, as indicated by a high calculated approximate entropy (8,9). This relative suppression of GH release seems to occur only after a certain duration of the critical condition rather than being dependent on the type or severity of the underlying disease. Secondly, the reduced amount of GH that is released in pulses was shown to be related to the low circulating levels of IGF-I, IGFBP-3 and acid-labile subunit (8,9). Indeed, it has been shown that when pulsatile GH secretion falls below a critical threshold during the chronic phase of illness, circulating IGF-I and acid-labile subunit progressively decrease over time (9). As low serum IGF-I and — even more so — low levels of acid-labile subunit are markers of protein wasting in this condition (17,23), these findings suggest that a relative hyposomatotropism may participate in the pathogenesis of the wasting syndrome distinctively in the chronic phase of protracted critical illness.

GH-SECRETAGOGUES IN PROTRACTED CRITICAL ILLNESS

Hypothalamic GH-releasing hormone (GHRH) is currently still considered to be the major endogenous and specific secretagogue for GH (24). Recently however, a series of synthetic peptides (GH-releasing peptides or GHRPs) and non-peptide agents have been shown to potently and specifically release GH (25–27), through a specific G-protein coupled receptor located in the hypothalamus and the pituitary (28). It now appears plausible that the — still unknown — endogenous ligand of the GHRP-receptor is another key factor in the physiological regulation of GH secretion.

Studying the GH responses to the administration of GH-secretagogues (GHRH and GHRP) enables — to a certain extent — differentiation between a primarily pituitary or hypothalamic origin of the relatively impaired GH release present in the chronic, intensive care-dependent phase of critical illness.

In a first prospective, controlled, cross-over study involving 40 patients (7), the responses to the IV administration of boluses of GHRH (1 μg/kg), GHRP-2 (1 μg/kg) and the combination of both secretagogues have been evaluated (7). A striking GH response to GHRP-2 was observed, consisting of a mean ± SEM peak GH of 51 ± 9 μg/L in older (66 ± 2 years) patients and of 102 ± 26 μg/L in younger (28 ± 3 years) patients (Figure 2). The mean GH response to GHRP-2 was more than 4-fold higher than to GHRH, the latter being subnormal in the majority of the patients. In addition, the response to GHRH administered together with GHRP-2 was found to be synergistic, as it was another 2.5-fold higher than to GHRP-2 alone.

These high GH responses to bolus injections of GHRP-2 and of GHRH + GHRP-2 in critical care conditions are remarkable, as they appear to overcome a number of normally inhibiting factors, namely, a non-fasting state; elevated serum cortisol, high glucose, insulin and fatty acid levels; low serum testosterone in males; and sometimes high age and obesity. Moreover, they credibly exclude a lack of pituitary capacity to *synthetise* GH as the ultimate mechanism underlying the blunted GH secretion during protracted critical illness. In fact, they provide indirect evidence for sufficient availability of endogenous GHRH to maintain GH synthesis (29–31).

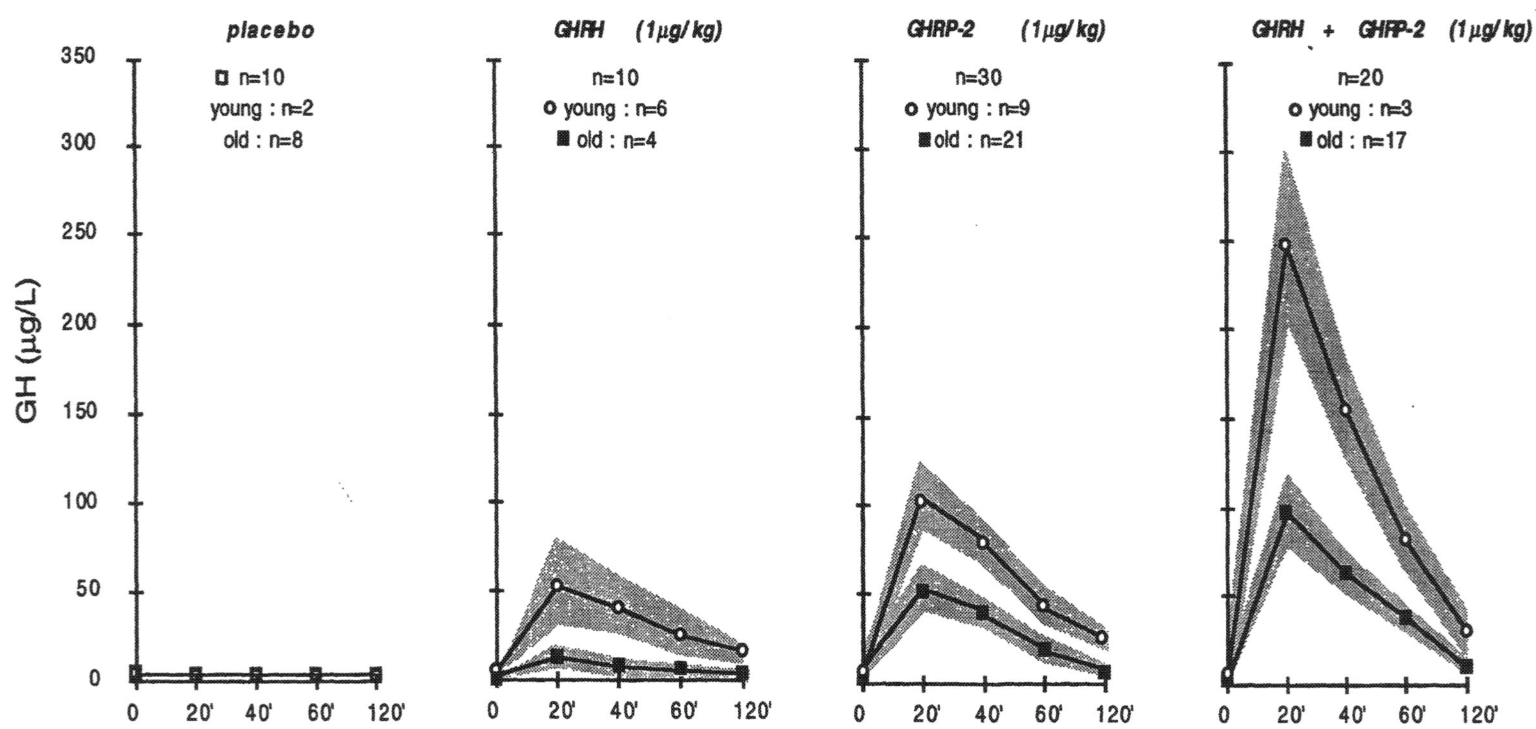

Figure 2. The somatotropes appear to readily release large amounts of GH upon stimulation with GH-secretagogues. Reproduced with permission from Ref. (7). Serum peak GH responses to placebo, GHRH (1 µg/kg IV), GHRP-2 (1 µg/kg IV) and GHRH + GHRP-2 (1 + 1 µg/kg IV) obtained in 11 younger (age 28 ± 2.6 years, open circles) and 29 older, critically ill patients (age 66 ± 2 years, squares). Lines interlink mean values, shaded area indicates mean ± SEM.

GHRPs are thought to reflect the actions of a still unidentified endogenous GHRP-like ligand and to act complementary with GHRH to release GH, in part through a functional SRIH antagonism, either directly or indirectly through the release of a hypothalamic U-factor (31–32). Experimentally, GHRP-induced GH release is GHRH-dependent and can be inhibited by SRIH (32). Consequently, one of the mechanisms possibly underlying the pronounced GH responsiveness in critical illness is low endogenous SRIH activity, which is perhaps in part a consequence of low circulating IGF-I (32–34). However, this mechanism is not evidently reconciled with reduced spontaneous GH pulse amplitude and sometimes virtually complete disappearance of GH pulses during critical illness (8,9). Moreover, in the face of merely a reduced SRIH tone, a higher GH response to GHRH would be expected (32).

Hence, if an endogenous GHRP-like ligand exists, a lack of this putative compound would explain the present findings. Indeed, a high GH response to GHRP-2 is conceivable in a constellation of low endogenous GHRP-like ligand and sufficient GHRH secretion (32). Ultimately, the combination of reduced availability of SRIH and endogenous GHRP-like ligand emerges as a plausible mechanism that clarifies the reduced GH burst amplitude, the increased frequency of spontaneous GH secretory bursts and the elevated interpulse levels as well as the striking responsiveness to GHRP-2 alone or in combination with GHRH, and this without markedly increased responsiveness to GHRH alone (7,18,35).

Two consecutive, controlled and randomized studies involved a total of 46 carefully selected patients suffering from protracted critical illness for at least several weeks (8,9). These patients were studied with overnight blood sampling every 20 minutes for 9 hours during 2 or 3 nights and the effects of continuous IV infusion of GHRH (1 μg/kg/h), GHRP-2 (1 μg/kg/h) and GHRH + GHRP-2 (1 + 1 μg/kg/h) were evaluated. Infusing GHRP-2, and even more so GHRH + GHRP-2, constantly over 21 h was found to substantially amplify pulsatile GH secretion (>6-fold and >10-fold respectively) in this condition of protracted critical illness, without altering the relatively high burst frequency, an effect that lasted up to 45 h (8,9) (Figure 3). The responses to GHRP-2 infusion in these critically ill patients were pronounced compared to the responses to GHRP-6 and to GHRH infusion reported in healthy volunteers (36–38). These high responses to the continuous infusion of hypothalamic releasing factors further support a hypothalamic origin of the relative hyposomatotropism in the chronic phase of critical illness.

In view of the high doses of continuously administered GHRH and GHRP-2, these observations also suggest that the pacemaker for pulsatile GH secretion in this condition is neither endogenous GHRH nor the presumed endogenous GHRP-like ligand, but rather an intermittent pituitary sensitivity to either or both of these releasing factors. SRIH is the most plausible candidate to determine the timing of the GH pulses. A reduced SRIH inhibitory action on the somatotropes is thought to be involved in the elevated pulsatile and basal GH secretion during starvation (39,40). However, in contrast to fasting, critical illness is associated with smaller, not larger, GH secretory bursts. Thus, more frequent withdrawal of SRIH may determine the observed high GH burst frequency, but it is unlikely that alterations in SRIH secretion exclusively explain the complete pattern of low amplitude GH

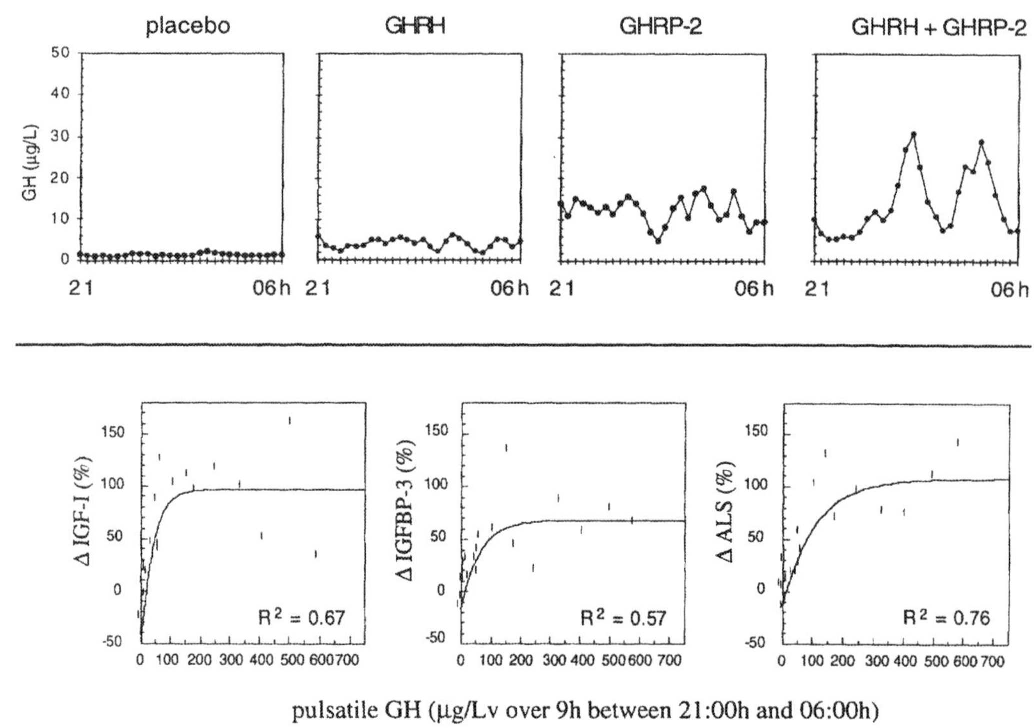

Figure 3. Upper part: Nocturnal serum GH profiles in the prolonged phase of critical illness illustrating the effects of continuous infusion of placebo, GHRH (1 µg/kg/h), GHRP-2 (1 µg/kg/h), or GHRH + GHRP-2 (1 + 1 µg/kg/h). Age range of the patients was 62–85 years; duration of illness was between 13 and 48 days; infusions were started 12 hours before the onset of the respective profiles. Reproduced with permission from Refs. (8) and (9). Lower part: Exponential regression lines have been reported between pulsatile GH secretion and the changes in circulating IGF-I, ALS and IGFBP-3 obtained with 45 h infusion of either placebo, GHRP-2 or GHRH + GHRP-2. They indicate that the parameters of GH responsiveness increase in proportion to GH secretion up to a certain point, beyond which further increase of GH secretion has apparently little or no additional effect. It is noteworthy that the latter point corresponds to a pulsatile GH secretion of approximately 200 µg/Lv over 9 h, or less, a value that can be evoked by the infusion of GHRP-2 alone. In the chronic, non-thriving phase of critical illness, GH-sensitivity is clearly present, in contrast to the acute phase of illness, which is thought to be primarily a condition of GH resistance. Reproduced with permission from Ref. (9).

secretion and pronounced responsiveness in particular to GHRP-2 during protracted critical illness. The observed GH-releasing effect of GHRP-2 infusion on pulsatile GH secretion is in line with the above suggested availability of endogenous GHRH to maintain the synthesis of GH (19,23), and with a lack of the endogenous GHRP-like ligand in the chronic phase of critical illness. In combination with a more frequent withdrawal of the suppressive SRIH tone, these neuroendocrine changes provide an explanation for the described findings, and may represent a pathophysiological basis for not secreting stored GH during critical illness, in contrast to what is observed in acute stress and starvation.

Infusion of GHRP-2 (1 µg/kg/h) and of GHRH + GHRP-2 (1 + 1 µg/kg/h) for 45 h during protracted critical illness evoked a proportionate rise in serum IGF-I (66% and

106%), IGFBP-3 (50% and 56%) and ALS (65% and 97%) indicating peripheral GH-responsiveness (8,9) (Figure 3). Exponential regression lines between pulsatile GH secretion on one hand and markers of GH-responsiveness on the other hand underline the presence of substantial GH-sensitivity, particularly in the range of pulsatile GH secretion that is normally observed during the initial phase of critical illness, a level which apparently can be obtained in the chronic phase by infusing GHRP-2 alone. The presence of considerable responsiveness to reactivated endogenous GH secretion clearly delineates the distinct pathophysiological paradigm of the chronic phase of critical illness, as opposed to the acute phase, which is thought to be primarily a condition of GH-resistance.

Recent data revealed that the responsiveness to GH-secretagogues in this condition is maintained for at least 5 days, with active feed-back inhibition preventing overtreatment, restoring (near) normal levels of IGF(BP)s and evoking an increase in biochemical markers of anabolism in bone tissue (41).

The remarkable observation that a constant IV infusion of a GHRP during *protracted* critical illness suffices to restore the pattern of pulsatile GH secretion observed during the *acute* phase of severe illness, and to evoke considerable IGF-I(BP) responses as long as acute intercurrent complications are absent, suggest a crucial role for the endogenous GHRP-like ligand in the stress response. The initial activation of GH release may be evoked by an increased activity of this endogenous GHRP-like ligand in the presence of more frequent SRIH withdrawal, whereas the relative suppression of GH secretion in the protracted phase of critical illness may be mediated by a deficiency (exhaustion of activation process?; depletion?) of the endogenous GHRP-like ligand within this constellation of low SRIH tone.

From a therapeutic perspective, the aforementioned data provide a sound pathophysiological basis to further explore the safety and efficacy of GH-secretagogue administration as a strategy to counter the overall "wasting syndrome" and, consequently, to actually accelerate the process of recovery from prolonged critical illness. The respect for active feedback inhibition loops and the observation that this strategy allows for peripheral adjustment of metabolic pathways according to the needs determined by the disease process, suggests that the infusion of GH-secretagogues may be a safer strategy than the administration of (high-doses) GH and/or IGF-I at a time when it is impossible to determine the optimal circulating levels of these hormones, particularly in the vulnerable elderly (42).

GH-SECRETAGOGUES AND OTHER ANTERIOR PITUITARY HORMONES IN PROTRACTED CRITICAL ILLNESS

The concept of a biphasic response of the anterior pituitary gland to sustained stress, as observed within the somatotropic axis, was shown to be equally applicable to prolactin (PRL) release, the thyrotropic axis and the gonadotropic axis (10,43,44). In general, the initial response to stress consists of an activation or at least maintenance of anterior pituitary hormone release together with an immediate inactivation of peripheral anabolic pathways. The chronic phase of protracted critical illness is characterized uniformly by a relative impairment of pulsatile release of anterior pituitary hormones, essentially of hypothalamic origin, in relation to reduced release of hormones by target organs.

Although the releasing capacity of GHRPs is thought to be specific for GH, interaction with other anterior pituitary axes has been documented during protracted critical illness.

Bolus injections of GHRP-2 as well as of GHRH in this condition evoked a minute but significant increase in serum cortisol and PRL, whereas GHRH, but not GHRP-2, paradoxically increased TSH and thyroid hormone levels (7).

Continuous infusion of GHRH was found to increase pulsatile thyrotropin (TSH) secretion and circulating levels of thyroid hormones, whereas GHRP-2, when continuously infused alone, slightly suppressed the pulsatile fraction of TSH release in prolonged critical illness (8).

The minute increase in PRL observed with bolus injections of GHRP-2 was maintained with continuous infusion, whereas the small rise in serum cortisol observed with boluses of GH-secretagogues appeared transient, as it was not present when the GH-secretagogues were infused continuously over 21–45 h (8,9,43).

Although continuous infusion of thyrotropin-releasing hormone (TRH) was found to increase nocturnal TSH release and circulating levels of thyroid hormones in the chronic phase of critical illness, the addition of GHRP-2 to the TRH infusion appeared necessary to also activate the pulsatile fraction of TSH secretion and to avoid a rise in serum concentrations of reverse T3 (9) (Figure 4). In addition, PRL secretion was found to be paradoxically suppressed when TRH was infused alone, while co-infusion of GHRP-2 tended to normalize the PRL response to TRH (9). These findings suggest that the endogenous GHRP-like ligand may play a role in the complex orchestration of anterior pituitary hormone release. In line with this reasoning is the observed absence of synchrony among GH, TSH and PRL release during prolonged critical illness, which is clearly abnormal. Continuous GHRP-2 infusion was found to re-synchronize GH, TSH and PRL secretion in this condition, which further supports the hypothesis of a relative deficiency of the putative endogenous GHRP-like ligand during protracted critical illness (45).

CONCLUSION

Administration of GHRP in a continuous infusion is able to transform the endogenously blunted pattern of pulsatile GH release present in the chronic phase of critical illness into a pattern that mimics the one observed in the acute phase of severe illness. These findings support the concept that increased activity of the putative endogenous GHRP-like ligand may be involved in the initial GH release in response to stress whereas a depletion of GHRP could explain the blunting of GH release observed when severe stress is sustained beyond the timeframe foreseen by nature.

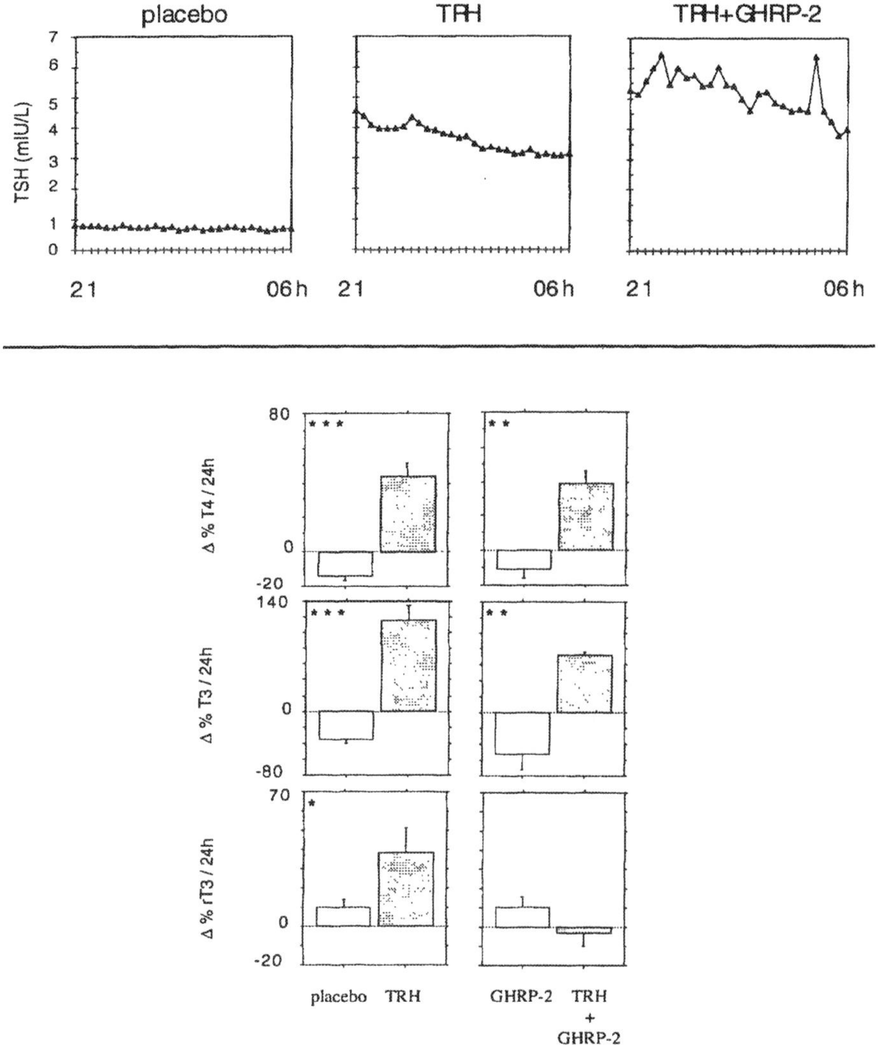

Figure 4. Upper part: Nocturnal serum TSH profiles in the prolonged phase of illness (duration of illness 15–18 days, patient's age 69–80 years) illustrating the effects of continuous infusion of placebo, TRH (1 μg/kg/h), TRH + GHRP-2 (1 μg/kg/h). Although TRH elevated TSH secretion, the addition of GHRP-2 to the TRH infusion appeared necessary to increase its pulsatile fraction. Reproduced with permission from Ref. (9). Lower part: Continuous administration of TRH (1 μg/kg/h), infused alone or together with GHRP-2 (1 + 1 μg/kg/h), induces a significant rise in serum T4 and T3 within 24 h. Reverse T3 is increased after infusion of TRH alone, but not if TRH is coinfused with GHRP-2. Studied patients were ill for 12 to 59 days; age range was 32–87 years. $^{*}P < 0.05$, $^{**}P < 0.001$, $^{***}P < 0.0001$. Reproduced with permission from Ref. (9).

The preserved peripheral GH and IGF-I responsiveness to reactivated GH release in the chronic phase of intensive care-dependent critical illness, at least as long as intercurrent (infectious) complications are absent, indicates that the hypothalamic component of the hyposomatotropism may contribute to the low IGF-I(BP) levels and to the feeding-resistant wasting syndrome in the "long-stay" patients on ICUs. These findings may have important therapeutic consequences. Moreover, infusion of GH-secretagogues may reflect a safer strategy compared to the administration of GH and/or IGF-I, as the presence of feed-back inhibition loops and the possibility of peripheral shifts in hormone metabolism and activity according to the needs determined by intercurrent complications, protects from side effects. It remains to be determined whether treatment with GH-secretagogues selectively in the chronic phase of critical illness will result in the expected beneficial clinical effects and will, ultimately, accelerate the recovery of those patients who need it most.

ACKNOWLEDGEMENTS

This work has been supported by research grants from the Fund for Scientific Research Flanders, Belgium (G.0162.96 and G.3c05.95N) and the Research Council of the University of Leuven (OT 95/24).

REFERENCES

1. Cuthbertson, D.P. (1932) Observations on the disturbance of metabolism produced by injury to the limbs. Q. J. Med. 25, 233–246.
2. Kinney, J.M., Duke, J.H., Long, C.L. and Gump, F.E. (1970) Tissue fuel and weight loss after injury. J. Clin. Pathol. 23, 65–72.
3. Streat, S.J., Beddoe, A.H. and Hill, G.L. (1987) Aggressive nutritional support does not prevent protein loss despite fat gain in septic intensive care patients. J. Trauma. 27, 262–266.
4. Gamrin, L., Essén, P., Forsberg, A.M., Hultman, E. and Wernerman, J. (1996) A descriptive study of skeletal muscle metabolism in critically ill patients: free amino acids, energy-rich phosphates, protein, nucleic acids, fat, water, and electrolytes. Crit. Care Med. 24, 575–583.
5. Döcke, W.-D., Randow, F., Syrbe, U., Krausch, D. et al. (1997) Monocyte deactivation in septic patients: restoration by INF-γ treatment. Nature Med. 3, 678–681.
6. Van den Berghe, G., de Zegher, F., Lauwers, P. and Veldhuis, J.D. (1994) Growth hormone secretion in critical illness: effect of dopamine. J. Clin. Endocrinol. Metab. 79, 1141–1146.
7. Van den Berghe, G., de Zegher, F., Bowers, C.Y. et al. (1996) Pituitary responsiveness to growth hormone (GH) releasing hormone, GH-releasing peptide-2 and thyrotropin releasing hormone in critical illness. Clin. Endocrinol. 45, 341–351.
8. Van den Berghe, G., de Zegher, F., Veldhuis, J.D. et al. (1997) The somatotropic axis in critical illness: effect of continuous GHRH and GHRP-2 infusion. J. Clin. Endocrinol. Metab. 82, 590–599.
9. Van den Berghe, G., de Zegher, F., Baxter, R.C. et al. (1998) Neuroendocrinology of prolonged critical illness: effect of continuous thyrotropin-releasing hormone infusion and its combination with growth hormone-secretagogues. J. Clin. Endocrinol. Metab. 83, 309–319.
10. Van den Berghe, G., de Zegher, F. and Bouillon, R. (1998) Acute and prolonged critical illness as different neuroendocrine paradigms. J. Clin. Endocrinol. Metab. 83, 1827–1834.
11. Noel, G.L., Suh, H.K., Stone, S.J.G. and Frantz, A.E. (1972) Human prolactin and growth hormone release during surgery and other conditions of stress. J. Clin. Endocrinol. Metab. 35, 840–851.
12. Ross, R., Miell, J., Freeman, E. and et al. (1991) Critically ill patients have high basal growth hormone levels with attenuated oscillatory activity associated with low levels of insulin-like growth factor-1. Clin. Endocrinol. 35, 47–54.
13. Voerman, H.J., Strack van Schijndel, R.J.M., de Boer, H., van der Veen, E.A. and Thijs, L.G. (1992) Growth hormone: secretion and administration in catabolic adult patients, with emphasis on the critically ill patient. Neth. J. Med. 41, 229–244.

14. Timmins, A.C., Cotterill, A.M., Cwyfan Hughes, S.C. et al. (1996) Critical illness is associated with low circulating concentrations of insulin-like growth factors-I and -II, alterations in insulin-like growth factor binding proteins, and induction of an insulin-like growth factor binding protein-3 protease. Crit. Care Med. 24, 1460–1466.
15. Hermansson, M., Wickelgren, R.B., Hammerqvist, F. et al. (1997) Measurement of human growth hormone receptor messenger ribonucleic acid by a quantitative polymerase chain reaction-based assay: demonstration of reduced expression after elective surgery. J. Clin. Endocrinol. Metab. 82, 421–428.
16. Baxter, R.C. (1994) Insulin-like growth factor binding proteins in the human circulation: a review. Horm Res. 42, 140–144.
17. Baxter, R.C. (1997) Acquired growth hormone insensitivity and insulin-like growth factor bioavailability. Endocrinol. Metab. 4(suppl.B), 65–69.
18. Ross, R.J.M., Miell, J.P., Holly, J.M.P et al. (1991) Levels of GH-binding activity, IGFBP-1, insulin, blood glucose and cortisol in intensive care patients. Clin. Endocrinol. 35, 361–367.
19. Ghahary, A., Fu, S., Shen, Y.J., Shankowsky, H.A. and Tredget, E.E. (1994) Differential effects of thermal injury on circulating insulin-like growth factor binding proteins in burn patients. Molec. Cell Biochem. 135, 171–180.
20. Hartman, M.L., Veldhuis, J.D., Johnson, M.L. et al. (1992) Augmented growth hormone secretory burst frequency and amplitude mediate enhanced GH secretion during a two day fast in normal men. J. Clin. Endocrinol. Metab. 74, 757–765.
21. Thissen, J.P., Ketelslegers, J.-M., Underwood, L.E. (1994) Nutritional regulation of insulin-like growth factors. Endocr. Rev. 15, 80–101.
22. Souba, W.W. (1997) Nutritional support. N. Engl. J. Med. 336, 41–48.
23. Hawker, F.H., Steward, P.M., Baxter, R.C. et al. (1987) Relationship of somatomedin-C/insulin-like growth factor-I levels to conventional nutritional indices in critically ill patients. Crit. Care Med. 15, 732–736.
24. Frohman, L.A. and Jansson, J.O. (1986) Growth hormone-releasing hormone. Endocr. Rev. 7, 223–253.
25. Bowers, C.Y., Momany, F.A., Reynolds, G.A. and Hong, A. (1984) On the *in vitro* and *in vivo* activity of a new synthetic hexapeptide that acts on the pituitary to specifically release growth hormone. Endocrinology 114, 1537–1545.
26. Korbonits, M., Grossman, A.B. (1995) Growth hormone-releasing peptide and its analogues. Novel stimuli to growth hormone release. TEM 6, 43–49.
27. Chapman, I.M., Bach, M.A., Van Cauter, E. et al. (1996) Stimulation of the growth hormone (GH)–insulin-like growth factor I axis by daily oral administration of a GH-secretagogue (MK-677) in healthy elderly subjects. J. Clin. Endocrinol. Metab. 81, 4249–4257.
28. Howard, A.D., Feighner, S.D., Cully, D.F. et al. (1996) A receptor in pituitary and hypothalamus that functions in growth hormone release. Science 273, 974–977.
29. Barinaga, M., Yamonoto, G., Rivier, G., Vale, W., Evans, R. and Rosenfeld, M.G. (1983) Transcriptional regulation of growth hormone gene expression by growth hormone-releasing factor. Nature 306, 84–85.
30. Holl, R.W., Thorner, M.O. and Leong, D.A. (1988) Intracellular calcium concentration and growth hormone secretion in individual somatotropes: effect of growth hormone-releasing factor and somatostatin. Endocrinology 122, 2927–2932.
31. Goth, M.I., Lyons, C.E., Canny, B.J. and Thorner, M.O. (1992) Pituitary adenylate cyclase activating polypeptide, growth hormone (GH)-releasing peptide and GH-releasing hormone stimulate GH release through distinct pituitary receptors. Endocrinology 130, 939–944.
32. Bowers, C.Y., Sartor, A.O., Reynolds, G.A. and Badger, T.M. (1991) On the actions of the growth hormone-releasing hexapeptide, GHRP. Endocrinology 128, 2027–2035.
33. Thorner, M.O., Vance, M.L., Hartman, M.L. et al. (1990) Physiological role of somatostatin on growth hormone regulation in humans. Metabolism 39, 40–42.
34. Berelowitz, M., Szabo, M., Frohman, L.A., Firestone, S., Chu, L. and Hintz, R.L. (1985) Somatomedin-C mediates growth hormone negative feedback by effects on both the hypothalamus and the pituitary. Science 230, 461–463.

35. Reichlin, S. (1992) Neuroendocrinology. In: Williams Textbook of Endocrinology. Wilson & Foster (eds). W.B. Saunders, Philadelphia, USA, pp. 135–219.
36. Vance, M.L., Kaiser, D.L., Evans, W.S. et al. (1985) Pulsatile growth hormone secretion in normal man during a continuous 24 h infusion of human growth hormone releasing factor (1-40): evidence for intermittent somatostatin secretion. J. Clin. Invest. 75, 1584–1590.
37. Jaffe, C.A., Ho, P.J., Demott-Friburg, R., Bowers, C.Y. and Barkan, A.L. (1993) Effects of a prolonged growth hormone (GH)-releasing peptide infusion on pulsatile GH secretion in normal man. J. Clin. Endocrinol. Metab. 77, 1641–1647.
38. Huhn, W.C., Hartman, M.L., Pezzoli, S.S. and Thorner, M.O. (1993) Twenty-four-hour growth hormone (GH)-releasing peptide (GHRP)-infusion enhances pulsatile GH secretion and specifically attenuates the response to a subsequent GHRP bolus. J. Clin. Endocrinol. Metab. 76, 1202–1208.
39. Ho, K.Y., Veldhuis, J.D., Johnson, M.L. et al. (1988) Fasting enhances growth hormone secretion and amplifies the complex rhythms of growth hormone secretion in man. J. Clin. Invest. 81, 968–975.
40. Hartman, M.L., Veldhuis, J.D., Johnson, M.L. et al. (1992) Augmented growth hormone secretory burst frequency and amplitude mediate enhanced GH secretion during a two day fast in normal men. J. Clin. Endocrinol. Metab. 74, 757–765.
41. Van den Berghe, G., Wouters, P., Weekers, F., Mohan, S., Baxter, R.C., Veldhuis, J.D., Bowers, C.Y., Bouillon, R. (1999) Reactivation of pituitary hormone release and metabolic improvement by infusion of growth hormone releasing peptide and thyrotropin-releasing hormone in patients with protracted critical illness. J. Clin. Endocrinol. Metab. 84, 1311–1323.
42. Johannsson, G., Rosén, T. and Bengtsson, B.-A. (1997) Individualized dose titration of growth hormone (GH) during GH replacement in hypopituitary adults. Clin. Endocrinol. 47, 571–581.
43. Van den Berghe, G., de Zegher, F., Veldhuis, J.D. et al. (1997) Thyrotropin and prolactin release in prolonged critical illness: dynamics of spontaneous secretion and effects of growth hormone secretagogues. Clin. Endocrinol. 47, 599–612.
44. Van den Berghe, G., de Zegher, F., Lauwers, P., Veldhuis, J.D. (1994) Luteinizing hormone secretion and hypoandrogenemia in critically ill men: effect of dopamine. Clin. Endocrinol. 41, 563–569.
45. Van den Berghe, G., de Zegher, F., Bouillon, R., Veldhuis, J.D., Bowers, C.Y. (1998) Anterior pituitary function in prolonged critical illness: effects of continuous infusion of growth hormone-secretagogues. In: Growth Hormone Secretagogues in Clinical Practice. Barry B. Bercu and Richard F. Walker (eds). Marcel Dekker, New York, pp. 209–219.
46. Van den Berghe, G., de Zegher, F. and Bouillon, R. (1998) The somatotropic axis in critical illness: effects of growth hormone secretagogues. Growth Hormone and IGF Research 8, 96–98.

Growth Hormone Secretagogues
Edited by E. Ghigo, M. Boghen, F.F. Casanueva and C. Dieguez

Chapter 19

Growth Hormone Secretagogues in Catabolic Illness

RICHARD C. JENKINS and RICHARD J.M. ROSS

Department of Medicine, University of Sheffield, Clinical Sciences Centre, Northern General Hospital, Sheffield, U.K.

ABSTRACT

Growth hormone secretagogues (GHS) have a potent action on the anterior pituitary gland to promote the secretion of primarily GH, and also to a lesser extent ACTH (and thus cortisol) and prolactin. In some catabolic conditions, where treatment with exogenous GH has been shown to have useful anabolic effects, GHS may offer a convenient alternative form of treatment. Studies to date are few but the acute GH secreting action of GHS has been shown to be preserved in fasting, critical illness, end-stage renal failure, exogenous and endogenous Cushing's syndrome and hyperthyroidism. The acute effects are maintained after prolonged infusions in critical illness and after recurrent treatment in diet-induced catabolism. Concomitant acute actions on ACTH and prolactin secretion seem to be attenuated with longer duration of treatment. Thus, initial studies support the potential utility of GHS as agents to treat selected catabolic states but further, long-term, studies with clinical end-points are required.

INTRODUCTION

Growth hormone secretagogues (GHS) are an expanding class of peptide and non-peptide molecules which stimulate the pituitary gland to secrete growth hormone (GH) through their own specific receptor (1). They act at both the hypothalamus and pituitary gland and thus their efficacy requires these structures to be intact. The effects of GHS are modulated by GH-releasing hormone (GHRH) and somatostatin but GHS do not act via these hormones or via other agents which can effect GH secretion such as thyrotrophin-releasing hormone (TRH).

GHS in healthy subjects have been found to stimulate not only GH secretion but also secretion of ACTH (and thus cortisol) and prolactin. Clearly, this raises the concern that increased cortisol concentrations could counteract the anabolic effects of elevated GH concentrations.

The use of GHS, as an alternative anabolic therapy to exogenous GH, can only be anticipated to be successful in conditions in which GH has been shown to be efficacious. To date these include patients undergoing some forms of surgery (2), patients taking glucocorticoids (3), patients with burns (4), the acquired immuno-deficiency syndrome (AIDS) (5,6), renal failure (7,8) and chronic obstructive airways disease (9). These conditions are chronic and an ideal anabolic agent should be active orally — the more recently developed GHS, such as MK-677, fit this criterion. Studies of GHS in catabolic states have predominantly involved single injections which can either provide data that these agents are active in the conditions studied or illustrate GH physiology in catabolism. Long-term studies will be required to determine whether acute efficacy can be translated into long-term anabolic benefits. As with the early trials of GH treatment, small metabolic studies must be followed by longer-term clinical studies with clinically relevant end-points such as length of hospital stay, functional performance or mortality. The results of published studies are discussed below according to catabolic state (Table 1).

EFFECT OF GHS ON GH SECRETION IN SPECIFIC CATABOLIC STATES

Diet-induced catabolism (including anorexia nervosa)

Many catabolic states are characterised by poor nutritional intake or a period of fasting. Fasting results in a picture of GH resistance with low IGF-I and high GH levels (10,11) as seen in many catabolic states. For these reasons, fasting in normal volunteers has frequently been used as a model to study the effects of anabolic agents.

Eight healthy subjects fasted on two occasions each lasting 2 weeks (12). During the second week of each period they either received oral MK-677 (25 mg daily) or placebo (Figure 1). Daily nitrogen balance was similarly negative in each group during the first study week but during the second week was 0.31 g/day with MK-677 treatment and –1.48 g/day with placebo. The GH response to MK-677 was reduced to 40% of the baseline value after 7 days treatment. Serum IGF-I and IGFBP-3 concentrations increased only with MK-677 treatment. Serum cortisol and prolactin concentrations increased significantly after MK-677 at the start of the treatment week but by the end of the week these responses had become attenuated and were not different to placebo.

Patients with anorexia nervosa respond to hexarelin (peak GH 64.8 mU/l) similarly to healthy controls (77.5 mU/l) (13). Women with voluntary weight loss (for aesthetic reasons) and consequent amenorrhoea and women on a short-term hypocaloric diet also respond similarly (60.3 and 99.6 mU/l respectively). Interestingly, hexarelin blunted the response to subsequent GHRH administration in normal women, women with voluntary weight loss and women on a short-term diet but not in women with anorexia nervosa.

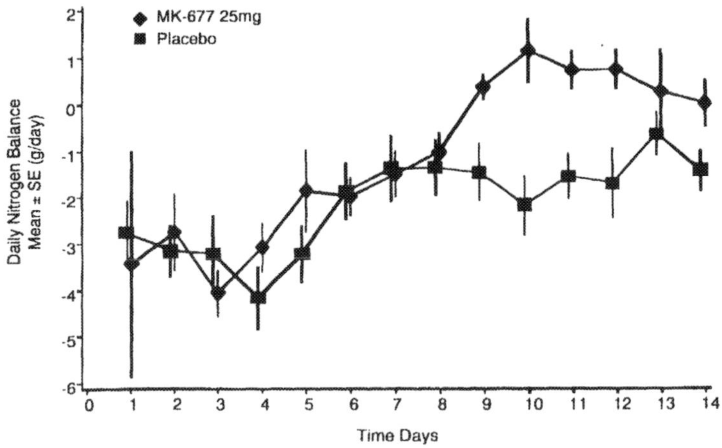

Critical illness

Studies on patients with critical illness usually include subjects with heterogeneous diseases whose pathophysiology will vary. Nevertheless, this area is one of the largest in which acute studies of GHS in catabolic patients have been performed. Van Den Berghe and colleagues have published a series of complex studies examining this patient group. Their work has focused on the effects of either intravenous boluses or prolonged intravenous infusions of GHRP-2, GHRH and TRH, either alone, in combination or sequentially. The first of these studies (14) demonstrated that GHRP-2 (1 μg/kg iv bolus) produced profound GH secretion (peak GH 43 μg/l) when compared to placebo (3.7 μg/l). GHRH (1 μg/kg iv bolus) was less effective (peak GH 20.4 μg/l) but when both GHRP-2 and GHRH were given together the effect was synergistic (134.6 μg/l). Addition of TRH (200 μg) to the combined treatment reduced the response slightly (110.2 μg/l). The GH response in the younger subjects (<40 years) was twice that in older subjects (>40 years).

The second study (15) gave patients two 21-hour infusions, separated by a three-hour gap, of placebo, GHRH (1 μg/kg/h), GHRP-2 (1 μg/kg/h) or GHRH and GHRP-2 together. GHRP-2 increased mean GH concentration, GH secretory pulse amplitude and basal GH secretion. Both peptides together again exhibited synergism. IGF-I concentrations rose within 24 hours and were higher after the combined treatment.

The most recent study (16) was similar to the last but also incorporated TRH into the infusion regimens. This study again found that GHRP-2 and combined GHRP-2 and GHRH increased IGF-I concentrations but also reported that two other GH-dependent proteins, IGFBP-3 and ALS, increased within 48 hours. TRH had no effect on GH secretion.

TABLE 1

SUMMARY OF PUBLISHED STUDIES OF THE EFFECTS OF GHS IN CATABOLIC CONDITIONS

Condition	Secretagogue	Dose	Route	GH secretion	ACTH secretion	Prolactin secretion	IGF-I conc.	IGFBP-3 conc.	Cortisol conc.	Refs.
Diet-induced catabolism	MK-677	25 mg	Oral	↑↑↑	NT	↑	↑	↑	↑	(12) *
	Hexarelin	1 μg/kg	IV	↑↑↑	NT	NT	NT	NT	NT	(13)
Anorexia	Hexarelin	1 μg/kg	IV	↑↑↑						
Critical illness	GHRH	1 μg/kg	IV	↑	NT	↔/↑	↔	NT	↑	(14)
	GHRP-2	1 μg/kg	IV	↑↑/↑↑↑		↑	↔		↑	
	GHRH + GHRP-2	As above	IV	↑↑↑↑↑↑		↔/↑	↔		↑	
	GHRH	1 μg/kg/hr	IVI	↑	NT	↑	↔/↑	NT	↔	
	GHRP-2	1 μg/kg/hr	IVI	↑↑		↔/↑	↑	↑	↔	(15,16,30)
	GHRH + GHRP-2	As above	IVI	↑↑↑↑		↔/↑	↑↑	↑	↔	
Renal failure	Hexarelin	1 μg/kg	IV	↑↑	↑	↔	NT	NT	↑	(17,18)
		2 μg/kg	IV	↑↑↑	↑	↔			↑	
Short-term exogenous glucocorticoids	L 692,429	0.2 mg/kg	IV	↑↑	NT	↑	NT	NT	↔	(21)
		0.75 mg/kg	IV	↑↑↑		↑↑			↔	
Long-term exogenous glucocorticoids	GHRH	100 μg	IV	↔	NT	NT	NT	NT	NT	(22)
	Hexarelin	100 μg	IV	↑↑						
	GHRH + Hexarelin	As above	IV	↑↑↑↑						
	GHRH	100 μg	IV	↔	NT	NT	↔	NT	NT	(23)
	GHRP-6	1 μg/kg	IV	↑↑						

Condition	Secretagogue	Dose	Route	GH secretion	ACTH secretion	Prolactin secretion	IGF-I conc.	IGFBP-3 conc.	Cortisol conc.	Refs.
	GHRH + GHRP-6	As above	IV	↑↑↑↑↑						
Cushing's — pituitary	GHRH	100 μg	IV	↔	NT	NT	↔	NT	NT	(23)
	GHRP-6	1 μg/kg	IV	↔/↑						
	GHRH + GHRP-6	As above	IV	↑/↑↑						
	GHRH	100 μg	IV	↔/↑	NT	NT	NT	NT	NT	(19)
	GHRP-6	100 μg	IV	↑						
	GHRH + GHRP-6	As above	IV	↑/↑↑						
	Hexarelin	2 μg/kg	IV	↑↑	↑↑↑	NT	NT	NT	↑↑	(20)
Cushing's — adrenal or ectopic	Hexarelin	2 μg/kg	IV	↑↑	↔				↔	
Hyperthyroidism	GHRH	100 μg	IV	↔/↑	NT	NT	NT	NT	NT	(25)
	GHRP-6	1 μg/kg	IV	↑↑↑						
	GHRH + GHRP-6	As above	IV	↑↑↑						

*The paper found different effects of MK-677 on days 1 and 7 of treatment; these are elaborated in the text. ↑, increased; ↓, decreased; ↔ unchanged; NT, not tested.

Renal failure

We have examined the acute effects of hexarelin on GH, ACTH, cortisol and prolactin secretion in patients with chronic renal failure on haemodialysis. An initial intravenous dose response study (17) in 20–40 year old well-nourished haemodialysis patients demonstrated exuberant GH secretion which was greater, albeit statistically similar, after 2 μg/kg than 1 μg/kg. The 2 μg/kg dose was then given to healthy volunteers (age, sex and nutritionally matched to the patient group) who produced indistinguishable responses to the patients suggesting that hexarelin efficacy is normal in haemodialysis-treated renal failure.

To examine the factors which affect hexarelin efficacy in renal failure we gave young (20–40 year old) and old (50–70 year old) poorly-nourished haemodialysis patients hexarelin (2 μg/kg iv) (18). Although the results did not reach statistical significance there were trends towards greater GH secretion in malnourished young patients in comparison to well-nourished young patients and lesser GH secretion in the old group compared to the young group ($p = 0.08$).

Glucocorticoid excess

Patients with Cushing's syndrome due to endogenous or exogenous glucocorticoid excess are catabolic and treatment with exogenous GH can be beneficial (3). Previous work has indicated that GH secretion is generally suppressed by glucocorticoids and a number of studies have examined whether GHS action is also suppressed in patients with glucocorticoid excess. Patients with endogenous Cushing's syndrome (predominantly Cushing's disease) were found to have markedly blunted responses to GHRH, GHRP-6 and combined GHRH/GHRP-6 compared to healthy controls (19). GH secretion in patients with Cushing's syndrome after GHRP-6 administration was about 25% of control values and although GHRH and GHRP-6 were synergistic the response was only about 12% that of the controls. Similarly, another group of patients with endogenous Cushing's syndrome (again mainly Cushing's disease) had a 50% reduction in peak GH, compared with controls, after hexarelin (20). The GHS, L 692,429 (0.2 mg/kg iv), was given to healthy subjects who had taken prednisolone (60 mg daily for 4 days) and this was found to reduce peak GH by over 50% (21). Increasing the dose of L 692,429 to 0.75 mg/kg improved the response to approximately 80% of the control value. In contrast, patients who had been on longer term prednisolone treatment (5–12.5 mg daily for 12–115 months) had an unaltered GH response to hexarelin alone but an attenuated response to combined hexarelin and GHRH (22). A comparison between the effects of endogenous (20–60 mg prednisolone daily for more than 6 months) and exogenous Cushing's syndrome (23) found that endogenous Cushing's syndrome attenuated the GH response to either GHRP-6 or combined GHRP-6/GHRH whilst exogenous Cushing's syndrome had no effect. The latter result is in contrast to previous studies and may be explained by an effect of duration or severity of Cushing's syndrome. In a similar vein, hexarelin was found to abolish the inhibitory effect of hydrocortisone on GH secretion in acromegaly (24).

In summary, the effects of GHS are reduced but still present in all groups studied with endogenous Cushing's syndrome. Exogenous Cushing's syndrome either has no effect on

GHS efficacy or reduces responses although this can be partially overcome by using a higher dose of GHS. GHRH and GHS were synergistic in all groups studied.

Hyperthyroidism

Acute administration of GHRP-6 (1 μg/kg iv) to hyperthyroid patients (25) produced similar peak GH concentrations (31.9 μg/l) to healthy controls (23.2 μg/l). The response to GHRH (100 μg iv) was reduced in hyperthyroidism (9.0 vs 27.0 μg/l in healthy controls) and the usual synergism between GHRH and GHRP-6 was not seen (22.5 vs 83.7 in healthy controls). A further study by the same group found a lower peak GH of 12.6 μg/l in hypothyroid individuals compared to 22.1 μg/l in controls after GHRP-6 (26).

EFFECTS OF GHS ON OTHER HORMONES IN CATABOLIC STATES

Effects on ACTH and cortisol

The anabolic use of GHS could potentially be offset by a concomitant increase in serum ACTH and thus cortisol. That GHS have this action in normals has been known for some time (27,28) and it is important to determine if it also occurs in catabolism where there are already disturbances of ACTH and cortisol. A number of studies have measured ACTH or cortisol response to GHS in catabolic states. Bolus GHRP-2 increased cortisol concentration by 35% in critical illness (14) whilst 21-hour duration overnight GHRP-2 infusion had no effect on mean cortisol concentration (15,16). We found that patients with renal failure on haemodialysis had increased ACTH and cortisol secretion following bolus intravenous hexarelin (18). In diet-induced catabolism, MK-677 had a significant effect to increase AUC cortisol and urinary free cortisol after the first dose but by the seventh dose no effect was seen (12). Prednisolone pre-treatment abolished the GHRP-6 effect on cortisol concentration (21). Hexarelin had a dramatic effect to increase ACTH and cortisol secretion in patients with Cushing's disease but had no effect in patients with Cushing's syndrome due to adrenal adenomas (20). The acute action on serum cortisol concentration was attenuated after 16 weeks of hexarelin treatment in elderly men (29).

This incomplete body of evidence seems to indicate that acute administration of GHS in catabolism does induce ACTH and cortisol secretion but this effect is lost with prolonged infusions or recurrent administration. The acute action on ACTH secretion is abolished by exogenous glucocorticoid excess, absent in adrenal Cushing's syndrome and exaggerated in Cushing's disease.

Effects on prolactin

Normal subjects secrete prolactin following acute administration of GHS (27,28). Acute treatment in diet-induced catabolism increased prolactin concentration but the effect was attenuated after 7 days of treatment (12). In critical illness, prolactin increased after bolus GHRP-2 (14) but a 21-hour GHRP-2 infusion has, in one study, apparently had no effect on prolactin (16) whilst in another study, prolactin levels increased by 21% (30). The prolactin-secreting effect was partially attenuated by prednisolone pre-treatment (21) and completely absent in renal failure (18).

SUMMARY

The acute action of GHS on GH- secretion has been shown to be present in patients with critical illness, end-stage renal failure, exogenous and endogenous Cushing's syndrome and hyperthyroidism. The acute effects on GH secretion are maintained after prolonged infusions in critical illness and after recurrent treatment in diet-induced catabolism. The concomitant acute actions on ACTH and prolactin secretion seem to be attenuated with longer duration of treatment. Thus, initial studies support the potential utility of GHS as agents to treat selected catabolic states but long-term studies of orally active agents with clinical end-points are required.

ACKNOWLEDGEMENTS

RCJ is supported by the University of Sheffield, the Northern General Hospital NHS Trust and Pharmacia and Upjohn Limited.

REFERENCES

1. Howard, A.D., Feighner, S.D., Cully, D.F. et al. (1996) A receptor in pituitary and hypothalamus that functions in growth hormone release. Sci 273, 974–7.
2. Ward, H.C., Halliday, D., Sim, A.J.W. (1987) Protein and energy metabolism with biosynthetic human growth hormone after gastrointestinal surgery. Ann. Surg. 206, 56–61.
3. Horber, F.F. and Haymond, M.W. (1990) Human growth hormone prevents the protein catabolic side effects of prednisone in humans. J. Clin. Invest. 86, 265–72.
4. Gilpin, D.A., Barrow, R.E., Rutan, R.L., Broemeling, L. and Herndon, D.N. (1994) Recombinant human growth hormone accelerates wound healing in children with large cutaneous burns. Ann. Surg. 220, 19–24.
5. Krentz, A.J., Koster, F.T., Crist, D.M., Finn, K., Johnson, L.Z., Boyle, P.J. and Schade, D.S. (1993) Anthropometric, metabolic, and immunological effects of recombinant human growth hormone in AIDS and AIDS-related complex. J. Acq. Imm. Def. Syn. 245, 251.
6. Mulligan, K., Grunfeld, C., Hellerstein, M.K., Neese, R.A. and Schambelan, M. (1993) Anabolic effects of recombinant human growth hormone in patients with wasting associated with human immunodeficiency virus infection. J. Clin. Endocrinol. Metab. 77, 956–62.
7. Ikizler, T.A., Wingard, R.L., Breyer, J.A., Schulman, G., Parker, R.A. and Hakim, R.M. (1994) Short-term effects of recombinant human growth hormone in CAPD patients. Kid. Int. 46, 1178–83.
8. Koch, V.H., Lippe, B.M., Nelson, P.A., Boechat, M.I., Sherman, B.M. and Fine, R.N. (1989) Accelerated growth after recombinant human growth hormone treatment of children with chronic renal failure. Journal of Pediatrics 115, 365–71.
9. Pape, G.S., Friedman, M., Underwood, L.E. and Clemmons, D.R. (1991) The effect of growth hormone on weight gain and pulmonary function in patients with chronic obstructive lung disease. Chest 99, 1495–500.
10. Isley, W.L., Underwood, L.E. and Clemmons, D.R. (1983) Dietary components that regulate serum somatomedin-C concentrations in humans. J. Clin. Invest. 71, 175–82.

11. Hartman, M.L., Veldhuis, J.D., Johnson, M.L., Lee, M.M., Alberti, K.G.M.M., Samojlik, E. and Thorner, M.O. (1992) Augmented growth hormone (GH) secretory burst frequency and amplitude mediate enhanced GH secretion during a two-day fast in normal men. J. Clin. Endocrinol. Metab. 74, 757–65.
12. Murphy, M.G., Plunkett, L.M., Gertz, B.J., He, W., Wittreich, J., Polvino, W.M., Clemmons, D.R. (1998) MK-677, an orally active growth hormone secretagogue, reverses diet-induced catabolism. J. Clin. Endocrinol. Metab. 83, 320–5.
13. Popovic, V., Micic, D., Djurovic, M., Obradovic, S., Casanueva, F.F. and Dieguez, C. (1997) Absence of desensitization by hexarelin to subsequent GH releasing hormone-mediated GH secretion in patients with anorexia nervosa. Clin. Endocrinol. 46, 539–43.
14. Van den Berghe, G., De Zegher, F., Bowers, C.Y., Wouters, P., Muller, P., Soetens, F., Vlasselaers, D., Schetz, M., Verwaest, C., Lauwers, P. et al. (1996) Pituitary responsiveness to GH-releasing hormone, GH-releasing peptide-2 and thyrotrophin-releasing hormone in critical illness. Clin. Endocrinol. 45, 341–51.
15. Van den Berghe, G., De Zegher, F., Veldhuis, J.D., Wouters, P., Awouters, M., Verbruggen, W., Schetz, M., Verwaest, C., Lauwers, P., Bouillon, R. et al. (1997) The somatotropic axis in critical illness: Effect of continuous growth hormone (GH)-releasing hormone and GH-releasing peptide-2 infusion. J. Clin. Endocrinol. Metab. 82, 590–9.
16. Van den Berghe, G., De Zegher, F., Baxter, R.C., Veldhuis, J.D., Wouters, P., Schetz, M., Verwaest, C., Van Der Vorst, E., Lauwers, P., Bouillon, R. et al. (1998) Neuroendocrinology of prolonged critical illness: Effects of exogenous thyrotrophin-releasing hormone and its combination with growth hormone secretagogues. J. Clin. Endocrinol. Metab. 83, 309–19.
17. Jenkins, R.C., Jones, J., Wilkie, M.E.R., Ghigo, E., El-Nahas, A.M. and Ross, R.J.M. (1998) Hexarelin-stimulated growth hormone secretion in patients on maintenance haemodialysis. Proceedings of the 24th International Symposium on Growth Hormone and Growth Factors in Endocrinology and Metabolism Abstract C4.
18. Jenkins, R.C., El-Nahas, A.M., Wilkie, M.E.R., Jones, J., Ghigo, E. and Ross, R.J.M. (1998) The effects of nutrition and age on hexarelin-induced anterior pituitary hormone secretion in adult patients on maintenance haemodialysis. J. Endocrinol. 156, 187.
19. Leal-Cerro, A., Pumar, A., Garcia-Garcia, E., Dieguez, C. and Casanueva, F.F. (1994) Inhibition of growth hormone release after the combined administration of GHRH and GHRP-6 in patients with Cushing's syndrome. Clin Endocrinol 41:649-54.
20. Ghigo, E., Arvat, E., Ramunni, J., Colao, A., Gianotti, L., Deghenghi, R., Lombardi, G. and Camanni, F. (1997) Adrenocorticotropin- and cortisol-releasing effect of hexarelin, a synthetic growth hormone-releasing peptide, in normal subjects and patients with Cushing's syndrome. J. Clin. Endocrinol. Metab. 82, 2439–44.
21. Gertz, B.J., Sciberras, D.G., Yogendran, L., Christie, K., Bador, K., Krupa, D., Wittreich, J.M and James, I. (1994) L-692,429, a non-peptide growth hormone (GH) secretagogue, reverses glucocorticoid suppression of GH secretion. J. Clin. Endocrinol. Metab. 79, 745–9.
22. Giustina, A., Bussi, A.R., Deghenghi, R., Imbimbo, B.P., Licini, M., Poiesi, C. and Wehrenberg, W.B. (1995) Comparison of the effects of growth hormone-releasing hormone and hexarelin, a novel growth hormone-releasing peptide-6 analog, on growth hormone secretion in humans with or without glucocorticoid excess. J. Endocrinol. 146, 227–32.
23. Borges, M.S., DiNinno, F.B., Lengyel, A.J. (1997) Different effects of growth hormone releasing peptide (GHRP-6) and GH-releasing hormone on GH release in endogenous and exogenous hypercortisolism. Clin. Endocrinol. 46, 713–8.
24. Giustina, A., Bresciani, E., Bugari, G,. Bussi, G., Deghenghi, R., Imbimbo, B. and Giustina, G. (1995) Hexarelin, a novel GHRP-6 analog, counteracts the inhibitory effect of hydrocortisone on growth hormone secretion in acromegaly. Endocrine Research 21, 569–82.
25. Ramos-Dias, J.C., Pimentel-Filho, F., Reis, A.F. and Lengyel, A.J. (1996) Different growth hormone (GH) responses to GH-releasing peptide and GH-releasing hormone in hyperthyroidism. J. Clin. Endocrinol. Metab. 81, 1343–6.

26. Pimentel-Filho, F.R., Ramos-Dias, J.C., Ninno, F.B.D., Facanha, C.F.S., Liberman, B. and Lengyel, A.J. (1997) Growth hormone responses to GH-releasing peptide (GHRP-6) in hypothyroidism. Clin. Endocrinol. 46, 295–300.
27. Gertz, B.J., Barrett, J.S., Eisenhandler, R., Krupa, D.A., Wittreich, J.M., Seibold, J.R. and Schneider, S.H. (1993) Growth hormone response in man to L-692,429, a novel non-peptide mimic of growth hormone-releasing peptide-6. J. Clin. Endocrinol. Metab. 77, 1393–7.
28. Imbimbo, B.P., Mant, T., Edward, M., Amin, D., Froud, A., Lenaerts, U., Boutignon, F. and Deghenghi, R. (1994) Growth hormone releasing activity of hexarelin in humans: A dose–response study. Eur. J. Clin. Pharmacol. 46, 421–5.
29. Rahim, A., O'Neill, P.A. and Shalet, S.M. (1998) The effect of chronic hexarelin administration on the pituitary–adrenal axis and prolactin. J. Endocrinol. (Supplement), 200.
30. Van den Berghe, G., De Zegher, F., Veldhuis, J.D., Wouters, P., Gouwy, S., Stockman, W., Weekers, F., Schetz, M., Lauwers, P., Bouillon, R. et al. (1997) Thyrotrophin and prolactin release in prolonged critical illness: dynamics of spontaneous secretion and effects of growth hormone-secretagogues. Clin. Endocrinol. 47, 599–612.

Growth Hormone Secretagogues
Edited by E. Ghigo, M. Boghen, F.F. Casanueva and C. Dieguez

Chapter 20

Treatment of Children with Short Stature by Growth Hormone Secretagogues

ZVI LARON

Endocrine and Diabetes Research Unit, Schneider Children's Medical Center of Israel, Beilinson Campus, Petah Tikva and Sackler Faculty of Medicine, Tel Aviv University, Israel

After the documentation that children respond to the acute administration of GH secretagogues (GH-S) even when administered orally or intranasally (1–4) by raising the serum growth hormone (GH) levels, it was obvious that long-term trials should be initiated. The fact that GH-S increase the endogenous GH release, and possibly synthesis, only in the presence of functional pituitary secretagogues (2,5) and available growth hormone releasing hormone (GHRH) (6) restricted the possible use of this new class of drugs to non-hypopituitary subjects.

However, the increasing use of GH in so called "non-conventional" conditions i.e. without endogenous GH deficiency, raised hopes for the clinical use of GH-S which has the advantage of acting without the need of painful injections. Among these indications are children with idiopathic short stature (growth retardation), familial short stature, intrauterine growth retardation (IUGR), Turner Syndrome, chronic renal failure, in all of which many trials with GH have been or are still being performed (7–9).

Due to limitations in production on the one hand and precautions toward possible undesirable effects on the other hand, the number of studies on the effectiveness and tolerance of these drugs in long-term growth promotion are scant. To the best of our knowledge there are only two published studies. One is our own study using Hexarelin (10) which is well documented. The second investigation using GRP-2 for 6 to 24 months has been presented only as an abstract (11). Herewith the summary of the known data.

STUDIES

The drug employed by us was Hexarelin (Hex = His-D-2-methyl-Trp-Ala-Trp-D-Phe-Lys-NH2) which was supplied by Dr. R. Deghenghi, (Europeptides, Argenteuil,

TABLE 1

Effect of various doses of intranasal Hexarelin for 7 days in children with idiopathic short stature on serum IGF-I, and plasma alkaline phosphatase

Patient no.	Sex	CA y:m	IGF-I (nMol/l)				Alkaline phospatase (U/L)			
				20* b.i.d.	20-20-40*	40* t.i.d.		20* b.i.d.	20-20-40*	40* t.i.d.
			Basal	Day 8	Day 8	Day 8	Basal	Day 8	Day 8	Day 8
1	M	5:9	5.8	8.3	(–)	(–)	135	151	(–)	(–)
2	F	5:10	12.75	(–)	13.6	20.7	162	(–)	167	142
3	M	5:10	7.2	6.3	(–)	8.6	189	190	(–)	208
4	M	7:6	14.2	16.9	12.8	16.1	284	259	265	259
5	M	9:1	8.4	10.1	16.65	18.75	213	210	240	241
6	M	9:11	7.1	6.8	9.5	14.25	218	217	254	238
7	M	11:4	15.4	17.4	26.54	22.75	139	143	137	153
				$p < 0.05$				$p < 0.05$		

CA = chronological age; y = years; m = months. Modified from Frenkel et al. (13).
* Daily dose of Hexarelin (μg/kg).

France). After ascertaining that a bolus intranasal (i.n.) insufflation of 20 μg/kg has an identical efficacy in raising serum hGH as an intravenous (i.v.) bolus of 1μg/kg in children with familial short stature (12), we proceeded to find a daily dose which will raise and maintain an elevated serum IGF-I (insulin-like growth factor-I) as a marker of linear growth stimulation. Due to the short half-life of the drug, we found that one or two daily intranasal applications were insufficient to raise serum IGF-I. We thus administered the drug three times a day (t.i.d.) (13) (Table 1). As a dose of 40 μg/kg t.i.d. achieved significant but lower than aimed changes, we decided to use 60 μg/kg t.i.d by i.n. spray, a dose, which was well tolerated by the children (14).

SUBJECTS

Eight prepubertal children (7 boys and 1 girl) aged 5.4 to 11.7 years were included in the long-term study (10). All were short (height < –1.9 SDS height) but had a normal hGH response (> 20 mU/l) to clonidine (15) and/or insulin hypoglycemia tests (16). An additional inclusion criterion was a normal response of hGH (40–60 mU/l) to one i.n. dose (20 μg/kg) of Hexarelin (12). Pertinent clinical data of these children is summarized in Table 2. The reasons for the trial were psychosocial difficulties in adaptation at school compared to their peers. The children were treated for 9 to 10 months according to the protocol. The study was approved by the local Ethics Committee and by the Ministry of Health and written informed consent was obtained from all the parents.

TABLE 2

PERTINENT CLINICAL DATA OF 8 PREPUBERTAL SHORT CHILDREN BEFORE INTRANASAL HEXARELIN (60 μg/kg t.i.d.) TREATMENT

Patient no.	Sex	CA (yrs)	BA (yrs)	Height (cm)	Height SDS	Growth velocity (cm/yrs)	Weight (kg)	Skinfold (subscapular) (mm)	Head circumference (cm)
1	M	5.4	4	99.5	−2.3	6.5	14.2	6	51.0
2	M	6.2	4.5	97.8	−3.7	4.4	11.6	6	49.5
3	M	6.2	.8	107.2	−1.9	5.5	14.9	5	47.3
4	F	6.4	3.5	105.5	−2.2	6.0	15.5	6	49.5
5	M	7.9	6	114.9	−1.9	5.9	20.9	8	51.0
6	M	8.4	6	116.5	−2.1	5.7	18.0	5	51.0
7	M	10.2	6	121.8	−2.5	4.8	21.9	5	53.0
8	M	11.7	9	133.5	−2.3	4.1	28.6	7	51.4
Mean ± SD				112.1±12.1		5.4±0.8	8.12±5.4	6.0±1	50.5±1.7

CA = chronological age, BA = Bone (skeletal age).

INVESTIGATION AND METHODS

The patients were examined weekly during the first month of treatment; bi-weekly during the second month, and monthly thereafter. Harpenden stadiometers were used by the same person for recumbent or standing anthropometric measurements. A Harpenden caliper served to measure suprailiac, triceps and subscapular skinfolds. Bone age was estimated using the Greulich and Pyle Atlas (17) from a hand and wrist radiography, with separate readings for carpal and phalangeal bones. The height was plotted on Tanner, Whitehouse and Takaishi growth charts (18). Blood for blood count, routine chemistry including liver tests, serum IGF-I and insulin were drawn after an overnight fast before and every 2 to 3 months during Hexarelin administration, and monthly thereafter for 3 months. All sera for hormone determinations were estimated in the same assay.

Human serum GH was measured by a radioimmunoassay modification of the method described by Laron & Mannheimer (19) using rabbit polyclonal antiserum to recombinant hGH (# 20.4.94) and rhGH (# 3-08P-525). The sensitivity of the method is 0.4 μg/l serum. The within assay coefficient of variation for serum with 6.2 μg/l hGH is 3.2%. Plasma TSH, free T4 and total T3 were measured by RIA as described previously (12). Serum IGF-I was measured by RIA after acid ethanol extraction followed by cryoprecipitation as previously described (20). The sensitivity of the assay is 2 nMol/L, the within assay coefficient of variation for a concentration of 20 nMol/L of IGF-I was 4.7%. Blood chemistry was determined by autoanalyzer (Hitachi, Japan). Statistical analysis was performed using the Student's paired *t* test.

RESULTS

The three times daily intranasal Hexarelin administration (60μg/kg) was associated with significant biochemical changes (Tables 3 and 4), most of which had already been observed 3 months after the start of treatment (14). The mean growth velocity of the children before treatment was 5.3 ± 0.84 cm/year (m ± SD) and increased to 8.3 ± 1.7 cm/year during 9 to 10 months of Hexarelin treatment ($p < 0.0001$) (Figure 1). During the same period the body weight of the children increased by 1.3 ± 0.5 kg (m ± SD), with a concomitant decrease of

TABLE 3

SERUM IGF-I (nMol/l) DURING INTRANASAL HEXARELIN (60 μg/kg t.i.d.) TREATMENT

Patient no.	Days			Months				
	0	7	14	1	2	3	6	9–10
1	5.5	9.6	9.7	8.8	7.8	9.0	9.8	
2	7.2	7.5	7.6	6.2	8.0	7.1	10.2	9.6
3	5.8	12.0	7.7	9.8	11.2	8.3	8.7	
4	12.7	22.2	16.7	17.3	13.2	16.3	14.5	13.9
5	14.2	15.1	16.8	16.3	19.0	15.6	19.0	22.3
6	13.0	12.3	13.6	12.9	13.4	18.7		
7	9.4	15.2	16.4	16.3	13.1	13.5	16.8	21.0
8	15.4	22.4	24.1	23.3	22.6	24.3	20.0	32.2
Mean±SD	10.4±3.9	14.5±5.4	14.1±5.6	13.9±5.5	13.5±5.1	14.1±5.9	14.1± 4.6	19.8±8.7

All the values during Hexarelin treatment were significantly higher than the basal levels ($p < 0.001$–0.05). Modified from Laron et al. (10).

TABLE 4

SERUM PHOSPHATE AND ALKALINE PHOSPHATASE (MEAN ± SD) DURING INTRANASAL HEXARELIN (60 μg/kg t.i.d.) TREATMENT

	Months				
	0	1	2	3	9–10
Phosphate (mMol/l)	1.5 0.1	1.7* 0.2	1.7* 0.2	1.7* 0.2	1.8** 0.1
Alkaline phosphatase (U/l)	219 74	249 75	265** 76	255* 80	261 75

*$p < 0.05$, **$p < 0.004$.

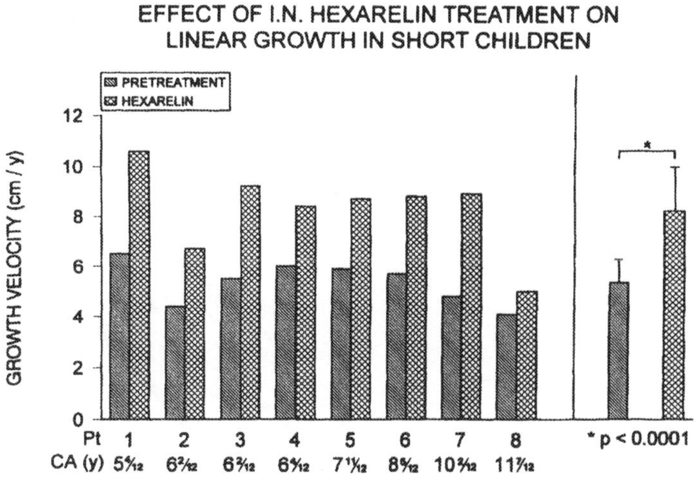

Figure 1. Growth velocity (cm/year) of 8 prepubertal boys with short stature before and at the end of 9–10 months intranasal Hexarelin treatment (60 μg/kg t.i.d.).

the subscapular subcutaneous tissue by a mean of 1.12 ± 0.8 mm ($p < 0.007$); denoting increase of muscle and bone mass and decrease in adipose tissue mass even in these slim children. A mean increase in the head circumference of 0.51 ± 0.3 cm ($p < 0.002$) was also observed (10).

Figures 2–6 illustrate the growth curves of 5 of the 8 children treated. It is seen that the growth varied, probably due to differences in the etiology of the short stature. Figure 2 shows that Hexarelin administration increased growth from under the 3rd centile to the 6th centile. The advancement of puberty kept the growth on the same centile also after discontinuation of treatment, there was, however, a concomitant diminution of the bone age retardation. Figure 3 presents a similar growth pattern, but in this boy subsequent treatment with hGH kept growth along the 10th percentile. The boy remained prepubertal and the bone age retardation remained the same. Figure 4 illustrates a small gain during Hexarelin treatment, but Figure 5 shows that after the interruption of Hexarelin, the growth returned to the hereditary centile and Figure 6 reveals lack of response to Hexarelin in a boy with marked familial short stature and IUGR (birth length = 45 cm). The different individual responses prove that presentation of mean values alone may lead to wrong conclusions.

Serum IGF-I rose from a mean of 10.4 ± 3.9 (SD) nMol/1 to 14.5 ± 5.4 nMol/1 ($p < 0.02$) already after 7 days of Hexarelin administration (13), and remained essentially constant throughout the first 6 months of treatment ($p < 0.0004$) (Table 3). In 3 of the 5 children there was a further increase in the following months. Serum inorganic phosphorus and alkaline phosphatase levels increased significantly from pretreatment levels of 1.53 ± 0.1 to 1.78 ± 0.1 mMol/1 ($p < 0.04$) and from 219 ± 74 to 260 ± 75 U/1 ($p < 0.05$) respectively. (Table 4). Serum cortisol and prolactin varied slightly (Table 5) but definitely did not increase with the doses of Hexarelin used. Despite the finding that TSH decreased

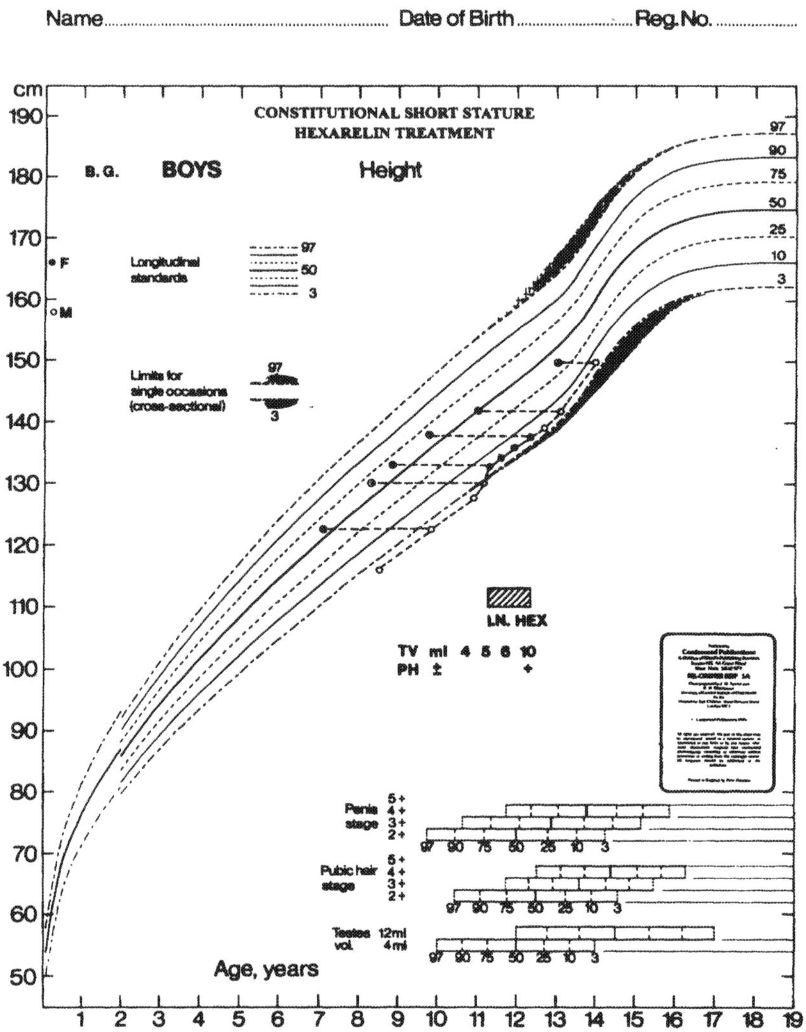

Figure 2. Growth chart of a boy with short stature before, during and after Hexarelin treatment (60 μg/kg t.i.d.).

TABLE 5

SERUM CORTISOL AND PROLACTIN (MEAN ± SD) BEFORE AND DURING INTRANASAL HEXARELIN (60 μg/kg t.i.d.) TREATMENT OF 8 PREPUBERTAL SHORT CHILDREN

	Months			
	0	3	6	9–10
Cortisol (nMol/L)	444±213	458±273	462±265	334±216
Prolactin (μg/L)	14.4±13.3	11.8±8.9	12.6±9.7	6.4±2.3

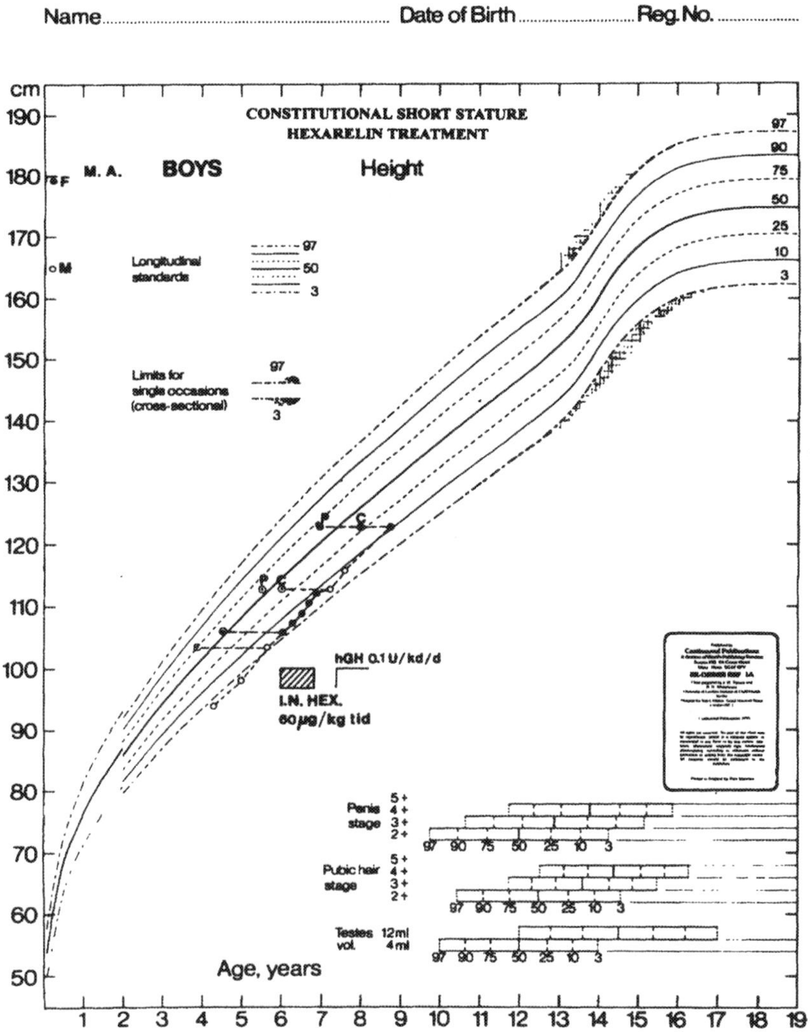

Figure 3. Growth chart of a boy with short stature before, during and after Hexarelin treatment (60 μg/kg t.i.d.).

significantly after an i.v. (1 μg/kg) or i.n. (20 μg/kg) bolus administration (12), no significant changes were observed during the long-term treatment in any of the thyroid function markers (serum TSH, fT_4, T_3) (10). Bone age advanced less or parallel with the chronological age. Blood glucose, electrolytes, creatinine and liver tests remained unchanged.

UNDESIRABLE EFFECTS

The drug was well tolerated, there was no local irritation, and no undesirable effects were reported. The fact that other investigators reported an increase in prolactin and cortisol is

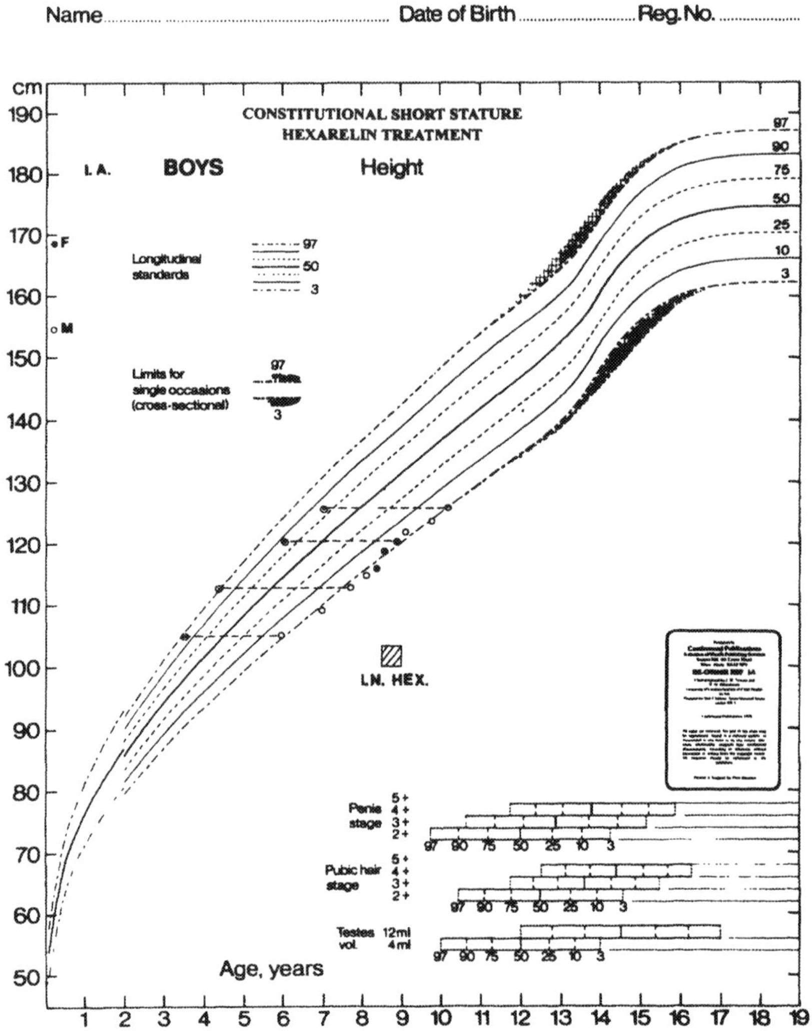

Figure 4. Growth chart of a boy with short stature before, during and after Hexarelin treatment (60 µg/kg t.i.d.)

probably due to the higher dose used (21,22,); but it may also be a transitory effect as that observed in our acute studies in which we registered a TSH suppression (2,12). It is of note that in *in vitro* studies we found that GHRP-1 but not GHRH inhibited TSH stimulated T_3 secretion and cAMP formation in cultured human thyroid follicles (23).

Knowing that long-term administration of GHRH induces desensitization to the drug (24), we tested this effect in the children treated by i.n. Hexarelin on a long term basis (25). Seven out of the 8 children included in the clinical trial underwent two types of test to determine the pituitary potential to secrete hGH before, at the end/or after interruption of the Hexarelin treatment.

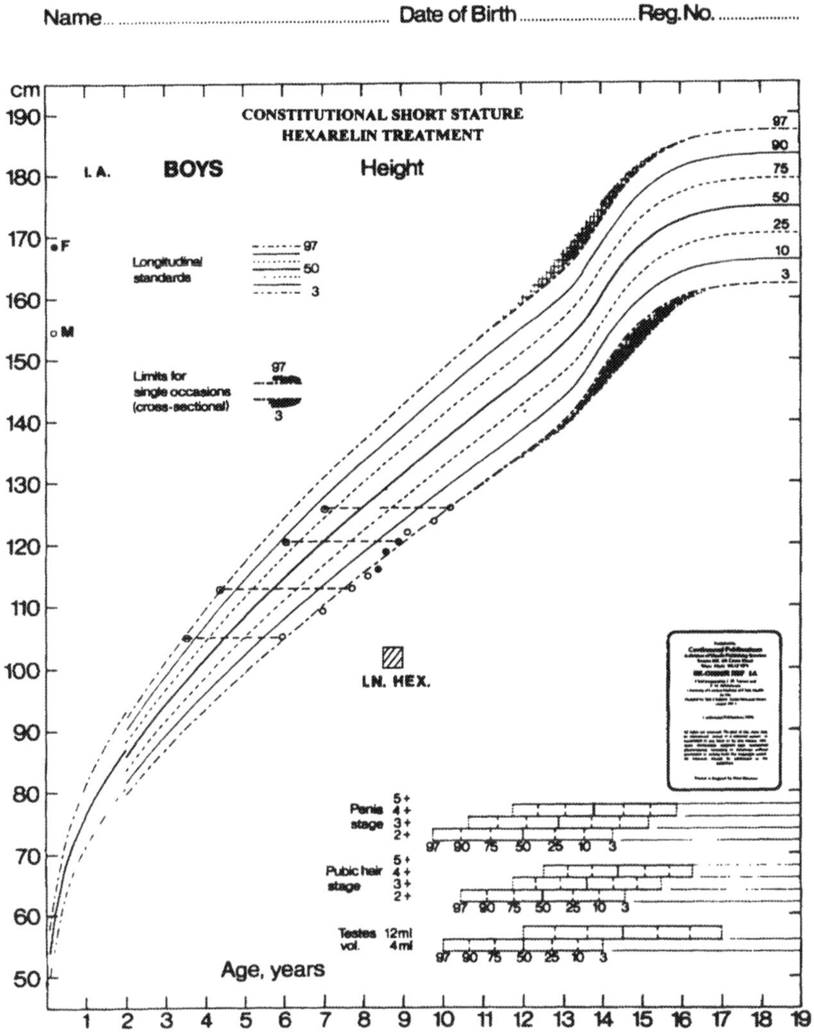

Figure 5. Growth chart of a boy with short stature before, during and after Hexarelin treatment (60 μg/kg t.i.d.).

Test 1: Serum hGH response to intranasal Hexarelin bolus

After an overnight fast, 20 μg/kg Hexarelin was administered intranasally, in the recumbent position, and blood for hGH determinations was drawn at 0, 15, 30, 45, 60, 90 and 120 minutes after drug administration. The test was performed at the start of the trial, after 7 days and after 6 months of continuous treatment.

Test 2: Serum hGH response to intravenous Hexarelin bolus

After an overnight fast, Hexarelin was administered intravenously (1 μg/kg), and blood was drawn at 0, 15, 30, 45, 60, 90 and 120 minutes after drug administration. The test was

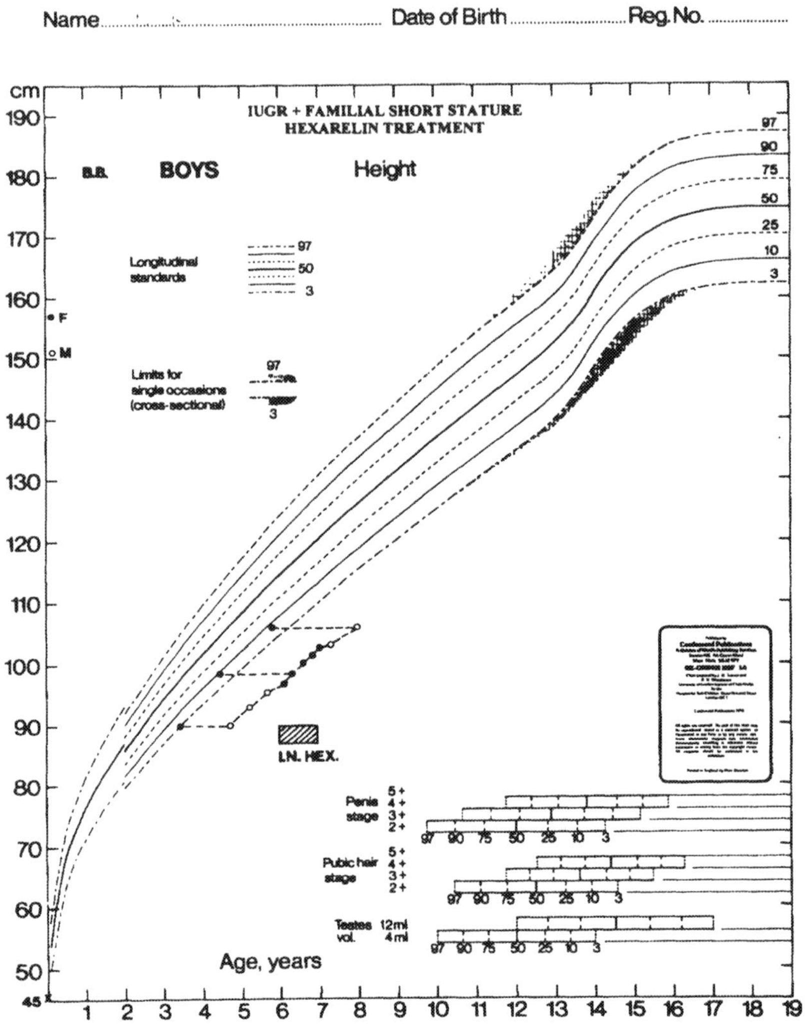

Figure 6. Growth chart of a boy with short stature before, during and after Hexarelin treatment (60 μg/kg t.i.d.)

performed before, at the end of the therapeutic trial, and 3 months after Hexarelin was stopped.

A definite decrease in the hGH response to both intravenous as well as intranasal Hexarelin bolus was registered. Figure 7 shows that already after 7 days there was a significant decrease in the response to an intranasal bolus from 70.6 ± 28.9 (m ± SD) to 34.1 ± 15.7 mU/L ($p < 0.002$). This level of suppression remained constant thereafter. Figure 8 shows the degree of desensitization observed after the intravenous Hexarelin bolus. The suppression of GH response by this test was 75% (from 84.8 ± 52 to 19.8 ± 10.9 mU/L; $p < 0.05$). Three months after discontinuation of Hexarelin the recovery of the pituitary somatrophs revealed a mean response of 42.1 ± 4.7 mU/L (50% of pretreatment response — $p < 0.005$).

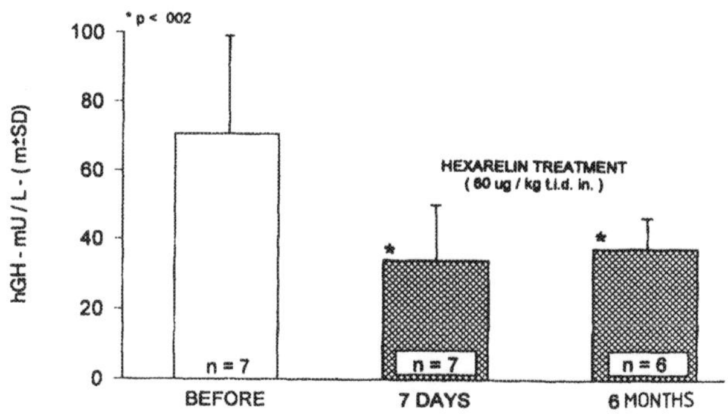

Figure 7. Mean serum hGH response (±SD) to one intranasal Hexarelin bolus (60 µg/kg t.i.d. i.n.) — before and during long-term Hexarelin treatment. Reproduced with permission from Klinger et al. (25).

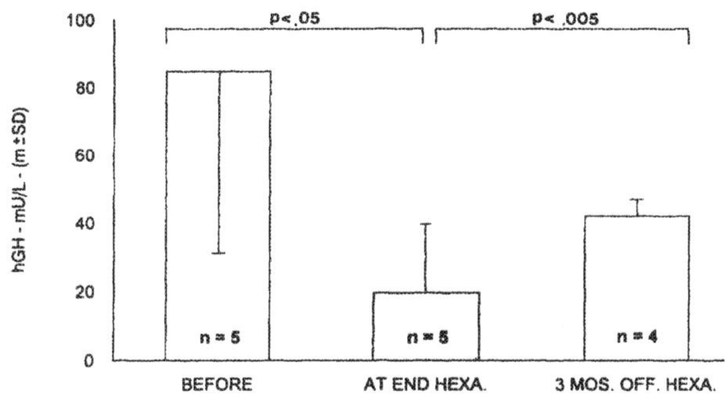

Figure 8. Peak response of serum hGH (m ± SD) to an intravenous bolus of Hexarelin (1 µg/kg) before, at the end of the long-term clinical trial and 3 months after stopping the treatment. Reproduced with permission from Klinger et al. (25).

Previous studies in adult men have also shown that a 6 hour infusion of GHRP-6 resulted in a dose-dependent desensitization to this peptide and a decline in hGH concentration (26). Jaffe et al. (27), infusing the same GHRP for 34 hours (1 µg kg^{-1} h^{-1}) to adult men, also observed a reduced response of hGH to acute GHRP-6 challenge compared to the pre-treatment response, but concluded that the pituitary remained responsive to stimuli. The mechanism of the desensitization is not known; however, it can be assumed that, similar to other hormones, it involves down-regulation at the receptor level.

The fact that most children grew faster during Hexarelin treatment than before while the endogenous GH secretion was suppressed, reveals that the latter process was only partial, as proven by circulating levels of serum IGF-I (Table 6) still higher than those at the initiation of treatment.

The only other long-term trial with a GH secretagogue is that reported in an abstract and poster by Pihoker et al (11). In that study 6 children with GH deficiency, aged 7–9 years were treated for 15 to 24 months three times daily by intranasal administration of GHRP-2 (10 to 15 µg/kg/dose). An increase in growth velocity from 4.1 to 6.0 cm per year was reported,

TABLE 6

Serum IGF-I lEVELS BEFORE, DURING AND AFTER LONG-TERM INTRANASAL HEXARELIN (60 μg/kg t.i.d.) TREATMENT OF 8 PREPUBERTAL BOYS WITH FAMILIAL SHORT STATURE

Patient no.	Serum IGF-I (nMol/L)		
	Basal	End of treatment	3 months off treatment
1	5.5	9.8	–
2	7.2	9.6	7.1
3	5.8	8.7	–
4	12.7	13.9	16.6
5	14.2	22.3	18.6
6	13.0	18.2	16.2
7	9.4	21.0	–
8	15.4	32.2	22.8
Mean ± SD	10.4±3.9	16.9±8.8	16.3±6.6
	$p < 0.02$		

without changes in serum IGF-I and IGFBP-3. One should look forward to the publication of the full results in order to understand how children with true GH deficiency could react to this type of drug and how growth velocity was accelerated without an increase in IGF-I or IGFBP-3.

Recent observations from our department have shown that Hexarelin, and probably all other GH-S, by increasing endogenous GH levels also raise serum insulin and Lp(a) the latter a recognized independent cardiovascular risk factor (28).

DISCUSSION

The limited clinical experience with the new class of drugs: "the GH secretagogues" whether small peptides or of non-peptide structure (29,30) has shown that their potency is similar to that of GHRH, but in contradistinction to the latter are also active when administered via the oral or intranasal route. The easy way of administration, even if at present it needs to be done three times a day for long acting action, and ready acceptability by the subjects, make the use of these drugs very attractive.

No local irritation has been reported by those using the intranasal insufflation nor have complaints been reported during the oral use (29,30).

The reports on a simultaneous increase in serum cortisol and prolactin are controversial (21,22) and seem to be linked to the dose used. The slight suppressive effect on TSH (12) seems to be of transitory nature (10). There is no doubt that more research has to be done on the specificity of the actions of the GH-S, especially with the new non-peptide oral preparations (31) which may act on gastrointestinal hormones as well. This is especially true considering the evidence that the GH-S acts by suppressing somatostatin (32,33), which is also present in the gastrointestinal tract and plays a role in the regulation of insulin and glucagon secretion from the pancreas and possibly also influences other local factors.

A definite undesirable effect is the desensitization effect observed already after a few days (25). However, our experience has shown that if the circulating IGF-I levels are maintained at pubertal levels the growth promoting action is not stopped (10). Whether the metabolic effects of the GH-S can be increased by using higher doses remains to be established, however, this could induce other undesirable effects overshadowing the anabolic actions and growth acceleration.

A clear limiting effect for the use of these drugs is the fact that they need functioning pituitary somatrophs and the presence of some GHRH; i.e. an intact hypothalamic pituitary axis (2,30). This excludes their use in patients with true GH deficiency (peak stimulated hGH of less than 3 to 5 μg). Whether these substances are active in partial GH deficiency remains to be established. The reported findings prove that GH-S can replace GH or IGF-I in conditions called "non-conventional" use of growth hormone, i.e. states in which one wishes to increase the secretion of endogenous hGH. Among those, in the pediatric age group, are short stature of various origins, familial, IUGR, chronic renal failure, Turner syndrome, etc., the last two indications being approved by insurance and health authorities in many countries.

The limited studies at present are insufficient to prove the optimal therapeutic approach, such as dose and routes of administration, or continuous versus intermittent administration to diminish the desensitization phenomenon.

Even if only of transitory benefit, as seems to be the case with the use of growth hormone (8), the use of these substances will be a useful addition to the drug armamentarium of the pediatric endocrinologist.

REFERENCES

1. Bowers, C.Y., Alster, D.K. and Frentz, J.M. (1992) The GH releasing activity of a synthetic heptapeptide in normal men and short stature children after oral administration. J. Clin. Endocrinol. Metab. 74, 292–298.
2. Laron, Z., Bowers, C.Y., Hirsch, D. et al. (1993) Growth hormone-releasing activity of growth hormone-releasing peptide-1 (a synthetic heptapeptide) in children and adolescents. Acta Endocrinol. 129, 424–426.
3. Loche, S., Cambiaso, P., Casini, M.R. et al. (1996) Effects of Hexarelin on growth hormone secretion in short normal children, in obese children, and in subjects with growth hormone deficiency. In: Growth Hormone Secretagogues. B.B. Bercu and R.F. Walker (eds). Springer-Verlag, New York, 347–358.

4. Bercu, B.B. and Walker, R.F. (1996) Evaluation of pituitary function in children using growth hormone secretagogues. J. Pediatr. Endocrinol. Metab. 9 (Suppl. 3), 325–332.
5. Pombo, M., Leal-Cerro, A., Barreiro, J. et al. (1996) Growth hormone releasing hexapeptide-6 (GHRP-6) test in the diagnosis of GH-deficiency. J. Pediatr. Endocrinol. Metab. 9 (Suppl. 3), 333–338.
6. Bowers, C.Y., Veeraragavan, K., Sethumadhvan, K. (1994) Atypical growth hormone releasing peptides. In: Growth Hormone II — Basic and Clinical Aspects. B.B. Bercu and R.F. Walker (eds). Springer Verlag, New York–Berlin, 203–222.
7. Lanes, R. (1995) Effects of two years of growth hormone treatment in short, slowly growing non-growth hormone deficient children. J. Pediatr. Endocrinol. Metab. 8 (Suppl.), 167–171.
8. Laron, Z., Klinger, B., Anin, S., Pertzelan, A. and Lilos, P. (1997) Growth during and 2 years after stopping GH treatment in prepubertal children with idiopathic short stature. J. Pediatr. Endocrinol. Metab. 10, 191–196.
9. Coste, J., Letrait, M., Claude, J. et al. (1997) Long term results of growth hormone treatment in France in children of short stature. Population register based study. B.M.J. 315, 708–713.
10. Laron, Z., Frenkel, J., Deghenghi, R., Anin, S., Klinger, B. and Silbergeld, A. (1995) Intranasal administration of the GHRP hexarelin accelerates growth in short children. Clin. Endocrinol. 43, 631–635.
11. Pihoker, C., Badger, T.M., Reynolds, G.A. and Bowers, C.Y. (1996) Intranasal growth hormone (GH) releasing peptide-2 (GHRP-2) in children with GH deficiency: Growth effects and GH secretion studies during treatment. Abstract P3-54 10th International Congress of Endocrinology, San Francisco, 1996, p. 768.
12. Laron, Z., Frenkel, J., Gil-Ad, I. et al. (1994) Growth hormone releasing activity by intranasal administration of a synthetic hexapeptide (hexarelin). Clin. Endocrinol. 41, 539–541.
13. Frenkel, J., Silbergeld, A., Deghenghi, R. and Laron, Z. (1995) Short term effect of intranasal administration of hexarelin ? synthetic growth hormone-releasing peptide. Preliminary communication. J. Pediatr. Endocrinol. Metab. 8, 43–45.
14. Laron, Z., Frenkel, J. and Silbergeld, A. (1996) Growth hormone-releasing peptide — Hexarelin — in children: Biochemical and growth promoting effects, Chapter 24. In: Growth Hormone Secretagogues. B.B. Bercu and R.F. Walker (eds). Springer Verlag, New York, pp. 379–387.
15. Laron, Z., Topper, E. and Gil-Ad, I. (1983) Oral clonidine — a simple, safe and effective test for growth hormone secretion. In: Evaluation of Growth Hormone Secretion. Z. Laron and O. Butenandt (eds). Pediatr. Adolesc. Endocrinol. 12, 103–115.
16. Josefsberg, Z., Kauli, R., Keret, R. et al. (1983) Tests for hGH secretion in childhood. Comparison of response of growth hormone to insulin hypoglycemia and to arginine in children with constitutional short stature in different pubertal stages. In: Evaluation of Growth Hormone Secretion. Z. Laron and O. Butenandt (eds). Pediatr. Adolesc. Endocrinol. 12, 66–74.
17. Greulich, W.W. and Pyle, S.I. (1959) Radiographic atlas of skeletal development of hand and wrist. Stanford University Press, Stanford, CA.
18. Tanner, J.M., Whitehouse, R.H. and Takaishi, M. (1966) Standards from birth to maturity for height, weight, height velocity and weight velocity. British children, 1965. Part I, II. Arch. Dis. Child 41, 454–71; 613–635.
19. Laron, Z. and Mannheimer, S. (1966) Measurement of growth hormone. Description of the method and its clinical applications. Isr. J. Med. Sci. 2, 115–119.
20. Laron, Z., Klinger, B., Jensen, L.T. and Erster, B. (1991) Biochemical and hormonal changes induced by one week of administration of rIGF-I to patients with Laron type dwarfism. Clin. Endocrinol. 35, 145–150.
21. Arvat, E., Di Vito, L., Maccagno, B. et al. (1997) Effects of GHRP-2 and hexarelin, two synthetic GH-releasing peptides, on GH, prolactin, ACTH and cortisol levels in man. Comparison with the effects of GHRH, TRH and hCRH. Peptides 18, 885–891.
22. Imbimbo, B.P., Mant, T., Edward, M. et al. (1994) Growth hormone releasing activity of Hexarelin in humans: a dose–response study. Eur. J. Clin. Pharmacol. 46, 421–425.
23. Kraiem, Z., Bowers, C.Y., Sobel, E. and Laron, Z. (1995) Growth hormone (GH)-releasing heptapeptide, but not GH-releasing hormone, inhibits thyrotropin-stimulated thyroid hormone

secretion and cAMP formation in cultured human thyroid follicles. Eur. J. Endocrinol. 133, 117–120.
24. Kirk, J.M.M., Trainer, P.J., Majrowski, W.J., Murphy, J. and Savage, M.O. (1994) Treatment with GHRH (1-29)NH2 in children with idiopathic short stature induces a sustained increase in growth velocity. Clin. Endocrinol. 41, 487–493.
25. Klinger, B., Silbergeld, A., Deghenghi, R., Frenkel, J. and Laron, Z. (1996) Desensitization from long-term intranasal treatment with hexarelin does not interfere with the biological effects of this growth hormone-releasing peptide in short children. Eur. J. Endocrinol. 134, 716–719.
26. De Bell, W.K., Pezzoll, S.S. and Thorner, M.O. (1991) Growth hormone (GH) secretion during continuous infusion of GH-releasing peptide: partial response attenuation. J. Clin. Endocrinol. Metab. 72, 1312–1316.
27. Jaffe, C.A., Ho, J., Demott-Friberg, R., Bowers, C.Y. and Barkan, A.L. (1993) Effects of a prolonged growth hormone (GH)-releasing peptide infusion on pulsatile GH secretion in normal men. J. Clin. Endocrinol. Metab. 77, 1641–1647.
28. Laron, Z., Wang, X.L., Silbergeld, A. and Wilcken, D.E.L. (1997) Growth hormone increases and insulin-like growth factor-I decreases circulating lipoprotein(a). Eur. J. Endocrinol. 136, 377–381.
29. Laron, Z. (1995) Growth hormone releasing peptides: clinical experience and therapeutic potential. Drugs 50, 595–601.
30. Ghigo, E., Arvat, E., Muccioli, G. and Camanni, F. (1997) Growth hormone-releasing peptides. Eur. J. Endocrinol. 136, 445–460.
31. Jacks, T., Smith. R., Judith, F. et al. (1996) MK-0677, a potent, novel, orally active growth hormone (GH) secretagogue: GH, insulin-like growth factor I, and other hormonal responses in beagles. Endocrinology 137, 5284–5289.
32. Smith, R.G., Van Der Ploeg, L.H.T., Howard, A.D. et al. (1997) Peptidomimetic regulation of growth hormone secretion. Endocrine Rev. 18, 621–645.
33. Conley, L.K., Teik, J., Deghenghi, R. et al. (1995) Mechanism of action of hexarelin and GHRP-6: analysis of the involvement of GHRH and somatostatin in the rat. Neuroendocrinology 61, 44–50.

Growth Hormone Secretagogues
Edited by E. Ghigo, M. Boghen, F.F. Casanueva and C. Dieguez

Chapter 21

Therapeutic Potential of GH Secretagogues in Adults

RALF M. NASS and MICHAEL O. THORNER

Division of Endocrinology and Metabolism, Department of Medicine, University of Virginia, Charlottesville, VA, 22908, U.S.A.

INTRODUCTION

Growth hormone releasing peptides (GHRP) were first discovered in 1976 (1). They are hepta- and hexapeptides which stimulate GH release in animals and humans. In addition, a series of non-peptidyl GH secretagogues have been developed. Further experiments showed that GHRPs act through a specific receptor, which was recently cloned (2). The existence of this receptor strongly suggests the existence of a natural GHRP-like ligand. This chapter will discuss some of the therapeutic applications of these GH secretagogues.

REGULATION OF GH SECRETION

Growth hormone is secreted by the pituitary, under complex hypothalamic control (3). The current concept of regulation of GH secretion includes that the hypothalamic peptides GHRH and somatostatin are key players and exert reciprocal effects on GH secretion. It is important to note that this concept neglects the fact that exact and reproducible measurements of these peptides in hypophysial portal blood are difficult (4). GH is also regulated by IGF-1 negative feedback at the pituitary (5) and by a GH short-loop negative feedback (6). Further parameters affecting the pattern of GH secretion include several metabolic components (7). As growth hormone is secreted physiologically in a pulsatile fashion, subcutaneous bolus injections of growth hormone constitute a non-physiologic intervention. The biologic significance of episodic secretion has been demonstrated in GH-deficient animals where GH administered in a pulsatile pattern was more efficient in improving growth than when given by constant infusion (8). Other animal studies have shown that the narrower GH pulses seen after GHRP, compared to GH administration, are less diabetogenic and have the same growth-promoting effects (9). Therefore the GHRPs and peptidomimetic GH secretagogues provide a more physiological approach to GH replacement, however subjects must have both an intact hypothalamic/pituitary axis and produce normal GH (10). Another advantage of treatment with GH secretagogues compared to GH treatment, is the moderating effect of insulin-like growth factor-I (IGF-I) feedback on the somatotroph, which could buffer against overtreatment.

EFFECTS OF AGING ON GH SECRETION

It is generally accepted that the secretion of growth hormone (GH) decreases with aging (11). In elderly subjects, the 24 hour integrated GH concentration is low and equal to levels observed in patients with GH deficiency (12). Different authors have described a decrease in GH secretory parameters from 15% to 70% in men and women over 60 years of age (13,14). Data in elderly men and women suggest that mean GH pulse amplitude and duration are reduced, while others describe a significant reduction in GH secretory burst frequency and half-life (14). However these observations were made before the advent of more sensitive GH assays which allow measurement of GH levels in all serum samples (15). In addition IGF-1 levels are significantly reduced (16).

It has been proposed that the relative GH deficiency of people over 60 years of age is partly responsible for the decrease in both muscle and bone mass and increase in adipose tissue seen in this population (17,18). In addition GH deficiency in adults is accompanied with changes in blood lipid profiles (19) that increase the risk for development of atherosclerotic vascular disease. Whether the age-dependent decline in peripheral GH levels also increases the risk of atherosclerotic vascular disease has yet to be determined.

Since the availability of unlimited supplies of recombinant DNA derived GH (20), several studies have been undertaken to investigate the benefits of GH therapy in different clinical settings, other than GH-deficient children. The results of these studies show that GH therapy has beneficial effects in GH-deficient adults with acquired GH deficiency during adulthood in terms of body composition, exercise capacity, blood lipid levels and quality of life (21–26). Additional studies in older patients show that the treatment with GH might have beneficial effects on bone mineral density, blood lipids, lean body mass and quality of life (27,28).

Data from several groups strongly support the hypothesis that hypothalamic impairment underlies GH insufficiency in aging (29). The mechanism responsible for this age-related impairment is not fully understood. The finding that arginine infusion, either alone or in combination with hexarelin, restores GH secretion to that observed in young adults (30), suggests a chronic increase in somatostatinergic tone as the underlying mechanism. Data investigating the effects of the combination of an arginine infusion and fasting support this concept (31). In this context it is interesting to note that GH secretagogues produce depolarization of somatotrophs and facilitate an influx of calcium, suggesting that they behave as functional antagonists of somatostatin at the level of the pituitary (32). Additional data (30,33) support the concept that the releasable pool of pituitary GH is preserved in aging, which is one requirement for the optimal effects of GH secretagogues.

Therefore, treatment with GH-releasing substances is potentially more favorable than that with exogenous GH in the elderly. As several GHRPs and non-peptidergic GH secretagogues have significant releasing effects after oral administration, several studies have investigated the clinical usefulness of these substances in the normal and GH insufficiency state.

THERAPEUTIC POTENTIAL OF GH SECRETAGOGUES IN AGING

Early clinical studies with GH secretagogues were done in healthy young normal volunteers to investigate the effects of a 24 h infusion of a GHRP on spontaneous pulsatile GH release (34). This study demonstrated that GH-releasing peptides can enhance GH secretion 8-fold compared to saline in normal young men and that the pattern of GH was pulsatile during the infusion. After a 24 h infusion of GHRP, the response to GHRH was not attenuated but enhanced, indicating that depletion of GH stores does not occur under these treatment conditions. A study performed in 24 healthy nonobese young male volunteers with L-692,429, a nonpeptide GH mimic (35), showed a significant GH peak response seen 30–45 min after initiation of a 15 min infusion. In addition the study proved the safety and tolerability of this compound.

A study by Aloi et al. (36) performed in healthy older men and women, suggested that acute administration of L-692,429 (L), in a dose of 0.75 mg/kg, causes a greater increase in GH secretion than GHRH. The integrated GH concentrations over 4 h after administration of this compound were significantly greater than after saline treatment.

Another study performed at the University of Virginia (37) investigated the effects of a continuous 24-h low dose infusion (0.05 mg/kg h) of (L) in older adults (four men and two women, aged 64–82 years) or a 12-h high dose infusion (0.1 mg/kg h). The 12-h high dose infusion resulted in a significant nearly 4-fold enhancement of pulsatile GH secretion (Figure 1). Secretion was increased by a greater mass of GH secreted per pulse, but not the number of secretion pulses. Both studies suggest that during continuous infusion of both GHRP and non- peptidergic GH secretagogues: (1) GH secretion remains pulsatile and (2) desensitization of the GH response does not occur.

Copinschi et al. (38) had reported that 7-day treatment of healthy normal young men (18–30 years) with 5 and 25 mg MK 677 resulted in an increase in pulse frequency of GH.

Ghigo and colleagues demonstrated that there is reduced somatotroph responsiveness to GHRPs in aging, which is most likely explained by a reduction of GHRH activity and an increase of somatostatinergic tone (30).

Another clinical trial (39) investigated the effects of an orally active GHRP mimetic (MK-677) in 32 older healthy subjects (15 women and 17 men, aged 64–81 years) in a randomized, double-blind, placebo-controlled trial. MK-677 was orally administered, once daily for 2 separate consecutive study periods of 14 and 28 days in doses of 2,10,25 mg or placebo. The twenty-five mg MK-677 dose increased the mean 24-h GH concentration significantly (97 ± 23%) (Figure 2). GH pulse height and interpulse nadir concentrations

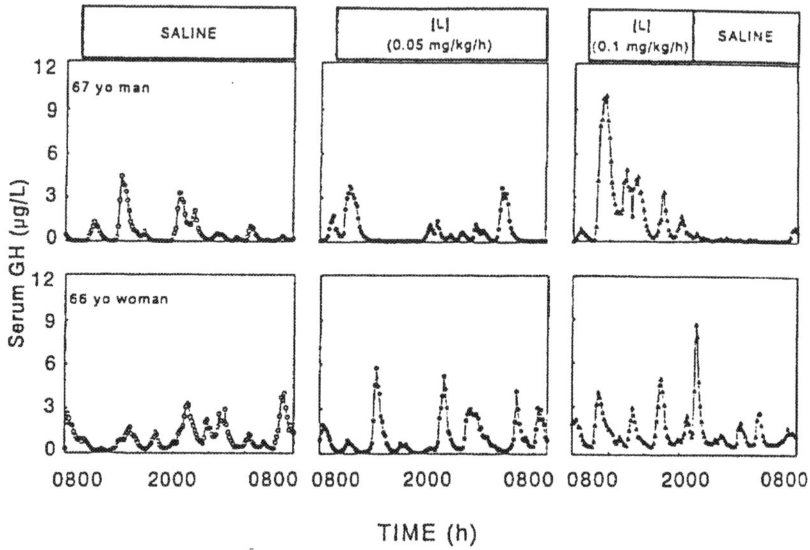

Figure 1. Individual 24-h GH concentration profiles for two representative older adults during the three treatment regimens, showing persistence of pulsatile GH release during [L] infusions: a 2-h baseline saline infusion (0800–1000 h) was followed by (1) a 24-h infusion (1000–1000 h) of saline (control: ○; *left panels*), (2) a 24-h infusion of low dose [L] (0.05 mg/kg h; ●; *middle panels*), or (3) a 12-h infusion (1000–2200 h) of high dose [L] (0.1 mg/kg h; ▲) followed by a 12-h infusion (2200–1000 h) of saline (*right panels*). Reproduced with permission from Chapman et al. (37).

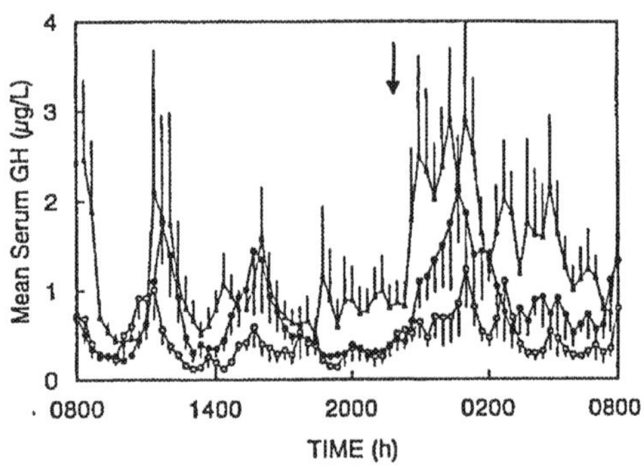

Figure 2. Mean (± SEM) serum GH concentrations (µg/L) in older subjects after 2 weeks of daily oral administration of placebo (○; n = 10), 10 mg/day MK-677 (●; n = 12), and 25 mg/day MK-677 (▲; n = 10). Evening treatment time (between 2200 and 2300 hours) is indicated by an arrow. Reproduced with permisssion from Chapman et al. (39). © The Endocrine Society.

also increased significantly without significant changes in the pulse number. The IGF-1 concentrations increased into the normal range for young adults (141 ± 21 µg/L at baseline, 219 ± 21 µg/L at 2 weeks and 265 ± 29 µg/L at 4 weeks) (Figure 3). Thyroid hormone levels, serum cortisol, diurnal cortisol pattern and urinary free cortisol concentrations did not change and prolactin concentrations remained within the normal range. Treatment with 10 and 25 mg MK-677 was associated with a significant increase in glucose and insulin, but as the expected enhancement of GH secretion produced by long-term MK-677 treatment will

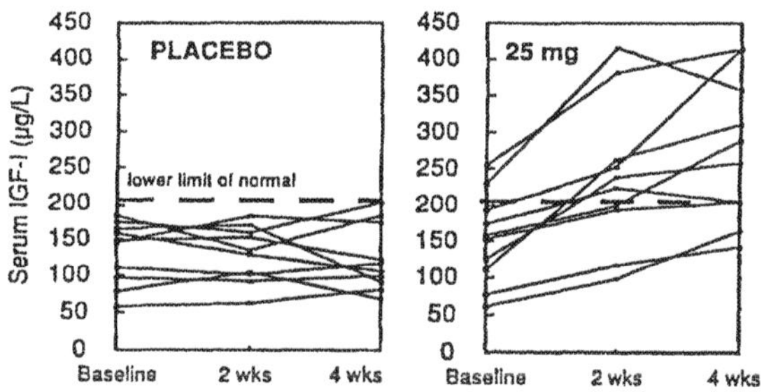

Figure 3. Serum IGF-I concentrations (µg/L) of individual older subjects at baseline and after 2 and 4 weeks of treatment with oral placebo (n = 10; left panel) or 25 mg MK-677 (n = 10; right panel) administered once daily in the evening between 2200 and 2300 hours. The dotted line represents the lower limit of the normal range for young adults aged 21–25 years (202–453 µg/L). Reproduced with permission from Chapman et al. (39). © The Endocrine Society.

result in reduction of body fat, it is possible that insulin sensitivity decreases and changes in glucose levels will improve. This study also confirmed that one month of treatment with MK-677 caused no desensitization of the hypothalamic–pituitary axis.

The authors have planned a two year, double-blind, placebo-controlled study of MK-677 in healthy elderly adults to investigate the effects on body composition and functional ability. This study should provide further information about safety and tolerance of MK-677 when given as long-term treatment.

THERAPEUTIC POTENTIAL OF GH SECRETAGOGUES IN CHILDREN

Loche and coworkers could demonstrate that hexarelin iv bolus infusions (2 µg/kg bw) can increase GH release in children with short stature and this response is further increased after priming with testosterone (40). In a trial performed by Mericq et al. (41), GHRP-2 was administered subcutaneously to six prepubertal children with growth hormone deficiency (GHD). GHD was defined as a GH response of less than 7 µg/L to at least two standard provocative tests and a growth velocity of 4 cm/year or less. The agents were administered for 6 months at increasing doses (0.3, 1.0, 3.0 µg/kg/day). At months 7 and 8 the children received GHRP-2 (3 µg/kg/day) together with GHRH (3 µg/kg/day). The maximal overnight GH and GH peak amplitude showed a progressive increase at the higher doses. The number of nocturnal GH peaks was unaffected. IGF-1 and IGFBP-3 levels did not increase significantly after this treatment period. Growth velocities were increased when compared to baseline. The study was not placebo-controlled. During the long treatment period the GHRP-2 injections were well tolerated.

Similar results have been reported by Laron et al. (42) using hexarelin administered 3 times daily intranasally to a group of prepubertal short children. However before more large scale studies with GHRPs can be conducted in children with GH deficiency, the pharmacokinetics and pharmacodynamics in pediatric patients must be characterized. First data about this issue have been published by Pihoker et al. (43).

THERAPEUTIC POTENTIAL OF GH SECRETAGOGUES IN OBESITY

The alterations in GH secretory pattern seen in both obesity and starvation underline the importance of nutritional conditions in GH secretion. Computer-assisted (deconvolution) analysis shows that the half-life of endogenous GH is significantly less in obese than in normal weight subjects (44). Furthermore, obese men had significantly fewer GH secretory bursts. Whereas GH secretion is enhanced during fasting, in obese individuals, spontaneous GH secretion is attenuated and the GH response to all tested stimuli (hypoglycemia, L-dopa, glucagon, exercise, clonidine and the secretagogues (GHRH and arginine) is significantly decreased (45,46). The fact that both exogenous GHRH- and GHRP-6-induced GH secretion are attenuated in obese volunteers eliminates a secretory deficit of either endogenous GHRH or the endogenous ligand for the GHRP-6-receptor as causative factors. This condition is reversed by weight loss or fasting.

The reason for the blunted GH response in obesity is unclear. Some data suggest that an enhanced somatostatinergic tone is the mechanism behind altered somatotrophic function in obesity (47). Other authors suggest that a chronic increase in somatostatin might lead to reduced function of somatotroph cells (48). Data published by Cordido et al. (49) showed that the combination of GHRH (100 μg iv) and GHRP-6 (100 μg iv) in obese subjects led to an increase in GH secretion, which did not differ greatly from that observed in normal subjects. Surprisingly, the combined administration of GHRP-6 plus GHRH seems to act independently of the somatostatinergic tone (50) as the effects of the combined administration of GHRP-6 and GHRH are not further affected by pretreatment with pyridostigmine. These authors also reported that the somatotrope cell in obesity has a considerable GH secretory capacity and that somatotrophs do not atrophy in the obese. Another interesting study suggested that free fatty acids were involved in the disrupted GH secretion of obesity (51).

L-692,429 has been investigated in two clinical situations in which GH secretion is attenuated: obese individuals, and non-obese individuals in the postprandial state (52). The doses administered were 0.2 mg/kg and 0.75 mg/kg over 15 min iv. The mean peak GH

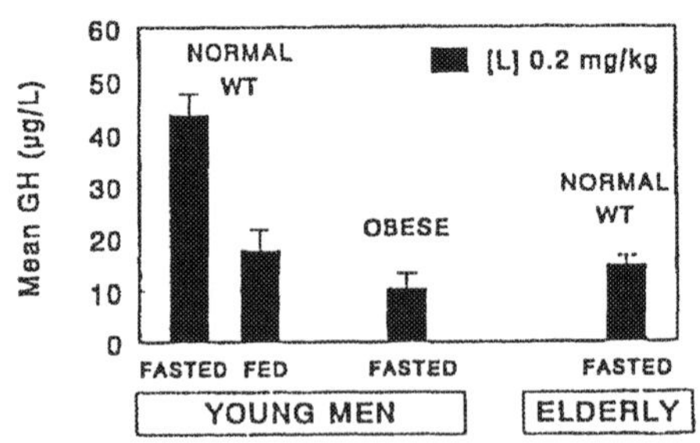

Figure 4. Mean (± SEM) peak GH response (μg/L) to 0.2 mg/kg L-692,429 in 10 non-obese, fasted (from midnight) and fed (standardized breakfast 45 minutes before treatment) young men and 12 obese, fasted (from midnight) young men ($P < 0.05$ compared to fasted, non-obese, young men). The mean peak GH response in normal-weight elderly subjects is shown in the right hand column for comparison. Reproduced with permission from Kirk et al. (52). © The Endocrine Society.

response to 0.2 mg/kg (L) in fasted non-obese individuals was approximately 4-fold higher than that in the fasted obese individuals (Figure 4). The mean peak GH response in fasted obese subjects was similar to fed non-obese subjects, which suggests that obesity may alter feedback to the hypothalamic–pituitary axis in a manner similar to food ingestion. The study confirmed once again that GH secretagogues are well tolerated and may have therapeutic benefits in this population as well.

A study by Svensson et al. (53) investigated the effects of 25 mg MK 677 daily for 8 weeks in 24 obese men. The IGF-1 levels increased approximately 40% as well as the IGFBP-3 levels, which were also significantly increased. Fat-free mass increased significantly, whereas total and visceral fat were not significantly changed. Interestingly, GH levels were increased by MK-677 treatment throughout the 8-week study period, even though the GH response to MK-677 was lower at 2 and 8 weeks compared to the initial response. The authors speculate that the observation period was too short or the dose too low to appreciate changes in visceral fat (Figure 5). Furthermore no women were included into the study. Johansson et al. (54) described that in obese males, GH treatment reduces the visceral fat mass over a 9 month period.

As obesity is an important risk factor in hypertension, diabetes and cardiovascular disease, the results of long-term studies investigating the effects of GH secretagogues on body composition may provide important information on possible therapeutic applications in other pathologic states as well.

THERAPEUTIC POTENTIAL OF GH SECRETAGOGUES IN THE CATABOLIC STATE

In the acute phase of critical illness the response of the somatotropic axis is characterized by an increase of the total amount of GH released. The interpulse concentrations of GH are relatively high whereas the serum concentrations of IGF-1 are low (55). These changes are probably explained by a state of GH resistance, which may be related to a decreased GH receptor expression (56). Whereas IGFBP-3 decreases, IGFBP-1 remains in normal or slightly elevated concentrations (57).

In prolonged critical illness GH secretion is characterized by a reduced pulse amplitude, elevated interpulse levels and the number of pulses is high (58). This condition is also associated with low levels of IGF-1,IGFBP-3 and acid-labile subunit (58). Growth hormone binding protein (GHBP) has been found to be low in critically ill patients; GHBP is positively regulated by GH (57).

Several studies have demonstrated that there are short-term benefits in nitrogen balance following the administration of pharmacological doses of GH to patients in an increased catabolic state. Vara-Torbeck and colleagues (59) investigated the effects of GH treatment (2.6 mg daily) in 180 patients undergoing cholecystectomy in a placebo-controlled study. The results suggest that there is a reduction in hospital stay and infection rate in the

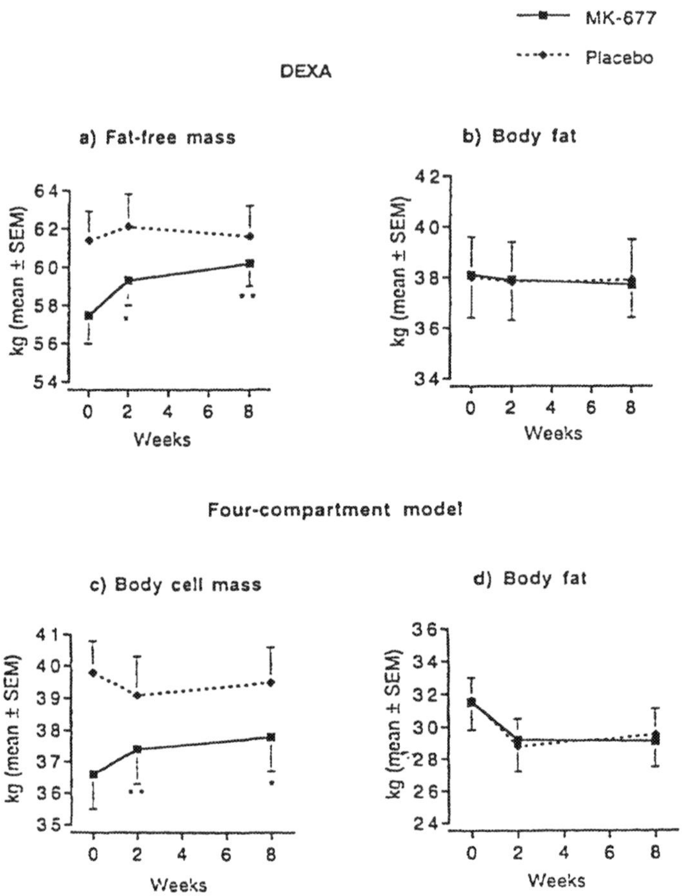

Figure 5. Body composition during 2-month treatment with MK-677 (25 mg) or placebo daily in obese subjects. (a) and (b) show FFM and body fat, respectively, as assessed by DEXA. (c) and (d) show body cell mass and body fat, respectively, as calculated using the four-compartment model. The vertical bars indicate the SE for the mean values shown. *P* values are based on a between-group analysis of the percent change from baseline. $^{*}P < 0.05$; $^{**}P < 0.01$. Reproduced with permission from Svensson et al. (53).

treatment group compared to those receiving placebo. Nevertheless, due to the state of GH resistance it is anticipated that ongoing trials with exogenous GH may be unable to demonstrate major benefit in the acute phase of illness (60). Recently a large multicenter study of GH treatment in intensive care unit patients was discontinued due to an increase in mortality of GH treated patients.

Short-term (<12 weeks) rhGH trials on the control of wasting in patients with AIDS demonstrated that rhGH increases LBM and decreases adipose tissue (61–68). The rhGH formulation of one company has recently been granted accelerated approval by the Food and Drug Administration (FDA) for the treatment of AIDS-related wasting syndrome.

The largest study so far has been performed by Schambelan et al. (61) with 178 HIV-positive patients with weight loss averaging 14%. The patients received either rhGH 0.1

mg/kg/d or placebo for 12 weeks and exhibited significant weight gain, significant difference in LBM and decreased fat mass with treatment.

Although in the later stage of the disease increased basal muscle protein degradation and decreased responsiveness of muscle protein synthesis to GH has been described (69).

Van den Berghe et al. (70) demonstrated that continuous infusion of GH secretagogues (GHRH and GHRP-2) amplifies pulsatile GH release and generates an increase of IGF-1 within 21 h in critical illness. An additional study in patients with prolonged critical illness showed that the coinfusion of GHRH and GHRP-2 increased pulsatile GH secretion 6- and 10-fold, and increased IGF-1 levels after 9 h infusion.

In conclusion in the chronic phase of critical illness, the whole somatotropic axis was found to be responsive to GH secretagogues. Initial attempts to investigate the effects of MK 677 in a catabolic study were reported by Clemmons et al. (71). This group demonstrated that 25 mg MK-677 administered daily for 7 days increased endogenous GH secretion and reversed nitrogen loss in normal volunteers made catabolic by caloric restriction (18 kcal/kg). All in all critically ill patients seem to be more resistant to the anabolic actions of GH. The role of synthetic GH and GH secretagogues as therapeutic agents in the catabolic state has yet to be established.

THERAPEUTIC POTENTIAL OF GH AND GH SECRETAGOGUES IN CHRONIC HEART FAILURE

Evidence exists that GH is a physiologic regulator of myocardial growth and performance (72). Growth hormone can activate cardiac cell growth and induce physiologic ventricular remodeling which is associated with enhanced contractile performance (73). Under circumstances of long-term excess of GH the force of cardiac contraction is increased despite a redistribution of the myosin heavy chain isoforms. This observation led to the hypothesis that GH may improve the thermodynamic efficiency of the contractile apparatus by reducing energy cost (74).

Studies published by two different groups (75,76) demonstrated that adults with GH deficiency have reduced left ventricular mass and impaired systolic function. The same authors describe six months after GH therapy a significant improvement of left ventricular mass, stroke volume and cardiac output and a reduction in peripheral vascular resistance. A sustained effect on cardiac performance, still present after 3 years' GH treatment, was reported by two open studies (77,78).

Similarly Frustaci et al. (79) reported an improvement in ventricular diameter, increase in wall thickness, ejection fraction and myofibrillar content in one patient with cardiac chamber dilatation after 3 months of GH therapy. Another study (80) performed in a small number of patients with dilated cardiomyopathy showed that three months of GH therapy with subcutaneous injections (50 μg–66 μg per kilogram of body weight per week) given every other day, resulted in an improvement of left-ventricular-wall thickness and reduced chamber size.

Animal data suggest that GH given during the early phase of remodeling after experimental myocardial infarction in rats attenuates the early pathologic LV remodeling and improves LV function (81). These data are supported by Isgaard et al. (82), who showed that GH in a physiological dose improves systolic function in rats with experimental heart failure. These studies emphasize the concept that GH has a positive impact on cardiac function in the pathologic conditions of heart failure and myocardial infarction (83) but no adequate placebo controlled studies are published yet. The only published data that addressed the question whether GH secretagogues have a beneficial impact on cardiac function have been published recently in animal data. DeGennaro et al. (84) described the administration of hexarelin over 15 days in GH-deficient animals treated with an anti-GH-releasing hormone (GHRH). Impaired somatotropic function was restored to normal which both counteracted ischemic damage and improved postischemic ventricular function. Interestingly, the vasopressor activity of angiotensin II reverted to those of controls. Although the role of GH in the regulation of cardiac function and myocardial structure has not yet been fully defined, the potential role of GH secretagogues in the therapy of cardiac pathology is an important area which deserves further investigation.

Finally, the authors would like to emphasize that the difference in specificity relating to the somatotropic axis and variable effects on the corticotropin axis of the GH secretagogues require that the results of the above mentioned studies in this chapter are interpreted based on the specific agent used. Some appear to affect appetite (85) which adds further complexity in addition to the differences in bioavailability and duration of action of these compounds.

REFERENCES

1. Momany, F.A., Bowers, C.Y., Reynolds, G.A., Hong, A. and Newlander, K. (1984) Conformational energy studies and *in vitro* and *in vivo* activity data on growth hormone-releasing peptides. Endocrinology 114, 1531–1536.
2. Howard, A.D., Feighner, S.D., Cully, D.F., Arena, J.P. and Liberator, P.A. et al. (1996) A receptor in pituitary and hypothalamus that functions in growth hormone release. Science 273, 974–977.
3. Thorner, M.O., Vance, M.L., Horvath, E. and Kovacs, K. (1998) The anterior pituitary. In: Williams Textbook of Endocrinology, 9th edit. J.D. Wilson and D.W. Foster (eds). W.B. Saunders & Co., Philadelphia, USA.
4. Frohman, L.A., Downs, T.R., Clarke, I.J. and Thomas, G.B. (1990) Measurement of growth hormone-releasing hormone and somatostatin in hypothalamic-portal plasma of unanesthetized sheep. Spontaneous secretion and response to insulin-induced hypoglycemia. J. Clin. Invest. 86, 17–24.
5. Hartman, M.L., Clayton, P.E., Johnson, M.L. et al. (1993) A low dose euglycemic infusion of recombinant human insulin-like growth factor I rapidly suppresses fasting-enhanced pulsatile growth hormone secretion in humans. J. Clin. Invest. 91, 2453–2462.
6. Tannenbaum, G.S. (1980) Evidence for autoregulation of growth hormone secretion via the central nervous system. Endocrinology 107, 2117–2120.
7. Ho, K.Y., Veldhuis, J.D., Johnson, M.L., Furlanetto, R., Evans, W.S., Alberti, K.G. and Thorner, M.O. (1988) Fasting enhances growth hormone secretion and amplifies the complex rhythms of growth hormone secretion in man. J. Clin. Invest. 81, 968–975.

8. Gevers, E.F., Wit, J.M. and Robinson, I.C.A.F. (1996) Growth, growth hormone (GH)-binding protein, and GH receptors are differentially regulated by peak and trough components of the GH secretory pattern in the rat. Endocrinology 137, 1013–1018.
9. Saenger, P. (1996) Editorial: Oral growth hormone secretagogues — Better than Alice in Wonderland's growth elixir? JCE&M 81, 2773–2775.
10. Mallo, F., Alvarez, C.V., Benitez, L., Burguera, B., Coya, R., Casanueva, F.F. and Dieguez, C. (1993) Regulation of His-D-Trp-Ala-Trp-D-Phe-Lys-NH_2 (GHRP-6)-induced GH secretion in the rat. Neuroendocrinology 57, 247–256.
11. Corpas, E., Harman, S.M. and Blackman, M.R. (1993) Human growth hormone and human aging. Endocrine Reviews 14, 20–39.
12. Zadik, Z., Chalew, S.A., McCarter, R.J., Meistas, M. and Kowarski, A.A. (1985) The influence of age on the 24-hour integrated concentration of growth hormone in normal individuals. JCE&M 60, 513–16.
13. Finkelstein, J.W., Roffwarg, H.P., Boyar, R., Kream, J. and Hellman, L. (1972) Age-related changes in the twenty-four hour spontaneous secretion of growth hormone. JCE&M 35, 665–70.
14. Ho, K.Y., Evans, W.S., Blizzard, R.M. et al. (1987) Effects of sex and age on 24-hour profile of growth hormone secretion in men: importance of endogenous estradiol concentrations. JCE&M 64, 51–8.
15. Chapman, I.M., Hartman, M.L., Straume, M., Johnson, M.L., Veldhuis, J.D. and Thorner, M.O. (1994) Enhanced sensitivity growth hormone (GH) chemiluminescence assay reveals lower postglucose nadir GH concentration in men than women. J. Clin. Endocrinol. Metab. 78, 1312–1319.
16. van Coevorden, A., Mockel, J., Laurent, E. et al. (1991) Neuroendocrine rhythm and sleep in aging men. Am. J. Physiol. 260, E651–E661.
17. Rudman, D. (1985) Growth hormone, body composition, and aging (Review). J. Am. Geriatr. Soc. 33, 800–807.
18. Rudman, D., Feller, A.G., Nagraj, H.S., Gergans, G.A., Lalitha, P.Y., Goldberg, A.F., Schlenker, R.A., Cohn, L., Rudman, I.W. and Mattson, D.E. (1990) Effects of human growth hormone in men over 60 years old. N. Engl. J. Med. 323, 1–6.
19. Rosen, T., Bengtsson, B.A. (1990) Premature mortality due to cardiovascular disease in hypopituitarism. Lancet 336, 285–288.
20. Seeburg, P.H., Shine, J., Martial, J.A., Ivarie, R.D., Morris, J.A., Ullrich, A., Baxter, J.D. and Goodman, H.M. (1978) Synthesis of growth hormone by bacteria. Nature 276, 795–98.
21. Attanasio, A.F., Lamberts, S.W.J., Matranga, A.M.C. et al. (1997) Adult growth hormone-deficient patients demonstrate heterogeneity between childhood onset and adult onset before and during human GH treatment. JCE&M 82, 82–88.
22. O'Halloran, D.J., Tsatsoulis, A., Whitehouse, R.W., Holmes, S.J., Adams, J.E. and Shalet, S.M. (1993) Increased bone density after recombinant human growth hormone therapy in adults with isolated GH deficiency. JCE&M 76, 1344–1348.
23. Johannsson, G., Rosen, T., Bosaeus, I., Sjostrom, L. and Bengtsson, B.A. (1996) Two years of growth hormone (GH) treatment increases bone mineral content and density in hypopituitary patients with adult-onset GH deficiency. JCE&M 81, 2865–2873.
24. Nass, R., Huber, R.M., Klauss, V., Muller, O.A., Schopohl, J. and Strasburger, C.J. (1995) Effect of growth hormone (hGH) replacement therapy on physical work capacity and cardiac and pulmonary function in patients with hGH deficiency acquired in adulthood. JCE&M 80, 552–557.
25. McGauley, G.A. (1989) Quality of life assessment before and after growth hormone treatment in adults with growth hormone deficiency. Acta Paediatr. Scand. 356(Suppl), 70–74.
26. Cuneo, R.C., Salomon, F., Watts, G.F., Hesp, R. and Sonksen, P.H. (1993) Growth hormone treatment improves serum lipids and lipoproteins in adults with growth hormone deficiency. Metabolism 42, 1519–1523.
27. Khorram, O., Yeung, M., Vu, L. and Yen, S.S. (1997) Effects of (norleucine27) growth hormone-releasing hormone (GHRH) (1-29)-NH2 administration on the immune system of aging men and women. JCE&M 82, 3590–96.

28. Vittone, J., Blackman, M.R., Busby-Whitehead, J., Tsiao, C., Stewart, K.J., Tobin, J., Stevens, T., Bellantoni, M.F., Rogers, M.A., Baumann, G., Roth, J., Harman, S.M. and Spencer, R.G. (1997) Effects of single nightly injections of growth hormone-releasing hormone (GHRH 1-29) in healthy elderly men. Metabolism: Clinical & Experimental 46, 89–96.
29. Thorner, M.O., Chapman, I.M., Gaylinn, B.D., Pezzoli, S.S. and Hartman, M.L. (1997) Growth hormone-releasing hormone and growth hormone-releasing peptide as therapeutic agents to enhance growth hormone secretion in disease and aging. Recent Progress in Hormone Research 52, 215–246.
30. Arvat, E., Gianotti, L., Grottoli, S., Imbimbo, B.P., Lenaerts, V., Deghenghi, R. et al. (1994) Arginine and growth hormone-releasing hormone restore the blunted growth hormone-releasing activity of hexarelin in elderly subjects. JCE&M 79, 1440–43.
31. Nass, R.M., Chapman, I.M., Skiles, E.H., Pezzoli, S.S., Vance, M.L., Hartman, M.L. and Thorner, M.O. (1997) Evidence for enhanced somatostatin release in the elderly: Effects of arginine and fasting in growth hormone release in young and older men. Abstr. P1–398, 79th Annual Meeting of the Endocrine Society, Minneapolis, USA.
32. Koch, B.D., Blalock, J.B. and Schonbrunn, A. (1988) Characterization of the cyclic AMP-independent actions of somatostatin in GH cells. An increase in potassium conductance is responsible for both the hyperpolarization and the decrease in intracellular free calcium produced by somatostatin. J. Biol. Chem. 263, 216–225.
33. Micic, D., Popovic, V., Doknic, M., Macut, D., Dieguez, C. and Casanueva, F.F. (1998) Preserved growth hormone (GH) secretion in aged and very old subjects after testing with the combined stimulus GH-releasing hormone plus GH-releasing hexapeptide-6. JCE&M 83, 2569–72.
34. Huhn, W.C., Hartman, M.L., Pezzoli, S.S. and Thorner, M.O. (1993) Twenty-four-hour growth hormone (GH)-releasing peptide (GHRP) infusion enhances pulsatile GH secretion and specifically attenuates the response to a subsequent GHRP bolus. JCE&M 76, 1202–1208.
35. Gertz, B.J., Barret, J.S., Eisenhaendler, R., Krupa, D.A., Wittreich, J.M., Seibold, J.R. and Schneider, S.H. (1993) Growth hormone response in man to L-692,429, a novel nonpeptide mimic of growth hormone-releasing peptide-6. JCE&M 77, 1393–1397.
36. Aloi, J.A., Gertz, B.J., Hartman, M.L., Huhn, W.C., Pezzoli, S.S., Wittreich, J.M., Krupa, D.A. and Thorner, M.O. (1994) Neuroendocrine responses to a novel growth hormone secretagogue, L-692,429, in healthy older subjects. JCE&M 79, 943–949.
37. Chapman, I.M., Hartman, M.L., Pezzoli, S.S. and Thorner, M.O. (1996) Enhancement of pulsatile growth hormone secretion by continuous infusion of a growth hormone-releasing peptide mimetic, L-692,429, in older adults: a clinical research center study. JCE&M 81, 2874–2880.
38. Copinschi, G., Van Onderbergen, A., L'Hermite-Baleriaux, M., Mendel, C.M., Caufriez, A., Leproult, R., Bolognese, J.A., DeSmet, M., Thorner, M.O. and Van Cauter, E. (1996) Effects of a 7-day treatment with a novel, orally active, growth hormone (GH) secretagogue, MK-677, on 24-hour GH profiles, insulin-like growth factor I, and adrenocortical function in normal young men. JCE&M 81, 2776–82.
39. Chapman, I.M., Bach, M.A., Van Cauter, E., Farmer, M., Krupa, D., Taylor, A.M., Schilling, L.M., Cole, K.Y., Skiles, E.H., Pezzoli, S.S., Hartman, M.L., Veldhuis, J.D., Gormley, G.J. and Thorner, M.O. (1996) Stimulation of the growth hormone (GH)-insulin-like growth factor I axis by daily oral administration of a GH secretagogue (MK-677) in healthy elderly subjects. JCE&M 81, 4249–4257.
40. Loche, S., Cambiaso, P., Casini, M.R., Imbimbo, B.P., Carta, D., Borrelli, P. and Cappa, M. (1996) Effects of hexarelin on growth hormone secretion in short normal children, in obese children, and in subjects with growth homone deficiency. In: Growth Hormone Secretagogues. B.B. Bercu and R.F. Walker (eds). Springer-Verlag, pp. 347–358.
41. Mericq, V., Cassorla, F., Salazar, T., Avila, A., Iniguez, G., Bowers, C.Y., Merriam, G.R. (1998) Effects of eight months treatment with graded doses of a growth hormone (GH)-releasing peptide in GH-deficient children. JCE&M 83, 2355–2360.

42. Laron, Z., Freckel, J., Deghenghi, R., Anin, S., Klinger, B. and Silbergeld, A. (1995) Intranasal administration of the GHRP hexarelin accelerates growth in short children. Clin. Endocrinol. 43, 631–635.
43. Pihoker, C., Kearns, G.L., French, D., Bowers, C.Y. (1998) Pharmacokinetics and pharmacodynamics of growth hormone-releasing peptide-2: a phase I study in children. JCE&M 83, 1168–1172.
44. Veldhuis, J.D., Iranmanesh, A., Ho, K.K.Y., Waters, M.J., Johnson, M.L. and Lizzaralde, G. (1991) Dual defects in pulsatile growth hormone secretion and clearance subserve the hyposomatotrophism of obesity in man. JCE&M 72, 51–59.
45. Williams, T., Berelowitz, M., Joffe, S.N. et al. (1984) Impaired growth hormone responses to growth hormone-releasing factor in obesity: a pituitary defect reversed with weight reduction. N. Engl. J. Med. 311, 1403–1407.
46. Ghigo, E., Procopio, M., Boffano, G.M. et al. (1992) Arginine potentiates but does not restore the blunted growth hormone response to growth hormone-releasing hormone in obesity. Metabolism: Clinical & Experimental 41, 560–563.
47. Kopelman, P.G., Noonan, K. (1986) Growth hormone response to low dose intravenous injections of growth hormone releasing factor in obese and normal weight women. Clin. Endocrinol. (Oxf.) 24, 157–164.
48. Cordido, F., Dieguez, C. and Casanueva, F.F. (1990) Effect of central cholinergic neurotransmission enhancement by pyridostigmine on the growth hormone secretion elicited by clonidine, arginine or hypoglycemia in normal and obese subjects. JCE&M 70, 1361–70.
49. Cordido, F., Penalva, A., Dieguez, C. and Casanueva, F.F. (1993) Massive growth hormone (GH) discharge in obese subjects after the combined administration of GH-releasing hormone and GHRP-6: evidence for a marked somatotroph secretory capability in obesity. JCE&M 76, 819–23.
50. Cordido, F., Penalva, A., Peino, R., Casanueva, F.F. and Dieguez, C. (1995) Effect of combined administration of growth hormone (GH)-releasing hormone, GH-releasing peptide-6, and pyridostigmine in normal and obese subjects. Metabolism: Clinical & Experimental 44, 745–48.
51. Cordido, F., Peino, R., Penalva, A., Alvarez, C.V., Casanueva, F.F. and Dieguez, C. (1996) Impaired growth hormone secretion in obese subjects is partially reversed by acipimox-mediated plasma free fatty acid depression. JCE&M 81, 914–918.
52. Kirk, S.E., Gertz, B.J., Schneider, S.H., Hartman, M.L., Pezzoli, S.S., Wittreich, J.M., Krupa, D.A., Seibold, J.R. and Thorner, M.O. (1997) Effect of obesity and feeding on the growth hormone (GH) response to the GH secretagogue L-692,429 in young men. JCE&M 82, 1154–59.
53. Svensson, J., Loenn, L., Jansson, J.O., Murphy, G., Wyss, D., Krupa, D., Cerchio, K., Polvino, W., Gertz, B., Bosaeus, I., Sjoestroem, L. and Bengtsson, B.A. (1998) Two-month treatment of obese subjects with the oral growth hormone (GH) secretagogue MK-677 increases GH secretion, fat-free mass, and energy expenditure. JCE&M 83, 362–369.
54. Johansson, G., Marin, P., Loenn, L. et al. (1997) Growth hormone treatment of abdominally obese men reduces abdominal fat mass, improves glucose and lipoprotein metabolism, and reduces diastolic blood pressure. JCE&M 82, 727–734.
55. Ross, R., Miell, J., Freeman, E. et al. (1991) Critically ill patients have high basal growth hormone levels with attenuated oscillatory activity associated with low levels of insulin-like growth factor-I. Clin. Endocrinol. (Oxf.) 35, 47–54.
56. Ross, R.J.M. and Chew, S.L. (1995) Acquired growth hormone resistance. European Journal of Endocrinology 132, 655–60.
57. Ross, R.J.M., Miell, J.P., Holly, J.M.P. et al. (1991) Levels of GH binding activity, IGFBP-1, insulin, blood glucose and cortisol in intensive care patients. Clin. Endocrinol. 35, 361–67.
8. Van den Berghe, G., de Zegher, F., Veldhuis, J.D. et al. (1997) The somatotropic axis in critical illness: effect of continuous GHRH and GHRP-2 infusion. J. Clin. Endocrinol. Metab. 82, 590–99.
9. Vara-Thorbeck, R., Guerrero, J.A., Rosell, J. et al. (1993) Exogenous growth hormone: effects on the catabolic response to surgically produced acute stress and on postoperative immune function. World J. Surg. 17, 530–38.

60. Van den Berghe, G., de Zegher, F., Bouillon, R. (1998) Acute and prolonged critical illness as different neuroendocrine paradigms JCE&M 83, 1827–1834.
61. Schambelan, M., Mulligan, K., Grunfeld, C., Daar, E.S., LaMarca, A., Kotler, D.P., Wang, J., Bozzette, S.A. and Breitmeyer, J.B. (1996) Recombinant human growth hormone in patients with HIV-associated wasting. A randomized, placebo-controlled trial. Serostim Study Group. Ann. Intern. Med. 125(11), 873–82.
62. Waters, D., Danska, J., Hardy, K., Foster, F., Qualls, C., Nickell, D. et al. (1996) Recombinant human growth hormone, insulin-like growth factor I, and combination therapy in AIDS-associated wasting-A randomized, double-blind, placebo controlled trial. Ann. Intern. Med. 125, 865–72.
63. Krentz, A.J., Koster, F.T., Crist, D.M. et al. (1993) Anthropometric, metabolic, and immunological effects of recombinant human growth hormone in AIDS and AIDS-related complex. J. AIDS 6, 245–251.
64. Mulligan, K., Grunfeld, C., Hellerstein, M.K., Neese, R.A. and Schambelan, M. (1993) Anabolic effects of recombinant human growth hormone in patients with wasting associated with human immunodeficiency virus infection. JCE&M 77, 965–62.
65. Daar, E.S., LaMarca, A., Schambelan, M., Grunfeld, C., Mulligan, K. and Breitmeyer, J. (1996) Effect of continuous growth hormone therapy on lean body mass (LBM) and AIDS-defining events in HIV-associated wasting (HIV-W). (Abstract.) Third Conference on Retroviruses and Opportunistic Infections, Washington, DC, p. 125.
66. Pinto, G., Brauner, R., Goulet, O., Clapin, A. and Blanche, S. (1997) Recombinant human growth hormone therapy for cachexia in HIV infected children. (Abstract.) Fourth Conference on Retroviruses and Opportunistic Infections, Washington, DC, p. 690.
67. Klaude, S., Fruhauf, L., Michels, B., Malkowski, J.P., Helm, E.B. and Althoff, P.H. (1996) HIV-associated wasting syndrome-therapy with mammalian cell-derived recombinant human growth hormone (r-hGH(m)). (Abstract.) XI International Conference on AIDS, Vancouver, British Columbia, Canada, p. 1384.
68. Berger, D.S., La Marca, A., Landy, H., Kauffman, R.S. and Breitmeyer, J. (1996) A Phase III study of recombinant human growth hormone (mammalian cell-derived) in patients with AIDS wasting. (Abstract.) XI International Conference on AIDS, Vancouver, British Columbia, Canada, p. 442.
69. McNurlan, M.A., Garlick, P.J., Steigbigel, R.T., DeCristofaro, K.A., Forst, R.A., Lang, C.H., Johnson, R.W., Santasier, A.M., Cabahug, C.J., Fuhrer, J. and Gelato, M.C. (1997) Responsiveness of muscle protein synthesis to growth hormone administration in HIV-infected individuals declines with severity of disease. Journal of Clinical Investigation 100 (8), 2125-32.
70. Van den Berghe, G., de Zegher, F., Baxter, R.C., Veldhuis, J.D., Wouters, P., Schetz, M., Verwaest, C., van der Vorst, E., Lauwers, P., Bouillon, R. and Bowers, C.Y. (1998) Neuroendocrinoloy of prolonged critical illness: effects of exogenous thyrotropin-releasing hormone and its combination with growth hormone secretagogues. JCE&M 83, 309–319.
71. Murphy, M.G., Plunkett, L.M., Gertz, B.J., He, W., Wittreich, J., Polvino, W.M. and Clemmons, D.R. (1998) MK-677, an orally active growth homone secretagogue, reverses diet-induced catabolism. JCE&M 83, 320–325.
72. Sacca, L., Cittadini, A., Fazio, S. (1994) Growth hormone and the heart. Endocr. Rev. 15, 555–73.
73. Yang, R., Bunting, S., Gillett, N., Clark, R. and Jin, H. (1995) Growth hormone improves cardiac performance in experimental heart failure. Circulation 92, 262–67.
74. Timsit, J., Riou, B., Bertherat, J. et al. (1990) Effects of chronic growth hormone hypersecretion on intrinsic contractility, energetics, isomyosin pattern, and myosin adenosine triphosphatase activity of rat left ventricle. J. Clin. Invest. 86, 507–15.
75. Amato, G., Carella, C., Fazio, S. et al. (1993) Body composition, bone metabolism, and heart structure and function in growth-hormone-deficient adults before and after GH replacement therapy at low doses. JCE&M 77, 1671–1676.

76. Merola, B., Cittadini, A., Colao, A. et al. (1993) Cardiac structural and functional abnormalities in adult patients, with growth hormone deficiency. JCE&M 77, 1658–1661.
77. Thuesen, L., Jorgensen, J.O.L., Muller, J.R. et al. (1994) Short and long-term cardiovascular effects of growth hormone therapy in growth hormone deficient adults. Clin. Endocrinol. (Oxf.) 41, 615–620.
78. Johannsson, G., Bengtsson, B.A., Andersson, B., Isgaard, J. and Caidahl, K. (1996) Long term cardiovascular effects of growth hormone treatment in GH-deficient adults. Preliminary data in a small group of patients. Clin. Endocrinol. (Oxf.) 45, 305–314.
79. Frustaci, A., Perrone, G.A., Gentiloni, N. and Russo, M.A. (1992) Reversible dilated cardiomyopathy due to growth hormone deficiency. Am. J. Clin. Path. 97, 503–511.
80. Fazio, S., Sabatini, D., Vigorito, C., Giordano, A., Guida, R., Pardo, F., Biondi, B. and Sacca, L. (1996) A preliminary study of growth hormone in the treatment of dilated cardiomyopathy. N. Engl. J. Med. 334, 809–814.
81. Cittadini, A., Grossman, J.D., Napoli, R., Katz, S.E., Stromer, H., Smith, R.J., Clark, R., Morgan, J.P. and Douglas, P.S. (1997) Growth hormone attenuates early left ventricular remodeling and improves cardiac function in rats with large myocardial infarction. J.Am. College Cardiol. 29, 1109–16.
82. Isgaard, J., Kujacic, V., Jennische, E., Holmang, A., Sun, X.Y., Hedner, T., Hjalmarson, A. and Bengtsson, B.A. (1997) Growth hormone improves cardiac function in rats with experimental myocardial infarction. Eur. J. Clin. Invest. 27, 517–25.
83. Caroll, P.V., Christ, E.R. et al. (1998) Growth hormone deficiency in adulthood and the effects of growth hormone replacement: A review. JCE&M 83, 382–395.
84. De Gennaro Colonna, V., Rossoni, G., Bernareggi, M., Muller, E.E. and Berti, F. (1997) Cardiac ischemia and impairment of vascular endothelium function in hearts from growth hormone-deficient rats: protection by hexarelin. Eur. J. Pharmacol. 334, 201–7.
85. Torsello, A., Luoni, M., Schweiger, F., Grilli, R., Guidi, M., Bresciani, E., Deghenghi, R., Muller, E.E., Locatelli, V. (1998) Novel hexarelin analogs stimulate feeding in the rat through a mechanism not involving growth hormone release. Eur. J. Pharmacol. 360, 123–9.

Growth Hormone Secretagogues
Edited by E. Ghigo, M. Boghen, F.F. Casanueva and C. Dieguez

Chapter 22

Growth Hormone Secretagogue Influences Feeding Behaviour in Experimental Animals

ICHIJI WAKABAYASHI, HITOSHI SUGIHARA and TAMOTSU SHIBASAKI

Nippon Medical School, Tokyo, Japan

A new class of pentapeptides or hexapeptides, growth hormone-releasing peptides (GHRPs), has been developed from met-enkephalin based on conformational energy calculations in conjunction with peptide chemistry and biological activity (1–3). GHRPs are potent and specific stimulators of GH secretion in multiple animal species as well as in humans when administered i.v., s.c., intranasally, or orally. Moreover, GHRPs potentiate the effect of native hypothalamic growth hormone-releasing hormone (GHRH) on GH secretion (1,2). However, GHRPs possess weak prolactin- and ACTH-releasing activities. Non-peptidyl GH secretagogues, including both spiroindane and benzolactam derivatives, whose mechanism of action mimic those of GHRPs, have also been developed (2,3). Both GHRPs and non-peptidyl GH secretagogues have been shown to share the same specific membrane receptor coupled to G protein in the pituitary gland and hypothalamus (3). Cloning of the GHRP receptor has contributed to the development of *in situ* hybridization and possible localization of the cells expressing the GHRP receptor mRNA, even though more studies are needed to accomplish the latter objective (4,5). In addition, it is strongly suggested that there is an endogenous ligand that could play an important role in the regulation of GH-axis.

Native hypothalamic GHRH has been shown to increase food intake in rodents by acting on the central nervous system (6,7). There is a complementary relationship between GHRPs and hypothalamic GHRH with respect to the effect on GH secretion (1–3). We hypothesized that such a complementary relationship should also exist for feeding behaviour. We present evidence that KP-102, a GHRP analogue, not only stimulates feeding behaviour, but also amplifies feeding behaviour induced by GHRH acting through the central nervous system (8–10).

Adult male rats were used throughout the studies and were housed in air-conditioned animal quarters, with the lights on between 08.00 and 20.00 h. Rat lab chow and water were available *ad libitum*. Prior to the study, a stainless steel cannula was implanted into the third

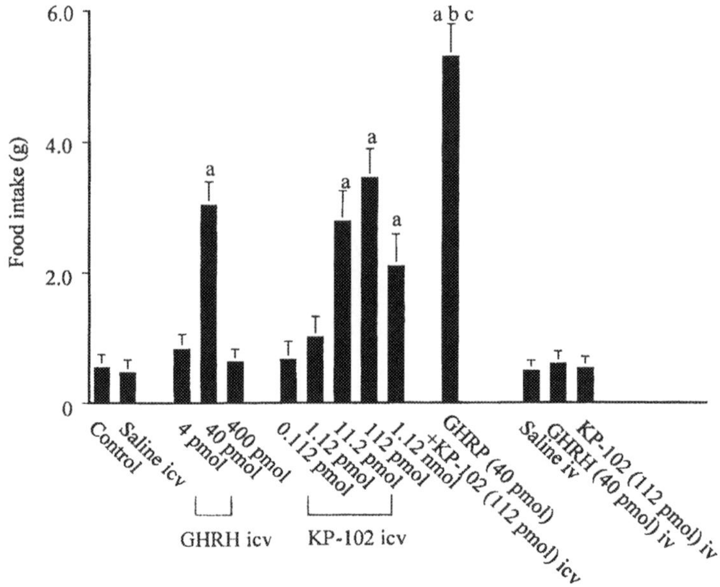

Figure 1. Effects of GHRH and KP-102 on food intake in free-feeding rats. Values are the mean 2-h food intake (13.00–15.00 h) after intracerebroventricular (i.c.v.) or intravenous (i.v.) administration of saline, human GHRH or KP-102. Each group consisted of 8 rats. Food intake was measured on 2 consecutive days from 13.00 to 15.00 h before treatments to obtain control values. Vertical bars indicate SEM. (a) $p < 0.05$ vs i.c.v. saline, (b) $p < 0.05$ vs 40 pmol of i.c.v. GHRH, (c) $p < 0.05$ vs 112 pmol of i.c.v. KP-102.

cerebral ventricle using a stereotaxic apparatus. In some studies, the stainless steel cannula was implanted into various regions of the hypothalamus as described later. The location of the cannula was verified histologically. All experiments were conducted between 13.00 and 15.00 h. Behavioural changes were also assessed by two independent researchers.

Intracerebroventricular (i.c.v.) administration of saline did not alter the food intake in free-feeding rats (Figure 1). I.c.v. injection of both GHRH and KP-102 significantly increased food intake in a dose-dependent manner, and the dose–response curves were bell-shaped. Although we confirmed the findings of Vaccarino et al., (6,7), larger doses of human GHRH were required to obtain a significant effect on food intake. Systemic administration of peptides did not increase food intake. When i.c.v. administration of KP-102 was tested in rats fasted for 24 h, the peptide did not significantly increase food intake as compared to that in control animals. It appeared that the stimulatory effect of i.c.v. KP-102 on food intake was masked by fasting-motivated feeding (9). Previous investigators observed that the ability of GHRP-6, a GHRP analogue, to stimulate eating was independent of its effect on GH-release (11), and found that systemic administration of rat GH did not influence food intake in rats (6). These findings suggested that KP-102 increases food intake by acting on the central nervous system and that its effect does not depend on the stimulation of GH release.

Systemic administration of GHRH derivatives has been demonstrated to induce expression of the c-*fos* gene or Fos protein, a marker of neuronal activity, in GHRH neurons

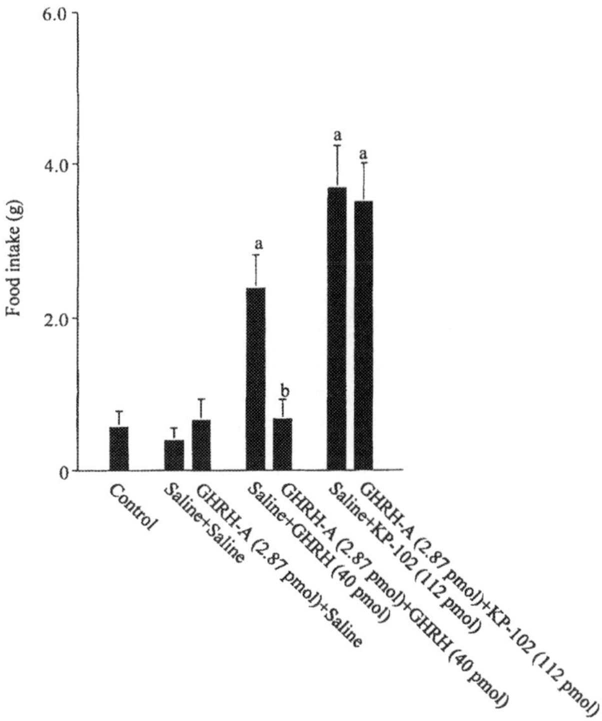

Figure 2. Effects of a GHRH antagonist (GHRH-A) on food intake stimulated by GHRH or KP-102 in free-feeding rats. Values are the mean 2-h food intake (13.00–15.00 h) after i.c.v. administration of saline, human GHRH or KP-102. Each group consisted of 6 rats. Control values represent the food intake measured on 2 consecutive days from 13.00 to 15.00 h. Vertical bars indicate the SEM. GHRH antagonist (GHRH-A) was administered centrally 15 min before i.c.v. administration of saline, GHRH or KP-102. (a) $p < 0.05$ vs saline + saline, (b) $p < 0.05$ vs saline + GHRH.

in the hypothalamic arcuate nucleus (ARC) in rats (12–14), and to stimulate GHRH release into the hypophysial portal vessels in sheep (15). These data suggest that GHRH neurons in the ARC are one of the central targets of GHRPs. In the present study, a combination of the maximally effective doses of GHRH and KP-102 exhibited an additive effect on the stimulation of food intake (Figure 1), suggesting that these agents stimulate food intake via different mechanisms. This interpretation is further supported by the finding that the increase of food intake induced by KP-102 was not inhibited by pretreatment with a GHRH antagonist, while such pretreatment completely blocked GHRH-induced food intake (Figure 2). These findings indicate that GHRH does not play a major role in KP-102-induced stimulation of food intake. Because neither i.c.v. nor i.v. administration of KP-102 or other peptides caused any observable general behavioural activation, the increase of eating was not considered to be due to a generalized behavioural activating property of the peptides.

The hypothalamic targets of GHRPs do not seem to be limited to GHRH neurons in the ARC, as suggested by following findings. First, synergistic effects of GHRH and GHRPs on GH secretion have been observed at maximal and supramaximal doses of these peptides (1). Second, in response to KP-102 or GHRP-6 administration, the c-*fos* gene is not only expressed by GHRH cells, but also by neuropeptide Y (NPY) cells in the ARC (13,14).

Third, *in situ* hybridization studies have revealed that the GHRP receptor gene is expressed in the ARC, the ventromedial nucleus (VMN), and other brain regions, even though the cells expressing this gene have not been characterized (4,5). The hypothalamic ARC and VMN are well recognized regions of the brain that participate in the regulation of feeding. NPY is known to be a potent orexigenic peptide. It is interesting to test whether stimulation of feeding by KP-102 is dependent on NPY via the Y5 receptor, which has been suggested as a 'feeding receptor'(16).

To obtain an insight into the site of action of KP-102 on feeding behaviour, the peptide was applied to the hypothalamic ARC, VMN, and lateral hypothalamic area (LAH). KP-102 stimulated food intake when it was applied to the VMH and ARC at doses ranging from 0.11 to 11.2 pmol. Drinking behaviour was also activated along with an increase in food intake (10). It has been reported that the stimulatory effect of GHRH on feeding was most sensitive when it was microinjected into the suprachiasmatic nucleus/medial preoptic area of the hypothalamus (17). Taken together, these findings suggest that the site of action on feeding differs between GHRPs and GHRH.

We have also examined the effects of somatostatin (SS), restraint stress, and corticotropin-releasing hormone (CRH) on KP-102-induced food intake (9). The mechanism by which GHRP-6 acts on the hypothalamus and stimulates GH secretion is sensitive to inhibition by SS (18). I.c.v. administration of SS did not influence food intake in freely feeding rats, but it partially inhibited the increase of food intake caused by i.c.v. application of KP-102. I.c.v. administration of CRH or restraint stress significantly inhibited food intake. Prior i.c.v. administration of KP-102 prevented the inhibition of food intake by CRH or restraint stress, but it did not increase beyond that in control animals. Both SS and CRH are likely mobilized in response to stress. These data may indicate that GHRP can counteract the decrease of feed intake under stress.

In conclusion, we demonstrated that i.c.v. administration of KP-102, a GHRP analogue, stimulated food intake at picomole doses and amplified the central effect of GHRH on feeding in rats. The effect of KP-102 on feeding does not appear to be dependent on GHRH, and seems to be mediated by a GHRP receptor that is expressed in the central nervous system.

ACKNOWLEDGMENTS

We thank Ms.R.Tokita for her technical assistance. This work was supported in part by grants from the Japanese Ministry of Education, Science, and Culture, Japan Private School Promotion Foundation and Foundation for Growth Science in Japan.

APPENDIX

KP-102 [D-Ala-D-β-Nal-Ala-Trp-D-Phe-Lys-NH2] was supplied by Kaken Pharmaceutical Co.Ltd.(Tokyo, Japan). Synthetic human GHRH 1-44 amide and [N-Ac-Tyr1, D-Arg2]-human GHRH 1-29 amide, a GHRH antagonist, were purchased from Bachem Feinchemikalien AG (Budendorf, Switzerland). A dose of GHRH antagonist was determined by the available literature (Lumpkin, M.D. and McDonald, J.K. (1989) Endocrinology 124, 1522–1531). Synthetic somatostatin 1-14 and human CRH were generously supplied by Dr. Nicholas Ling. Each of these peptides was dissolved in saline (0.9% NaCl).

REFERENCES

1. Bowers, C.Y. (1993) GH releasing peptides — structure and kinetics. J. Pediatr. Endocrinol. 6, 21–31.
2. Korbonits, M. and Grossman, A.B. (1995) Growth hormone-releasing peptide and its analogues. Trends Endocrinol. Metab. 6, 43–49.
3. Smith, R.G., Van der Ploeg, L.H.T., Howard, A.D., Feighner, S.D., Cheng, K., Hickey, G.J., Wyvratt, M.J., Fisher, M.H., Nargund, R.P. and Patchett, A.A. (1997) Peptidomimetic regulation of growth hormone secretion. Endrocr. Rev. 18, 621–646.
4. Guan, X.-M., Yu, H., Palyha, O.C., McKee, K.K., Feighner, S.D., Sirinathsinghji, D.J.S., Smith, R.G., Van der Ploeg, L.H.T. and Howard, A.D. (1997) Distribution of mRNA encoding the growth hormone secretagogue receptor in brain and peripheral tissue. Mol. Brain Res. 48, 23–29.
5. Bennet, P.A., Thomas, G.B., Howard, A.D., Feighner, S.D., Van der Ploeg, L.H.T., Smith, R.G. and Robinson, I.C.A.F. (1997) Hypothalamic growth hormone secretagogue-receptor (GHS-R) expression is regulated by growth hormone in the rat. Endocrinology 138, 4552–4557.
6. Vaccarino, F.J., Bloom, F.E., Rivier, J., Vale, W. and Koob, G.F. (1985) Stimulation of food intake in rats by centrally administered hypothalamic growth hormone-releasing factor. Nature 314, 167–168.
7. Vaccarino, F.J., Feifel, D., Rivier, J., Vale, W. and Koob, G.F. (1988) Centrally administered hypothalamic growth hormone-releasing factor stimulates food intake in free-feeding rats. Peptides 9 (Suppl. 1), 35–38.
8. Okada, K., Ishii, S., Minami, S., Sugihara, H., Shibasaki, T. and Wakabayashi, I. (1996) Intracerebroventricular administration of the growth hormone-releasing peptide KP-102 increases food intake in free-feeding rats. Endocrinology 137, 5155–5157.
9. Shibasaki, T., Yamauchi, N., Takeuchi, K., Ishii, S., Minami, S. and Wakabayashi, I. (1997) Effect of fasting, stress, corticotropin-releasing factor or somatostatin on growth hormone-releasing peptide KP-102-induced food intake in rats. Soc. Endocrine Abstr. (P1-66).
10. Suzuki, K., Ohata, H., Arai, K., Wakabayashi, I. and Shibasaki, T. (1998) Growth hormone-releasing peptide stimulates feeding behaviour through the ventromedial nucleus (VMH) and arcuate nucleus (ARC) of the hypothalamus in rats. Soc. Endocrine Abstr. (P2-196).
11. Locke, W., Kirgis, H.D., Bowers, C.Y. and Abdoh, A.A. (1995) Intracerebroventricular growth hormone-releasing peptide–6 stimulate eating without affecting plasma growth hormone response in rats. Life Sci. 56, 1347–1352.
12. Dickson, S.L., Leng, G., Robinson, I.C.A.F. (1993) Systemic administration of growth hormone-releasing peptide (GHRP-6) activates hypothalamic arcuate neurons. Neuroscience 53, 303–306.
13. Kamegai, J., Hasegawa, O., Minami, S., Sugihara, H. and Wakabayashi, I. (1996) The growth hormone-releasing peptide KP-102 induces c-*fos* expression in the arcuate nucleus. Mol. Brain Res. 39, 153–159.
14. Dickson, S.L., Luckman, S.M. (1997) Induction of c-*fos* messenger ribonucleic acid in neuropeptide Y and growth hormone (GH)-releasing factor neurons in the rat arcuate nucleus following systemic injection of the GH secretagogue, GH-releasing peptide-6. Endocrinology 138, 771–777.
15. Guillaume, V., Magnonan, E., Cataldi, M., Dutour, A., Sauze, N., Renard, M., Razafindraibe, H., Conte-Devolx, B., Deghenghi, R., Lenaerts, V. and Oliver, C. (1994) Growth hormone

(GH)-releasing hormone secretion is stimulated by a new GH-releasing hexapeptide in sheep. Endocrinology 135, 1073–1076.

16. Gerald, C., Walker, M.W., Criscione, L., Gustafson, E.L., Batz-Hartmann, C., Smith, K.E., Vaysse, P., Durkin, M.M., Laz, T.M., Linemeyer, D.L., Schaffhauser, A.O., Whitebread, S., Hofbauer, K.G., Taber, R.I., Branchek, T.A. and Weinshank, R.L. (1996) A receptor subtype involved in neuropeptide-Y-induced food intake. Nature 382, 168–171.
17. Vaccarino, F.J. and Hayward, M. (1988) Microinjections of growth hormone-releasing factor into the medial preoptic area/suprachiasmatic nucleus region of the hypothalamus stimulate food intake in rats. Regul. Pept. 21, 21–28.
18. Fairhall, K.M., Mynett, A., Robinson, I.C.A.F. (1995) Central effects of growth hormone-releasing hexapeptide (GHRP-6) on growth hormone release are inhibited by central somatostatin action. J. Endocrinol. 144, 555–560.

Growth Hormone Secretagogues
Edited by E. Ghigo, M. Boghen, F.F. Casanueva and C. Dieguez

Chapter 23

Growth Hormone Secretagogues and Sleep

AXEL STEIGER

Max Planck Institute of Psychiatry, Department of Psychiatry, Munich, Germany

INTRODUCTION

Peptides play a key role in sleep regulation. This has been shown by a host of studies in animals, in normal human controls and in patients with psychiatric and endocrine disorders [for review see Refs. (1,2)]. These studies demonstrated specific effects of various peptides on sleep. In particular, the influence of growth hormone-releasing hormone (GHRH) on sleep was extensively investigated. GHRH is the one endogenous substance for which sleep-promoting effects are best documented (2,3). This explains the recent interest of some researchers in the question of whether, similar to GHRH, growth hormone (GH) secretagogues modulate sleep. Indeed sleep promotion by GH secretagogues was reported by two independent laboratories, suggesting that ligands of the GH-secretagogue receptor participate in sleep regulation and that GH secretagogues may be useful in the treatment of sleep disorders, particularly insomnia. This chapter first gives a short introduction to the field of sleep endocrinology, then focuses on current knowledge about the effects of GH secretagogues on sleep and finally makes an attempt to integrate these data into the framework of what is known about the role of various peptides in sleep regulation.

SLEEP ENDOCRINOLOGY

Sleep in humans and animals is characterized by an electrophysiological component and a neuroendocrine component. The electrophysiological component refers to the cyclic occurrence of non rapid eye movement (nonREM) and rapid eye movement (REM) periods, which can be seen on a sleep electroencephalogram (EEG; polygraphy). The sleep EEG is analyzed conventionally by visual scoring, which divides polygraphic activity into the sleep stages REM, awake and 1–4 of nonREM sleep (including slow wave sleep [SWS], stages 3 and 4) (4). More recently, EEG spectral analysis is used in addition to investigate the amount and course of various frequency bands during sleep. The endocrine component of sleep is reflected by distinct patterns of secretion of various hormones. These can be investigated by blood sampling via long catheter simultaneously with sleep-EEG recordings.

In young normal human subjects the major amount of SWS, and correspondingly of EEG delta power, occurs during the first half of the night. This electrophysiological activity is closely but not absolutely linked to the nocturnal GH peak. During this interval cortisol levels reach their nadir. In contrast, during the second half of the night during nonREM sleep SWS is rare and stage 2 sleep preponderates, the major portion of REM sleep occurs, GH levels are low and cortisol secretion rises stepwise until awakening (5,6). Interestingly, during normal aging a similar pattern of sleep-endocrine activity develops. Common features of depression and aging include the occurrence of more shallow sleep, a decrease in SWS, disturbed sleep continuity, shortened REM latency (the interval between sleep onset and the first REM period) and elevated cortisol levels. The change in cortisol is more distinct in patients with depression, who frequently show hypercortisolism. Due to a flattened amplitude, cortisol secretion is higher during the first half of the night in the elderly than in young subjects (7–10) (see Figure 1).

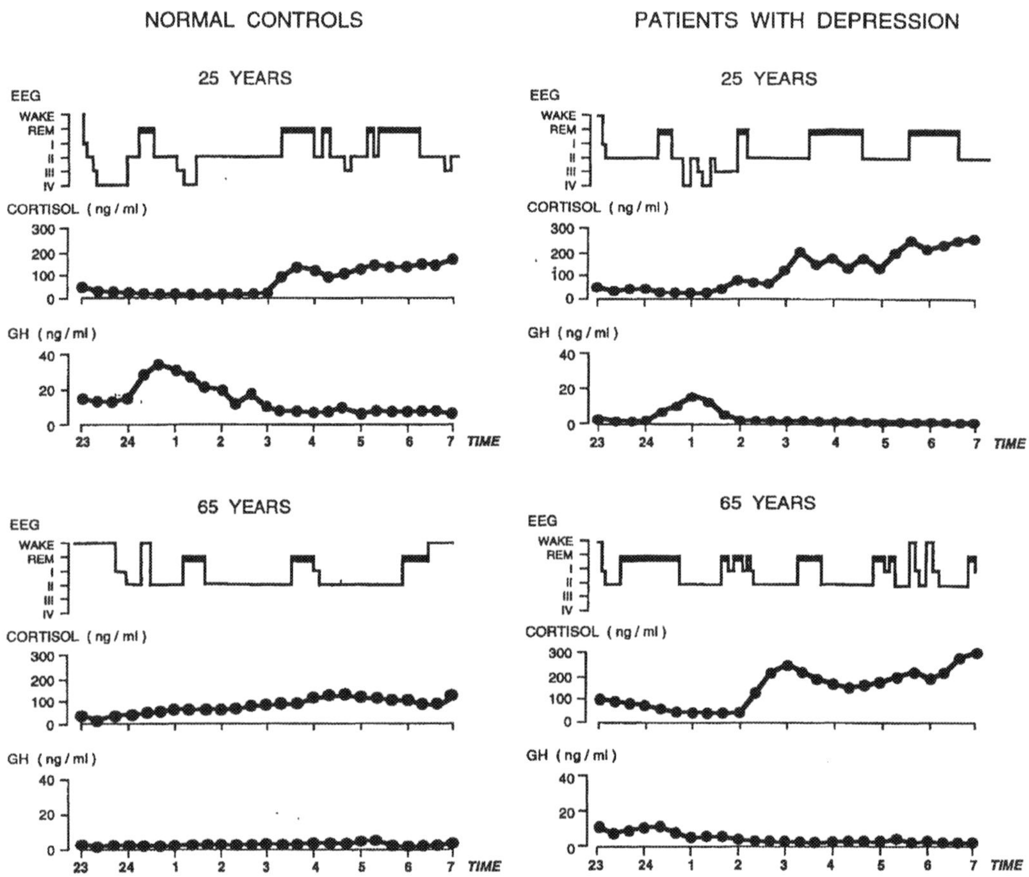

Figure 1. Sleep EEG and nocturnal secretion of cortisol and growth hormone (GH) in representative young and old normal subjects and patients with depression. REM = rapid eye movement sleep. I–IV indicate stages of nonREM sleep.

These observations suggest that the neurophysiological and neuroendocrine components of sleep have common regulators. As pointed out later on in more detail there is strong evidence that the neuropeptides GHRH and corticotropin-releasing hormone (CRH) are these regulators and that they have a reciprocal interaction.

EFFECTS OF GH SECRETAGOGUES ON THE SLEEP EEG AND HORMONE SECRETION

Studies on GHRP-6

Frieboes and colleagues (11) were the first to investigate the effect of a GH secretagogue, namely GH-releasing peptide 6 (GHRP-6), on sleep. Previous human and animal research had demonstrated that the hexapeptide GHRP-6 stimulates GH release in a dose-dependent manner (12). Most studies on normal controls and on short-statured children have investigated GH release after i.v., oral and intranasal administration of this substance during the daytime and have reported significant increases [for a review, see Ref. (11)]. Twenty-four-hour infusions of GHRP-6 failed to enhance nocturnal GH release in humans (12,13). Conflicting data were reported on the effects of GHRP-6 on cortisol secretion. Cortisol levels remained unchanged after oral GHRP-6 during the daytime in normal controls (14). Hayashi et al. (15), however, found a slightly but significantly elevated cortisol concentration after i.v. GHRP in the morning in healthy volunteers. The effect of GHRP-6 administration around sleep onset on the sleep EEG and sleep-associated hormone secretion was previously unknown. To clarify this issue, Frieboes et al. (11) chose a protocol with simultaneous investigation of the sleep EEG and hormone secretion under baseline conditions and after pulsatile i.v. administration of GHRP-6 in normal control subjects. Analogous protocols were used in various studies in our laboratory on sleep effects of several peptides in humans (16).

Seven young normal male controls with a mean age of 25.3 years (S.D. 1.3; range 21–30) participated in this study (11). It consisted of two sessions at an interval of one week in which placebo or active GHRP-6 was administered according to a randomized schedule. Each session consisted of two successive nights in the sleep laboratory. The first night of each session served for adaptation to the laboratory setting. On the second night an indwelling catheter was inserted into a forearm vein. The catheter was connected to a plastic tube that ran through a sound-proof lock into the adjacent room. Beginning at 2200 h blood was collected every 20 min until 0700 h through the long catheter for later analysis of the plasma concentrations of GH, cortisol and ACTH. From the adjacent room sleep-EEG recordings (EEG, electrooculogram, electromyogram and ECG) were monitored from 2300 h to 0700 h. Outside of this period sleeping was not permitted. To mimic the physiological release of neuropeptides GHRP-6 was given in a pulsatile fashion. At 2200, 2300, 0000 and 0100 h a single bolus of 50 μg of the substance (Clinalfa, Läufelfingen, Switzerland) or placebo was administered i.v.

After GHRP-6 time spent in sleep stage 2 increased significantly (245.4 ± 25.8 min after placebo vs. 270.1 ± 25.3 min after GHRP-6; $p < 0.02$). The amount of intermittent

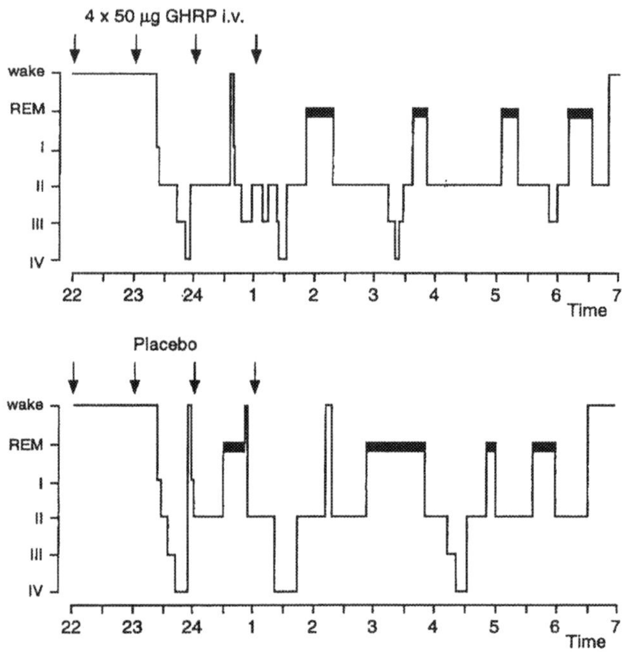

Figure 2. Hypnograms of a representative subject after 4 x 50 μg GHRP-6 and after placebo.

wakefulness showed a nonsignificant trend to decrease (19.9 ± 16.0 min after placebo vs. 12.2 ± 13.2 min after GHRP-6) (Figure 2). There were no changes in any other sleep-EEG variables, including the amount of SWS and REM sleep. Furthermore, EEG spectral analysis failed to show any significant differences between GHRP-6 and placebo.

With regard to endocrine effects, GHRP-6 prompted an elevation of GH that was significantly above the GH surge after placebo. Moreover, the HPA hormones cortisol and ACTH were stimulated by GHRP-6. Significant elevations of the ACTH concentration and area under the curve were found between 2200 and 0200 h, whereas the ACTH secretion for the total night remained unchanged. The mean cortisol concentration was significantly enhanced after GHRP-6 during the total night. During the first half of the night the mean concentration and area under the curve of cortisol were elevated. Moreover, the cortisol nadir was elevated after GHRP-6.

The effect of GHRP-6 on the sleep EEG appears to depend on dosage, route and time of administration. Recently Frieboes et al. (17) amplified their first study on i.v. GHRP-6 and used different routes of administration, namely intranasal, oral and sublingual administration. Administration of 300 μg/kg bodyweight GHRP-6 given orally at 2100 h was followed by a decrease in sleep stage 2 in the second half of the night. After sublingual administration of 30 μg/kg bodyweight GHRP-6 at 2245 h the sleep EEG remained unchanged, whereas intranasal administration of 30 μg/kg bodyweight GHRP-6 increased sleep stage 2 in the second half of the night by trend. Spectral analysis of total-night nonREM sleep revealed a decrease in delta power by trend.

Study on GHRP-2

In the most recent study in this field Moreno-Reyes et al. (18) failed to find major effects of another GH secretagogue, GHRP-2, on the sleep EEG. In their study 7 young normal male controls (24–30 years old) were examined in the sleep laboratory on 4 consecutive nights. On days 2 and 4 blood samples for GH and prolactin determinations were collected at 15-min intervals for 25 consecutive hours starting at 1500 h. During both nights sleep-EEG recordings were obtained and i.v. injections of GHRP-2 (1 μg/kg body weight) or saline, in randomized order, were given via a long catheter after 60 sec of the third REM period (ie, between 0323 and 0509 h). Except for a nonsignificant tendency to an increased amount of wakefulness during the first hour of the injections, the sleep EEG was not affected by GHRP-2. The GHRP-2 injections were followed by trend in prolactin elevation and by GH pulses within or around the upper limit of the physiological range.

Studies on MK-677

Sleep-promoting effects after prolonged oral treatment with a GH secretagogue were documented in the recent report by Copinschi et al. (19). These authors selected the novel, orally active GH secretagogue MK-677 for two separate studies in young and older normal control subjects. Nine healthy young men with a mean age of 27 years (SD 3; range 18–30) participated in the first study. The protocol was designed as a double-blind, placebo-controlled 3-period crossover study. Each subject participated in 3 treatment periods, presented in random order and separated by at least 14 days. Each period involved administration of the drug as a single oral dose at bedtime by 22.45 h for 7 consecutive days. Doses were 5 or 25 mg MK-677 and matching placebo. Prior to the beginning of the study all subjects spent an adaptation night in the sleep laboratory. Throughout the entire study the subjects were asked to maintain a regular sleep-wake cycle (2300–0700 h). On day 6 of each period they came to the sleep laboratory at 1900 h for a reacclimatization night. On the following day at 1700 h a catheter was inserted into a forearm vein and 1-ml blood samples were obtained at 15 min intervals for 25 consecutive hours starting at 1800 h. During both days 6 and 7 the sleep EEG was recorded from 2300 to 0700 h, according to standard guidelines (4).

For technical reasons the statistical calculations could be performed only for the nights with blood samplings and involved 8 of the 9 subjects. All sleep-EEG parameters were similar in the placebo and low-dose (5 mg) conditions. After high-dose (25 mg) treatment the duration of stage 4 was prolonged by nearly 50% (37 ± 7 min after placebo vs. 54 ± 10 min after MK-677; $p < 0.05$). The amount of REM sleep was more than 20% higher than after placebo (85 ± 7 min after placebo vs. 103 ± 3 min after MK-677; $p < 0.005$). The increases in stages 4 and REM reflected a nonsignificant trend towards smaller amounts of wake which decreased, on average, by 34% as compared to the placebo condition (77 ± 16 min after placebo vs. 52 ± 10 min after MK-677). Other sleep-EEG variables did not differ significantly between the three treatment conditions. The cumulative profiles of stages wake, 4 and REM indicate that the effects of high-dose treatment on stage 4 and on wake were mostly apparent during the beginning of the night, whereas the increase in REM sleep

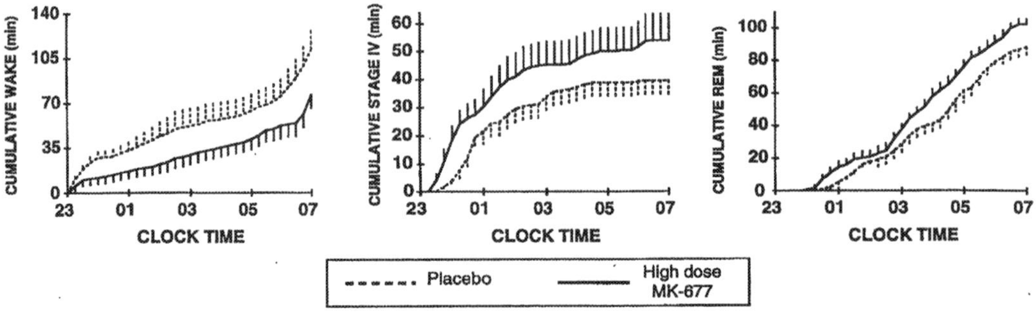

Figure 3. Cumulative minutes of sleep stages (awake, 4, REM) in young normal controls after 7 days po treatment with 25 mg MK-677 (solid lines) of placebo.

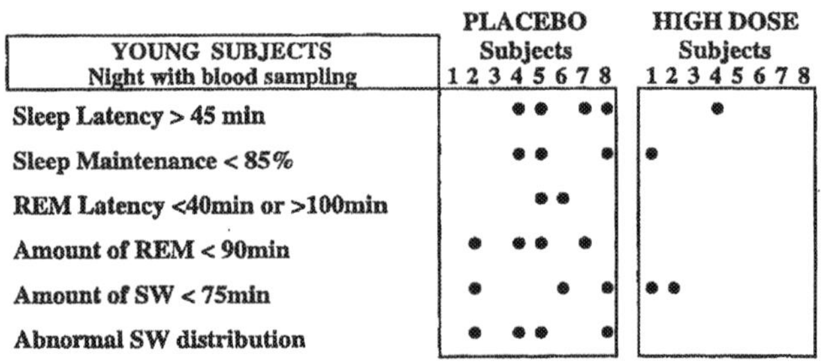

Figure 4. Schematic representation of sleep quality in young subjects during nights with placebo, 2 mg MK-677 and 25 mg MK-677.

occurred around the middle of the sleep period (see Figure 3). The frequency of deviation from normal sleep decreased from 42% under placebo to 8% under high-dose MK-677 ($p <$ 0.03) (see Figure 4). In young men the frequency of detectable GH pulses was higher with both dosages of MK-677 than with placebo. GH secretion, however, was not significantly increased following any of the dosages. In contrast, insulin-like growth factor I (IGF-I) levels increased in a dose-dependent manner.

In the second study 6 healthy fully self-sufficient older subjects (4 men, 2 women), ages 65–71 years were included. They participated in 2 successive treatment periods separated by a 14-day washout period. The treatment period involved administration of the drug as a single oral dose between 2200 and 2300 h for 14 consecutive days. Doses were 2 mg during the first treatment period and 25 mg MK-677 for the second treatment period. The sleep EEG was recorded for 2 consecutive nights at baseline (before the beginning of the first treatment period) and at the end of each treatment period. On each occasion a catheter was inserted into a forearm vein at 0700 h immediately after the first night and 1-ml blood samples were obtained at 20 min intervals for 25 consecutive hours. During both nights the sleep EEG was recorded from 2300 to 0700 h.

In the second study valid sleep-EEG recordings were obtained during all the re-acclimatization nights and on 17 of the 18 nights with blood sampling. Therefore statistical calculations were performed for all subjects for the nights without blood sampling and for 5 of 6 subjects for the nights with blood sampling. At baseline sleep-EEG variables were as expected in the elderly population, with increased amounts of wake and decreased amounts of stages 4 and REM. Under the treatment with the low MK-677 dose (2 mg) a significant increase in REM time was found (50 ± 10 min under placebo vs. 72 ± 9 min under MK-677; $p < 0.005$). During the nights with blood sampling REM latency was decreased following low-dose treatment (228 ± 43 min after placebo vs. 72 ± 8 min after MK-677; $p < 0.02$). The frequency of deviations from normal sleep was also decreased under low-dose treatment in the elderly ($p < 0.02$).

In the elderly the total amount of GH secreted during 24 hours remained unchanged following treatment with low-dose MK-677 but was significantly increased after 2 weeks of treatment with high-dose MK-677 (25 mg). Similarly, nocturnal GH secretion was significantly higher following high-dose treatment, but not following low-dose treatment, than at baseline.

DISCUSSION — SLEEP EFFECTS OF GH SECRETAGOGUES AND THE PHYSIOLOGY OF PEPTIDERGIC SLEEP REGULATION

The findings presented in the previous sections show that GH secretagogues are capable of promoting sleep, whereas the substances investigated so far exert different effects on the sleep EEG. I.v. GHRP-6 and by trend intranasal GHRP-6 increased one component of nonREM sleep, namely stage 2 sleep. After prolonged oral administration of MK-677, another component of nonREM sleep, stage 4 sleep, and also stage REM increased in young men, whereas in the elderly there was an increase in REM sleep only. After i.v. GHRP-6, GH, cortisol and ACTH secretion were stimulated, whereas in the young controls an increase in the frequency of GH pulses was found as the only endocrine effect of MK-677. I.v. GHRP-2, however, failed to affect sleep EEG, whereas it prompted a transient elevation of prolactin secretion.

The promotion of sleep by the GH secretagogues investigated and particularly the sleep-EEG changes after MK-677, is similar to that after acute central and systemic administration of GHRH to animals and systemic administration to humans. Various studies in rats and rabbits in several laboratories have demonstrated that sleep, and particular SWS, is promoted by central and also by i.v. GHRH administration. Conversely more shallow sleep occurred after experimental inhibition of GHRH [for review see Ref. (3)]. As found by three independent laboratories, i.v. GHRH also promotes sleep in humans.

We studied the effects of GHRH on human sleep by a protocol analogous to that of Frieboes et al. (11) with administration of GHRP-6. In our study repetitive hourly administration of 4 × 50 μg GHRH to young normal males from 2200 to 0100 h enhanced SWS and GH secretion and blunted cortisol release (20). The comparison of repetitive vs. continuous administration of GHRH underlined that it is crucial to administer neuropeptides in a pulsatile fashion mimicking the physiological secretion pattern (21). Increases in SWS and GH (and also in REM sleep) were found in this study only after a protocol identical to our initial study (20). The sleep EEG remained unchanged, however, when GHRH was continuously infused. The effects of repetitive GHRH in these studies (20,21) are opposite to the changes in pulsatile CRH seen in young normal males. After $4 \times 50\ \mu g$ CRH SWS decreased during the second half of the night and REM sleep was diminished throughout the night; cortisol levels were elevated and the GH surge was blunted (22). In a further study 0.3 μg/kg GHRH was administered in 3 different protocols: (1) after the onset of the first SWS period; (2) after 60 sec of the third REM period; and (3) after sleep deprivation lasting until 0400 h (23). These 3 protocols yielded the following results: (1) unchanged SWS but increased REM sleep; (2) an increase in SWS and a decrease in intermittent wakefulness; and (3) a decrease in intermittent wakefulness. Taken together, these data confirm that GHRH is a sleep-promoting substance in humans and suggest that CRH impairs sleep.

It is well documented that the changes in the sleep EEG after the peptides GHRH and CRH represent CNS effects which occur independently of the changes in peripheral hormone secretion. This is evident from the observation that SWS is enhanced by administration of cortisol to young and old normal human controls and is suppressed by administration of GH to normal controls and to animals, probably by feedback inhibition of CRH and of GHRH, respectively [for review see Ref. (2)]. Furthermore, systemic administration of GHRH to hypophysectomized rats selectively enhanced SWS, whereas after systemic GHRH in intact rats enhanced both SWS and REM sleep in this experiment (24). These data suggest that peripherally administered GHRH promotes SWS by central action, whereas GH stimulates REM sleep. However, since immunoneutralization of GH is followed by decreases in both nonREM and REM sleep in the rat, this suggests that GH may have nonREM sleep-promoting effects, particularly in chronic conditions (25). Finally, the effects of intracerebral injections of insulin-like growth factor-1 (IGF-1) in the rat were investigated recently (26). A small dose of IGF-1 modestly enhanced nonREM sleep. When the dose was increased a marked suppression of nonREM sleep was found. These results suggested that IGF-1 can promote nonREM sleep and may contribute to the mediation of the effects of GH on sleep. The acute sleep suppressive activity of the high dose of IGF-1 was attributed to inhibition of endogenous GHRH.

Kerkhofs et al. (23) concluded from their data that GHRH is most effective when given during intervals of shallow sleep. However, this hypothesis is not supported by 3 studies involving i.v. GHRH in 3 situations in which shallow sleep is a robust finding: (1) in young normal men in the early morning hours, (2) in patients with depression, and (3) in elderly normal controls. In the first study $4 \times 50\ \mu g$ GHRH was given i.v. between 0400 and 0700 h. The sleep-EEG effects were weak and the only significant change was a decrease in REM density (a measure for the amount of rapid eye movements during REM periods). SWS remained unchanged in this study. GH was significantly enhanced, whereas HPA hormones were not affected. These data suggest that GHRH is less effective in young normal men when it is given during the second half of the night, when the amounts of SWS and GH are low, but the activity of the HPA system is increasing. In contrast, GHRH given in a pulsatile fashion during the first 3 hours of the night (20) or by single boluses after sleep onset until the third REM period (23) exerted sleep-promoting effects lasting until the second half of the night. These observations support the hypothesis (27) that there is a time window near sleep onset when the physiological action of GHRH takes place, coinciding with relatively low activity of CRH. In the second study 4 × 50 μg GHRH was administered i.v. to patients of both sexes having an acute episode of depression (28). Again the sleep-EEG effects of GHRH were much weaker in depressed patients than in young normal subjects. There were several nonsignificant trends towards improved sleep quality with SWS and REM time tending to increase and intermittent wakefulness tending to decrease. Similar to the first study, however, the only significant change in the sleep EEG after GHRH was a decrease in REM density during the second half of the night. The GH peak was distinctly augmented, whereas cortisol and ACTH secretion remained unchanged. In the third study an analogous protocol was applied to healthy elderly men and women (29). Four × 50 μg i.v. GHRH prompted some sleep-promoting effects, but again these were much weaker than in young subjects. The number of intermittent awakenings decreased significantly and the first nonREM period was prolonged. The GH surge was enhanced significantly, but less than in young subjects. HPA hormone secretion remained unchanged. This study demonstrated a reduced potency of GHRH to modulate sleep-endocrine activity in the elderly. Similarly, during the daytime the response of GH to GHRH was much smaller in elderly than in young controls (30).

Recent data from our laboratory suggest that in the elderly GHRH and somatostatin exert opposite effects not only on GH release but also on the sleep EEG. Repetitive administration of 4 × 50 μg somatostatin to elderly normal controls of both sexes prompted impairment of sleep: we observed a decrease in total sleep time and REM time and an increase in intermittent wakefulness (31). This finding is similar to the observation that nonREM sleep decreased in rats treated with a long-acting somatostatin analogue (32). In contrast, in young normal men neither continuous (33), nor single (34), nor repetitive i.v. (20) administration of somatostatin exerted any effect on the sleep EEG. Our data suggest that the capacity of GHRH to counteract the impairment of sleep by somatostatin and (as discussed in more detail below of CRH) declines with increasing age. This hypothesis fits with preclinical findings showing that the activity of the GHRH system declines during aging, whereas the somatostatin system remains largely unchanged (35).

Clearly, there are similarities between the reduced efficacy of acute i.v. GHRH (29) and prolonged oral MK-677 (36) in elderly compared to young controls. Interestingly, a recent study by Murck et al. (37) showed even deterioration of sleep after subchronic GHRH administration. This finding is in contrast to the increase in REM sleep observed after subchronic treatment with MK-677 in healthy seniors. In their study Murck et al. (37) tested whether the effect of GHRH on sleep-endocrine activity is enhanced after "GHRH priming". This method was introduced by Iovino et al. (38). These authors had previously reported that the blunted capacity of GHRH to stimulate GH release during the daytime in the elderly can be revised by "priming" (100 μg GHRH i.v. every second day for 12 days). Murck et al. (37) studied the sleep-endocrine activity of two elderly men on four occasions. Either placebo or 4 × 50 μg GHRH was given between 2200 and 0100 h at baseline on two consecutive nights, followed by 12 days with GHRH priming in accordance with Iovino et al. (38). Then the effect of GHRH was retested, and finally after a placebo period of 12 days a second retest was performed. The efficacy of GHRH was not improved by priming in either subject. Sleep quality was impaired even in comparison to baseline after priming, and this effect persisted until the end of the protocol.

These data support the hypothesis of a reciprocal interaction between GHRH and CRH in sleep regulation, which was first proposed by Ehlers and Kupfer (39). Apparently GHRH triggers SWS and GH release and inhibits cortisol secretion through suppression of CRH. In contrast, CRH has opposite effects; it stimulates cortisol and decreases SWS and GH secretion. Thus CRH has activating and sleep-impairing effects. This is in accordance with its role as a key regulator of behavioral adaptation to exposure to stress. In contrast, GHRH is a sleep-promoting substance; high GHRH activity is associated with enhanced SWS. On the other hand, there is a link between low GHRH levels and shallow sleep. This is further illustrated by the observation that subjects with dwarfism show less SWS than age-matched normal controls (40), which is probably caused by a central GHRH deficiency. Hypothalamic GHRHmRNA levels peak in rats around light onset, when the animals tend to sleep, decline during the light phase and are low during their activity at the dark phase (41). Correspondingly, in young normal subjects endogenous GHRH levels are thought to be highest during the first half of the night resulting in the highest amounts of SWS and GH and in the nadir of cortisol. During the second half of the night, the influence of CRH predominates. Therefore cortisol is released and the amounts of SWS and GH are low. Changes in the GHRH:CRH ratio may result in changes in sleep-endocrine activity. Overactivity of CRH is well documented in depression (42), whereas the activity and/or efficiency of GHRH decline during aging (38). Consequently, in both conditions CRH predominates, prompting similar alterations in sleep-endocrine activity. During aging the increasing influence of somatostatin may act as an additional sleep-impairing factor (Figure 5). During the recovery night after sleep deprivation changes in sleep-endocrine activity were reported that were opposite to those found during depression and aging, namely increases in SWS and GH (43). A recent study by Toppila et al. (44) suggests accumulation of GHRH during sleep deprivation. It appears possible that GHRH is excessively produced during sleep deprivation and overrides the effects of CRH.

Beside GHRH, CRH and somatostatin other neuropeptides play a role in sleep regulation. Particularly for galanin, neuropeptide Y (NPY) and vasoactive intestinal polypeptide (VIP), sleep-EEG effects were documented in human studies. A role of galanin in sleep regulation appears likely since REM-sleep deprivation induced galanin gene expression (45).

After pulsatile i.v. administration of $4 \times 50\,\mu g$ galanin REM sleep was prolonged for the first 3 sleep cycles and stage 2 sleep decreased during the total night. EEG spectral analysis showed an increase in EEG delta power for the total night, whereas hormone secretion remained unchanged (46). In two separate studies 2 dosages of NPY ($4 \times 50\,\mu g$ and $4 \times 100\,\mu g$) were given to young normal men (47). The major effect of the lower dose was a blunting of cortisol and ACTH. After the higher dose only marginal suppression of HPA hormones

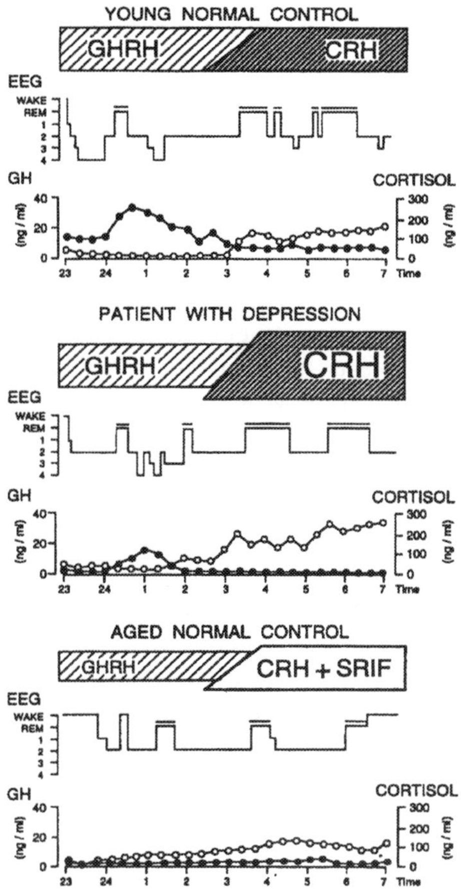

Figure 5. Model of peptidergic sleep regulation (© 1997, American Sleep Disorders Association and Sleep Research Society. Reprinted by permission).

was found, but sleep-onset latency was markedly shortened and EEG delta power was decreased during the second half of the night. The sleep-EEG findings in humans are similar to those after i.c.v. administration of NPY to rats (48). In animals REM sleep and nonREM sleep increased after VIP. After repetitive administration of 4 $\times$ 50 μg VIP to young normal males the sleep cycles were decelerated during the first 3 cycles due to increased duration of both REM- and nonREM-sleep periods. Furthermore, the cortisol nadir was advanced. These findings suggest an influence of VIP on circadian rhythms which is probably mediated at the suprachiasmatic nucleus (49).

CONCLUSIONS AND PERSPECTIVES

Two major conclusions can be drawn from the reported sleep studies with GH secretagogues: GH secretagogues are capable of promoting sleep, and it appears likely that this effect is centrally mediated. The most distinct sleep-EEG changes were found after acute i.v. GHRP-6 and after prolonged oral MK-677 in young normal male controls. After both substances sleep was enhanced, whereas GHRP-6 and MK-677 stimulated different components of nonREM sleep, namely either stage 2 or stage 4. After MK-677 in addition REM sleep increased.

The different effects of GHRP-6 and GHRH on nonREM sleep may be explained by their binding to different receptors. GHRP-6 and MK-677 bind to the GH-secretagogue receptor (50), which is distinct from the GHRH receptor. GHRH prompted an enhancement of SWS, whereas GHRP-6 elevated stage 2 sleep. The latter effect appears to be a unique finding in normal controls as no other endogenous or synthetic substance is known to share this effect in normal subjects. In a similar vein, however, in patients with depression Antonijevic et al. (I.A. Antonijevic, R.M. Frieboes, H. Murck, A. Steiger, unpublished) observed a normalization of the reduced amount of stage 2 sleep after GHRH. Because none of the hormones elevated after GHRP-6, GH, cortisol or ACTH, enhance stage 2 sleep it appears likely that this effect represents a central action of the substance. Similarly, Copinschi et al. (19) suggested that the sleep-EEG changes after MK-677 reflect a central action of the drug. This view was supported by the absence of a correlation between hormones and sleep-EEG effects particularly in the young subjects. Furthermore, the effects of MK-677 differ from those of i.c.v. IGF-1 on sleep (26). Whereas the effects of MK-677 on GH release in the study by Copinschi et al. (19) were only marginal, it can be argued that the increase in stage 4 sleep after MK-677 may be due to an enhanced release of central GHRH. This idea is in line with the hypothesis that MK-677 stimulates arcuate neurons containing GHRH and reduces somatostatin (51). In this model, however, it is difficult to explain the increase in REM sleep in young and old subjects after MK-677. Stimulation of REM sleep would be explained by an increase in GH (24,26) or by enhanced activity of somatostatin (52). The most marked effect of GHRP-6 was found after pulsatile i.v. administration. This supports the view that this is the most appropriate method to prompt acute sleep-EEG changes by peptides. Nevertheless, it is interesting that intranasal administration of the peptide also prompted a trend towards increased stage 2 sleep. In addition, EEG-delta power decreased by trend specifically after this route of administration. This effect may be explained by a release of NPY after GHRP-6, which is suggested by preclinical studies (53). In rats (48) and humans (47), NPY suppressed EEG-delta activity.

Opposite to GHRH, which suppressed cortisol (20), and in contrast to oral MK-677, which did not affect cortisol release (19) after GHRP-6, in addition to GH ACTH and cortisol were also stimulated. These findings are similar to those of Hayashi et al. (15), who found that 15–30 min after i.v. GHRP-6 in the morning cortisol levels were slightly but significantly increased.

The lack of an effect of i.v. GHRP-2 given as a single dose at the onset of the third REM period does not rule out an influence of this compound on sleep as long as no data are available on the influence of its pulsatile administration on the sleep EEG. Nevertheless, it appears remarkable, that i.v. GHRP-2 (18) did not share the effect of GHRH in the analogue protocol with i.v. GHRH during the third REM period (23).

Whereas the endogenous ligand of the GH-secretagogue receptor has not yet been identified, it appears to be a candidate for a further factor of the somatotropic system participating in sleep regulation. The capacity of GH secretagogues to promote sleep appears to be of clinical relevance. In most of the world today, the most frequently used hypnotics are benzodiazepines. These drugs are not capable of inducing natural sleep for they suppress SWS, EEG-delta power and REM sleep (54). In addition, at least after brief administration of benzodiazepines, blunted GH release and stimulation of prolactin were found (55,56). Furthermore, the risk of addiction in benzodiazepine treatment is well established. Therefore, there is a need for the development of hypnotics better related to human physiology. It appears possible that in the future GH secretagogues may be used as hypnotics. The study results with MK-677 are encouraging. It is of particular interest that sleep in the elderly was improved, since with increasing age sleep quality deteriorates and the use of hypnotics increases. Caution is called for in the therapeutic use of GH secretagogues because stimulation of the HPA system appears to be an unwanted side effect after some of the substances. Therefore, further studies with long-term administration of GH secretagogues are necessary including monitoring of HPA activity. The preliminary findings suggest a relatively low risk after intranasal and oral administration of these substances.

REFERENCES

1. Obál, Jr. F, Opp, M., Sary, G., Krueger, J.M. (1990) Endocrine mechanisms in sleep regulation. In: Inoué S., Krueger J.M. (eds.), Endogenous sleep factors. SPB Academic Publishing, The Hague, pp. 109–120.
2. Steiger, A., Holsboer, F. (1997) Neuropeptides and human sleep. Sleep 20, 1038–1052.
3. Krueger, J.M., Obál, Jr. F. (1993) Growth hormone-releasing hormone and interleukin-1 in sleep regulation. FASEB J 7, 645–652.
4. Rechtschaffen, A, Kales, A. (1968) A Manual of Standardized Terminology, Techniques and Scoring System for Sleep Stages of Human Subjects. US Department of Health, Education & Welfare, Neurological Information Network, Bethesda, MD.
5. Weitzman, E.D. (1976) Circadian rhythms and episodic hormone secretion in man. Annu. Rev. Med. 27, 225–243.
6. Steiger, A., Herth, T., Holsboer, F. (1987) Sleep-electroencephalography and the secretion of cortisol and growth hormone in normal controls. Acta Endocrinol. (Copenh.) 116, 36–42.
7. Van Coevorden, A., Mockel, J., Laurent, E. et al. (1991) Neuroendocrine rhythms and sleep in aging men. Am. J. Physiol. 260, E651–E661.
8. Bliwise, D.L. (1993) Sleep in normal aging and dementia. Sleep 16, 40–81.
9. Reynolds, 3d, C.F, Kupfer, D.J. (1987) Sleep research in affective illness: state of the art circa 1987. Sleep 10, 199–215.
10. Steiger, A., von Bardeleben, U., Herth, T., Holsboer, F. (1989) Sleep EEG and nocturnal secretion of cortisol and growth hormone in male patients with endogenous depression before treatment and after recovery. J. Affect. Disord. 16, 189–195.
11. Frieboes, R.-M., Murck, H., Maier, P., Schier, T., Holsboer, F., Steiger, A. (1995) Growth hormone-releasing peptide-6 stimulates sleep, growth hormone, ACTH and cortisol release in normal man. Neuroendocrinology 61, 584–589.
12. Huhn, W.C., Hartman, M.L., Pezzoli, S.S., Thorner, M.O. (1993) Twenty-four-hour growth hormone (GH)-releasing peptide (GHRP) infusion enhances pulsatile GH secretion and specifically attenuates the response to a subsequent GHRP bolus. J. Clin. Endocrinol. Metab. 76, 1202–1208.

13. Jaffe, C.A., Ho, P.J., DeMott-Friberg, R., Bowers, C.Y., Barkan, A.L. (1993) Effects of a prolonged growth hormone (GH)-releasing peptide infusion on pulsatile GH secretion in normal men. J. Clin. Endocrinol. Metab. 77, 1641–1647.
14. Bowers, C.Y., Alster, D.K., Frentz, J.M. (1992) The growth hormone-releasing activity of a synthetic hexapeptide in normal men and short statured children after oral administration. J. Clin. Endocrinol. Metab. 74, 292–298.
15. Hayashi, S., Okimura, Y., Yagi, H. et al. (1991) Intranasal administration of His-D-Trp-Ala-Trp-D-Phe-LysNH2 (growth hormone releasing peptide) increased plasma growth hormone and insulin-like growth factor-I levels in normal men. Endocrinol. Jpn. 38, 15–21.
16. Steiger, A., Antonijevic, I.A., Bohlhalter, S., Frieboes, R.M., Friess, E., Murck, H. (1998) Effects of hormones on sleep. Horm. Res. 49, 125–130.
17. Frieboes, R.-M., Murck, H., Antonijevic, I.A., Holsboer, F., Steiger, A. (1997) Sleep-endocrinological effects of GH-releasing peptide-6 in man after different routes of application. Exp. Clin. Endocrinol. Diabetes. 105 (Suppl. 1), 1.
18. Moreno-Reyes, R., Kerkhofs, M., L'Hermite-Balériaux, M., Thorner, M.O., Van Cauter, E., Copinschi, G. (1998) Evidence against a role for the growth hormone-releasing peptide axis in human slow-wave sleep regulation. Am. J. Physiol. 274, E779–E784.
19. Copinschi, G., Leproult, R., Van Onderbergen, A. et al. (1997) Prolonged oral treatment with MK-677, a novel growth hormone secretagogue, improves sleep quality in man. Neuroendocrinology 66, 278–286.
20. Steiger, A., Guldner, J., Hemmeter, U., Rothe, B., Wiedemann, K., Holsboer, F. (1992) Effects of growth hormone-releasing hormone and somatostatin on sleep EEG and nocturnal hormone secretion in male controls. Neuroendocrinology 56, 566–573.
21. Marshall, L., Mölle, M., Böschen, G., Steiger, A., Fehm, H.L., Born, J. (1996) Greater efficacy of episodic than continuous growth hormone releasing hormone (GHRH) administration in promoting slow wave sleep (SWS). J. Clin. Endocrinol. Metab. 81, 1009–1013.
22. Holsboer, F., von Bardeleben, U., Steiger, A. (1988) Effects of intravenous corticotropin-releasing hormone upon sleep-related growth hormone surge and sleep EEG in man. Neuroendocrinology 48, 32–38.
23. Kerkhofs, M., Van Cauter, E., Van Onderbergen, A., Caufriez, A., Thorner, M.O., Copinschi, G. (1993) Sleep-promoting effects of growth hormone-releasing hormone in normal men. Am. J. Physiol. 264, E594–E598.
24. Obál Jr., F., Floyd, R., Kapás, L., Bodos, B., Krueger, J.M. (1996) Effects of systemic GHRH on sleep in intact and in hypophysectomized rats. Am. J. Physiol. 270, E230–E237.
25. Obál Jr., F., Bodosi, B., Szilagyi, A., Kacsoh, B., Krueger, J.M. (1997) Antiserum to growth hormone decreases sleep in the rat. Neuroendocrinology 66, 9–16.
26. Obál Jr., F., Kapas, L., Bodosi, B., Krueger, J.M. (1998) Changes in sleep in response to intracerebral injection of insulin-like growth factor-1 (IGF-1) in the rat. Sleep Research Online 1, 87–91.
27. Ehlers, C.L., Reed, T.K., Henriksen, S.J. (1986) Effects of corticotropin-releasing factor and growth hormone-releasing factor on sleep and activity in rats. Neuroendocrinology 42, 467–474.
28. Steiger, A., Guldner, J., Colla-Müller, M., Friess, E., Sonntag, A., Schier, T. (1994) Growth hormone-releasing hormone (GHRH)-induced effects on sleep EEG and nocturnal secretion of growth hormone, cortisol and ACTH in patients with major depression. J. Psychiatr. Res. 28, 225–238.
29. Guldner, J., Schier, T., Friess, E., Colla, M., Holsboer, F., Steiger, A. (1997) Reduced efficacy of growth hormone-releasing hormone in modulating sleep endocrine activity in the elderly. Neurobiol. Aging 18, 491–495.
30. Shibasaki, T., Shizume, K., Nakahara, M. et al. (1984) Age-related changes in plasma growth hormone response to growth hormone-releasing factor in man. J. Clin. Endocrinol. Metab. 58, 212–214.
31. Frieboes, R.-M., Murck, H., Schier, T., Holsboer, F., Steiger, A. (1997) Somatostatin impairs sleep in elderly human subjects. Neuropsychopharmacology 16, 339–345.

32. Beranek, L., Obál Jr., F., Taishi, P., Bodosi, B., Laczi, F., Krueger, J.M. (1997) Changes in rat sleep after single and repeated injections of the long-acting somatostatin analog ocreotide. American Journal of Physiology — Regulatory Integrative & Comparative Physiology 273, R1484–R1491.
33. Parker, D.C., Rossman, L.G., Siler, T.M., Rivier, J., Yen, S.S., Guillemin, R. (1974) Inhibition of the sleep-related peak in physiologic human growth hormone release by somatostatin. J. Clin. Endocrinol. Metab. 38, 496–499.
34. Kupfer, D.J., Jarrett, D.B., Ehlers, C.L. (1992) The effect of SRIF on the EEG sleep of normal men. Psychoneuroendocrinology 17, 37–43.
35. Sonntag, W.E., Boyd, R.L., Booze, R.M. (1990) Somatostatinergic gene expression in hypothalamus and cortex of aging male rats. Neurobiol. Aging 11, 409–416.
36. Copinschi, G., Van Onderbergen, A., L'Hermite-Balériaux, M. et al. (1996) Effects of a 7-day treatment with a novel orally active nonpeptide growth hormone secretagogue, MK-677, on 24-hour growth hormone profiles, insulin-like growth factor-I and adrenocortical function in normal young men. J. Clin. Endocrinol. Metab. 81, 2776–2782.
37. Murck, H., Frieboes, R.-M., Schier, T., Steiger, A. (1997) Longtime administration of growth hormone-releasing hormone (GHRH) does not restore the reduced efficiency of GHRH on sleep endocrine activity in 2 old-aged subjects — a preliminary study. Pharmacopsychiatry 30, 122–124.
38. Iovino, M., Monteleone, P., Steardo, L. (1989) Repetitive growth hormone-releasing hormone administration restores the attenuated growth hormone (GH) response to GH-releasing hormone testing in normal aging. J. Clin. Endocrinol. Metab. 69, 910–913.
39. Ehlers, C.L., Kupfer, D.J. (1987) Hypothalamic peptide modulation of EEG sleep in depression: a further application of the S-process hypothesis. Biol. Psychiatry 22, 513–517.
40. Åström, C., Lindholm, J. (1990) Growth hormone-deficient young adults have decreased deep sleep. Neuroendocrinology 51, 82–84.
41. Bredow, S., Taishi, P., Obál Jr., F., Guha-Thakurta, N., Krueger, J.M. (1996) Hypothalamic growth hormone-releasing hormone mRNA varies across the day in rat. NeuroReport 7, 2501–2505.
42. Holsboer, F. (1995) Neuroendocrinology of affective disorders. In: Bloom, F.E., Kupfer, D.J. (eds.). Neuropsychopharmacology. 4th Generation of Progress. Raven Press, New York, pp. 957–970.
43. Davidson, J.R., Moldofsky, H., Lue, F.A. (1991) Growth hormone and cortisol secretion in relation to sleep and wakefulness. J. Psychiatry Neurosci. 16, 96–102.
44. Toppila, J., Asikainen, M., Alanko, L., Turek, F.W., Stenberg, D., Porkka-Heiskanen, T. (1996) The effect of REM sleep deprivation on somatostatin and growth hormone-releasing hormone gene expression in the rat hypothalamus. J. Sleep Res. 5, 115–122.
45. Toppila, J., Stenberg, D., Alanko, L. et al. (1995) REM sleep deprivation induces galanin gene expression in the rat brain. Neurosci. Lett. 183, 171–174.
46. Murck, H., Maier, P., Frieboes, R.-M., Schier, T., Holsboer, F., Steiger, A. (1995) Galanin promotes REM sleep in man. Pharmacopsychiatry 28, 201
47. Bohlhalter, S., Antonijevic, I.A., Brabant, G., Holsboer, F., Steiger, A. (1997) Short term pulsatile administration of neuropeptide Y suppresses ACTH and cortisol secretion, promotes sleep, but does not affect leptin secretion in man. Exp. Clin. Endocrinol. Diabetes 105 (Suppl. 1), 111 Abstract.
48. Ehlers, C.L., Somes, C., Lopez, A., Kirby, D., Rivier, J.E. (1997) Electrophysiological actions of neuropeptide Y and its analogs: new measures for anxiolytic therapy? Neuropsychopharmacology 17, 34–43.
49. Murck, H., Guldner, J., Frieboes, R.-M. et al. (1996) VIP decelerates non-REM–REM cycles and modulates hormone secretion during sleep in men. Am. J. Physiol. 271, R905–R911.
50. Howard, A.D., Feighner, S.D., Cully, D.F. et al. (1996) A receptor in pituitary and hypothalamus that functions in growth hormone release. Science 273, 974–977.

51. Smith, R.G., Pong, S.S., Hickey, G. et al. (1996) Modulation of pulsatile GH release through a novel receptor in hypothalamus and pituitary gland. Rec. Prog. Horm. Res. 51, 261–286.
52. Danguir, J. (1986) Intracerebroventricular infusion of somatostatin selectively increases paradoxical sleep in rats. Brain Res. 367, 26–30.
53. Korbonits, M., Little, J.A., Forsling, M.L. et al. (1998) The effect of growth hormone secretagogues on the release of growth hormone-releasing hormone, somatostatin, vasopressin, and corticotrophin-releasing hormone from the rat hypothalamus in vitro. In: Bercu, B.B., Walker, R.F. (eds.), Growth hormone secretagogues in clinical practice. Marcel Dekker, New York, pp. 231–249.
54. Borbély, A.A., Achermann, P. (1991) Ultradian dynamics of sleep after a single dose of benzodiazepine hypnotics. Eur. J. Pharmacol. 195, 11–18.
55. Copinschi, G., Van Onderbergen, A., L'Hermite-Balériaux, M. et al. (1990) Effects of the short-acting benzodiazepine triazolam, taken at bedtime, on circadian and sleep-related hormonal profiles in normal men. Sleep 13, 232–244.
56. Steiger, A., Guldner, J., Lauer, C., Meschenmoser, C., Pollmächer, T., Holsboer, F. (1994) Flumazenil exerts intrinsic activity on sleep EEG and nocturnal hormone secretion in normal controls. Psychopharmacology 113, 334–338.

Growth Hormone Secretagogues
Edited by E. Ghigo, M. Boghen, F.F. Casanueva and C. Dieguez

Chapter 24

Hexarelin, A Synthetic Growth Hormone Secretagogue, Exhibits Protectant Activity in Experimental Myocardial Ischemia and Reperfusion

FERRUCCIO BERTI, GIUSEPPE ROSSONI and VITO DE GENNARO COLONNA

Department of Pharmacology, Chemiotherapy and Medical Toxicology, University of Milan, Milan, Italy

INTRODUCTION

The observation that growth hormone (GH) administration to normal subjects induced positive inotropic effect was the first strong support to the rational use of this hormone in cardiovascular diseases. On the other hand, GH deficiency is a complex syndrome in which the cardiovascular involvement represents one of the most prominent features. In fact, it is now increasingly recognised that, together with decreased lean body mass and bone density and declined muscle power and exercise tolerance, hypopituitary subjects may face premature cardiovascular mortality which may be related to atherogenesis. In this respect, ultrasonographic study of Markussis et al. (1) showed a significant increase in the intima-media thickness of the carotid arteries with atheromotous plaques in carotid and femoral arteries of asymptomatic adult hypopituitary patients.

Rosen and Bengtsson (2) first reported that patients with hypopituitarism, when given the proper thyroid, adrenal and gonadal replacement therapy, without any specific GH replenishment, showed an increased mortality from cardiovascular events, especially myocardial infarction and cardiac failure. In adult subjects with hypopituitarism and severe GH deficiency, Shahi et al. (3) found a significant correlation between serum levels of insulin-like growth factor-1 (IGF-1), the mediator of most of the effects of pituitary GH, and left ventricular mass. This suggested that GH plays a role for the maintenance of cardiac size in adulthood. Furthermore, these authors reported that some of their patients showed left ventricular diastolic dysfunction and ischemic-like ST segment alterations during exercise testing indicating that the small coronary arteries of these patients were compromised in their function. According to the available data, peripheral action of GH are also dependent on IGF-1 which exerts its own effects such as enhancement of contractility on cardiac myocytes and increase in size of these cells. The evidence for direct and indirect effects of

GH on cardiac function is based on the observation of GH cardiac receptors and IGF-1 messenger RNA (mRNA) expression in animal models and an increase in cardiac tissue of IGF-1 mRNA expression in response to GH administration (4).

Cardioprotective effect in myocardial ischemia followed by reperfusion has also been reported with IGF-1 and the suggested mechanism was referred to inhibition of polymorphonuclear leukocyte (PMN)-induced cardiac necrosis and attenuation of reperfusion-induced apoptosis of cardiac myocytes (5). Furthermore IGF-1 has been shown to be a regulator of vascular function by stimulating nitric oxide (NO) production in cultured vascular endothelium. The effect of IGF-1 in ischemia may be due to a reduction of PMN infiltration during reperfusion via stimulation of the release of endothelial NO. In fact, it has been reported that NO inhibits neutrophyl adherence to the coronary endothelial lining and decreases their accumulation within the myocardium. Thus, NO by down-regulating neutrophyl-endothelium interaction at the coronary level, may inhibit transcellular metabolism and the consequent formation of cardiotoxic leukotrienes (6).

In a recent study of our group (7), isolated hearts obtained from GH-deficient rats, when subjected to low flow ischemia and reperfusion showed a marked ischemic tissue damage consisting in worsening of post-ischemic ventricular dysfunction. In this study the replacement therapy with GH resulted in a clear protection of the hearts. Similar protective data were also obtained with the GH-secretagogue hexarelin not only in GH-deficient rats (8) but also in aged rats (9) which are known to be prone to ischemic damage. In this chapter we will review our findings obtained with hexarelin in the light of the emerging evidence for beneficial effects of GH-secretagogues in ischemic heart diseases, underlying also the important role of GH/IGF-1 axis in the maintenance of a normal function of vascular endothelium.

THE HEXAPEPTIDE HEXARELIN

In recent years several synthetic peptides have been shown to be active in inducing GH secretion in different animal species. Among them a family of growth hormone releasing factors (GHRP-6, GHRP-1, GHRP-2) have demonstrated high potency and selectivity in stimulating GH secretion. In particular the hexapeptide GHRP-6, developed from an enkephalin analogue, has been shown to stimulate selectively GH release both *in vitro* and *in vivo*. The mechanism of action of GHRP-6 is not completely understood, however the compound seems to act at the pituitary site activating intracellular messenger pathways different from those utilized by GH releasing hormone (GHRH), and at the hypothalamic level where receptors for GHRP have been demonstrated.

Hexarelin (His-D-2Methyl-Trp-Ala-Trp-D-Phe-Lys-NH_2) belongs to the family of GHRP-6 analogues in which tryptophan (Trp) has been substituted by the more stable 2-Methyl-Trp. Hexarelin has been shown to be more potent and long lasting in stimulating GH release than the parent compound GHRP-6 in different animal species, man included (10).

HEXARELIN PROTECTS THE HEARTS OF GH-DEFICIENT RATS FROM ISCHEMIA REPERFUSION DAMAGE

Over the past few years, significant advances have been made in our understanding of the cellular and molecular mechanism involved in GH action, including its effect on cardiac tissue (11). Furthermore the experimental evidence points to a role of GH in cardiac pathophysiology. In line with previous experiments, where hearts excised from GH-deficient rats are more sensitive to the damaging effects of global flow reduction and reperfusion (7), we investigated whether hexarelin, like GH, was capable of reversing cardiac ventricular dysfunction in the same rat model of GH-deficiency. Briefly, the protocol of these experiments was the following. Male Sprague–Dowley rats of 20 days were randomly assigned to four experimental groups of 10 animals each: (a) controls (treated with normal rabbit serum); (b) GH-deficients (treated with rabbit anti GHRH-serum); (c) GH-deficients treated with hexarelin; (c) controls treated with hexarelin. Hexarelin (80 μg/kg, bid, sc) was given to the animals from postnatal day 25 to 40 in addition to normal rabbit serum or rabbit anti GHRH-serum. GH-deficients rats were killed 14 days after the last injection of hexarelin. Anterior pituitaries were removed for determination of GH mRNA level and blood was collected for evaluation of plasma IGF-1 concentration (8).

Electrically paced hearts in the four experimental groups were perfused retrogradely through the aorta with gassed Krebs-Henseleit solution and subjected to 20 min ischemia (flow rate 2 ml/min) and reperfusion (flow rate 12 ml/min) was followed for 30 min as previously described (7).

The results obtained with these experiments clearly demonstrate that rats given the rabbit anti GHRH serum were truly GH-deficients since their growth rate, pituitary GH mRNA and plasma IGF-1 levels were significantly reduced, as already demonstrated by Cella et al. (12) (Table 1). In these animals, administration of hexarelin restored the

TABLE 1

BODY AND HEART WEIGHTS AND MARKERS OF SOMATOTROPIC FUNCTION IN CONTROL AND GH-DEFICIENT RATS TREATED OR NOT WITH HEXARELIN

Treatment	Body weight (g)	Heart weight (mg)	Heart weight / Body weight (mg/g)	Pituitary GH mRNA (%)	Plasma IGF-1 (ng/ml)
NRS	193.1 ± 2.2	1475 ± 10.1	7.63	100	169 ± 5.0
GHRH-Ab	168.2 ± 2.1_	1295 ± 9.0_	7.69	–51.2 ± 1.7*	93 ± 2.4_
HEXA	190.0 ± 1.5	1451 ± 10.2	7.60	–2.7 ± 44	158 ± 5.2
GHRH-Ab + HEXA	192.8 ± 1.8	1480 ± 12.8	7.67	–7.0 ± 6.1	157 ± 2.4

Treatment legend as in Figure 1. Figures related to body and heart weights and plasma levels of IGF-1 are mean values ± SEM of 10 determinations. Figures related to pituitary GH mRNA are mean values ± SEM of 5 determinations. *$P < 0.01$ vs. NRS and GHRH-Ab + HEXA.

somatotropic function with normalisation of the above biological markers. Restoration of GH mRNA levels in GH-deficient adult male rats given at the same dose of hexarelin as that used in the present experiments was already reported by Torsello et al. (13).

The mechanism(s) underlying the action of hexarelin is not fully understood. This peptide may modulate GH secretion by acting directly on the pituitary (14) and/or at

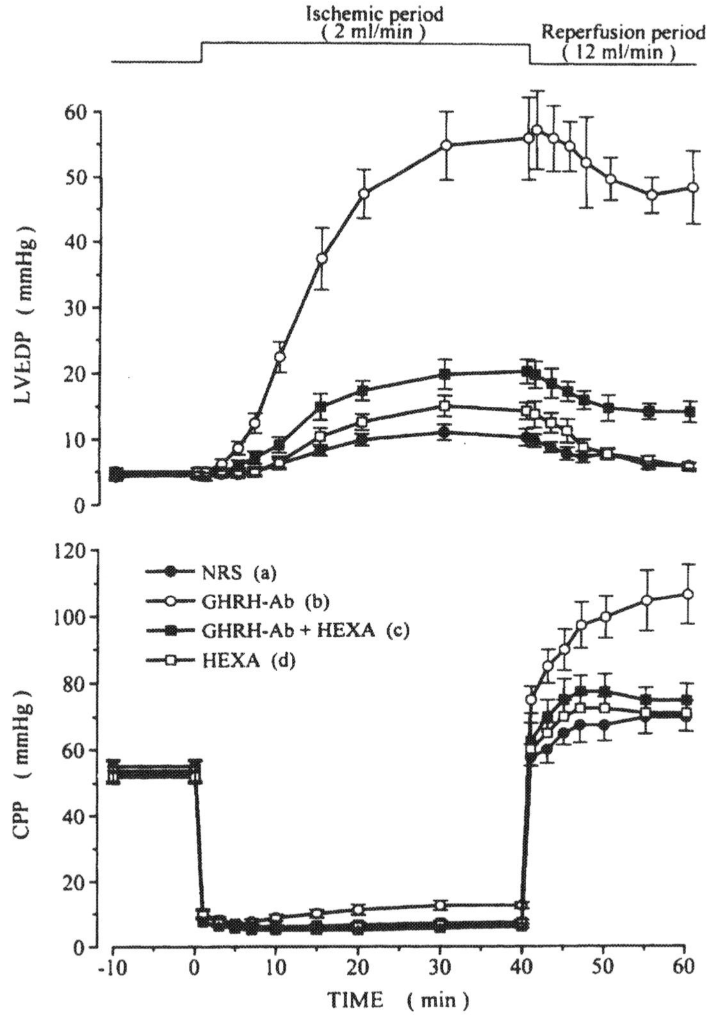

Figure 1. Perfusion experiments with paced isovolumic left heart preparations from normal rabbit serum (NRS, group a), anti-GHRH serum (GHRH-Ab, group b), anti-GHRH serum + hexarelin (GHRH-Ab + HEXA, group c) and hexarelin (HEXA, group d) treated rats. Each point represents the mean values and vertical bars the SEM from 10 hearts. LVEDP = left ventricular-end diastolic pressure; CPP = coronary perfusion pressure. The areas under the curve (AUCs) of LVEDP are: (a) 499 ± 55; (b) 2563 ± 197; (c) 916 ± 84 and (d) 622 ± 68. Statistical significance: b vs. a, c and d, $P < 0.001$; c vs. d and a, $P < 0.05$; d vs. a, $P > 0.05$. AUC was estimated according the trapezoid method: in ordinate, LVEDP in mmHg; in abscissa, time from 0 to 60 min.

hypothalamic level by modulating the release of somatostatin (15) and/or GHRH (16) and/or unknown factors.

Whatever the mechanism of action might be, hexarelin-induced restoration of somatotropic function appears to be instrumental to the striking improvement of the post-ischemic ventricular function recorded in the isolated hearts (Figures 1–2). However, the observation that hexarelin induced a clear-cut protection against myocardial damage also in control rats without modifying somatotropic function raises the important issue of its true mechanism of action and suggests that the peptide may also act directly on the heart. In support of this view, mRNA coding for a receptor related to growth hormone-releasing secretagogues, such as hexarelin, has been recently detected in rat cardiac tissue (17). Furthermore, the possibility cannot be ruled out and is currently being investigated that hexarelin, besides increasing plasma IGF-1 levels, may stimulate local IGF-1 biosynthesis or induce accumulation of the peptide in the heart by interfering with its degradation (18). In this connection, an increased responsiveness of the cardiac myofilaments to IGF-1 made locally available by hexarelin should be considered and this might explain the more pronounced improvement of the isolated heart contractility upon reperfusion. Reportedly, IGF-1 has positive inotropic effects in healthy male volunteers (19) and on rat papillary muscle, with increasing force development and rise in free peak calcium in isolated cardiac myocytes (20). Furthermore, the ability of IGF-1 to limit reperfusion injury in rat hearts subjected to ischemia (5) and in functionally impaired hearts of rats undergoing myocardial infarction (21) has been clearly demonstrated.

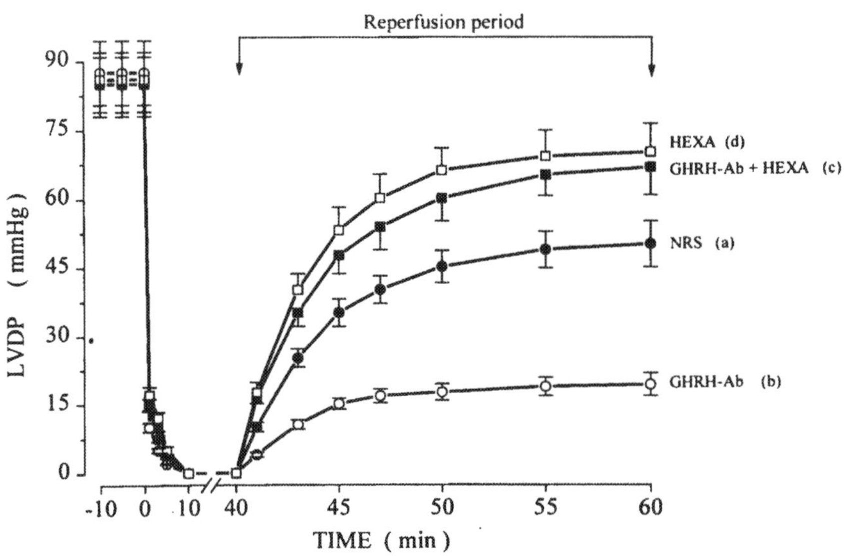

Figure 2. Left ventricular developed pressure (LVDP) in isovolumic left heart preparations subjected to global low-flow ischemia and reperfusion. Each point represents the mean values and vertical bars the SEM from 10 hearts. The areas under the curve of LVDP are: (a) 877 ± 54; (b) 386 ± 29; (c) 1154 ± 82 and (d) 1256 ± 108. Statistical significance: b vs. d, c and a, $P < 0.01$; d and c vs. a, $P < 0.05$; d vs. c, $P > 0.05$. For abbreviations see caption of Figure 1.

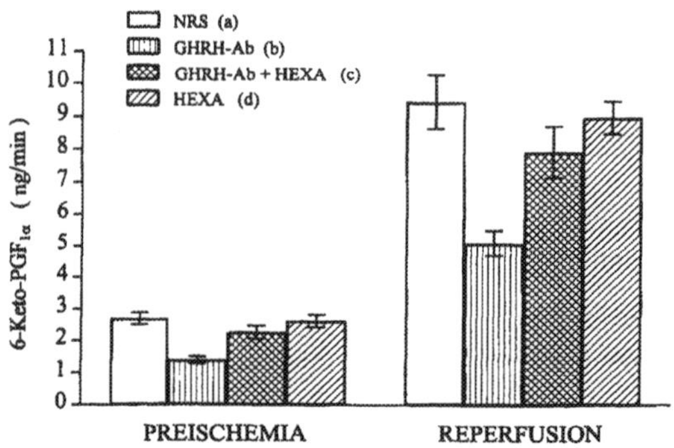

Figure 3. Rate of release of 6-keto-$PGF_{1\alpha}$ in perfusates of isovolumic left heart preparations from rats of the 4 experimental groups. Each column represents the mean values and vertical bars the SEM from 10 hearts. Perfusates were collected for 5 min before flow reduction (pre-ischemia) and during the first 10 min of reperfusion. Statistical significance: a vs. b, $P < 0.001$; a vs. c and d, $P > 0.05$. For abbreviations see caption of Figure 1.

Another interesting feature of GH deficiency that emerges from these isolated heart experiments was the reduced formation of the stable metabolite of prostacyclin (PGI_2), 6-keto-$PGF_{1\alpha}$, not only during the pre-ischemic phase but in particular during reperfusion (Figure 3). This may bear some relevance to the aggravation of the ischemic damage detected in the hearts of these animals. In fact, in hearts from GH-deficient hexarelin-treated rats, the attenuation of the ischemic damage was associated with a recovery of 6-keto $PGF_{1\alpha}$ release within the range of values of control preparations (Figure 3). It is well known that insufficient production of primary prostaglandins may be associated with further aggravation of tissue damage, in particular in early reperfusion (22). Along this line, it has been already reported that prostacyclin mimetics or PGI_2-releasers prevent ventricular contracture of ischemic hearts and improve heart mechanics at reperfusion (23).

HEXARELIN PREVENTS ALTERATIONS OF VASCULAR ENDOTHELIUM-DEPENDENT RELAXANT FUNCTION IN GH-DEFICIENT RATS

Another important finding of our study was the hyper-reactivity of coronary smooth muscles to angiotensin II in heart preparations from GH-deficient rats (Figure 4). This, especially when viewed in conjunction with a clear-cut reduction of PGI_2 generation by the cardiac tissues, not only denotes damage of the vascular endothelial-dependent relaxant mechanism but also emphasises the crucial role of somatotropic function in maintaining the integrity of the vascular endothelial cell lining. In fact, in heart preparations from GH-deficient rats treated with hexarelin, the recovery of somatotropic function was associated with normalisation of the vasopressor activity of angiotensin II and with preserved generation of PGI_2. The competence of the eicosanoid to modulate the vasopressor activity of endothelin-1 in the isolated perfused rabbit heart has already been demonstrated (24).

Changes in coronary perfusion pressure in response to acetylcholine have been already reported in isolated hearts from GH-deficient rats, in which replacement therapy with GH

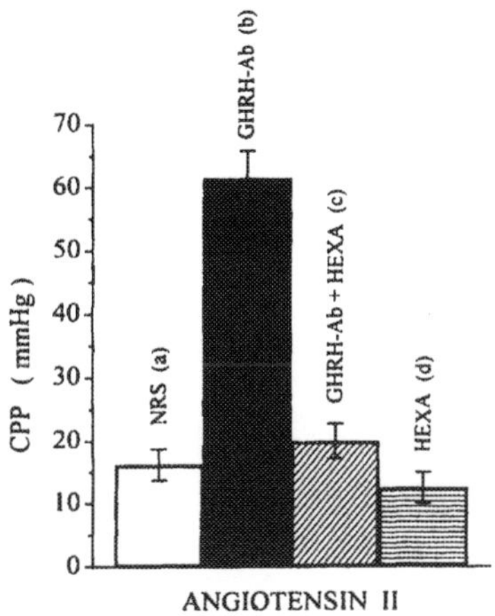

Figure 4. Vasopressor activity of angiotensin II (1 μg) injected in isovolumic left heart preparations during the pre-ischemic phase. CPP, coronary perfusion pressure. Each column represents the mean values and vertical bars the SEM from 10 hearts. Statistical significance: b vs. a, c and d, $P < 0.001$; a vs. c and d, $P > 0.05$. For abbreviations see caption of Figure 1.

restored the physiologic response of the coronary vasculature to the neuromediator (7). In spite of the complexity of the mechanism underlying the vasopressor response to acetylcholine in this animal species, these previous data and our present results support the idea that for a modulatory response of the vascular tissue to vasoconstriction, a preserved somatotropic function is needed.

It is also of relevance that the damage of the vascular endothelial cells in GH deficiency is a phenomenon most likely widespread in the circulation. In fact, using aortic rings from GH-deficient rats, the results obtained indicated changes of the two important systems regulating vascular tonus: generation of both PGI_2 and NO. Again, normalisation of the somatotropic function by hexarelin was linked to a restored generation of PGI_2 by vascular segments, coupled with preservation of acetylcholine and L-NMMA responses and reduced hyper-reactivity to endothelin-1 (Figures 5–6).

Collectively, the present results indicate that hexarelin given to rats with selective GH deficiency is capable of restoring somatotropic function to an extent similar to that induced by GH replacement therapy. However, even if the beneficial effect of hexarelin on the post-ischemic ventricular dysfunction appears to involve restoration of pituitary GH mRNA and plasma IGF-1 levels, a direct action of the hexapeptide on specific myocardial receptors should be considered.

Finally, the hyper-reactivity to vasoconstrictors of the coronary vasculature and aortic rings of GH-deficient rats would indicate that a normal function of the GH/IGF-1 axis is crucial for preventing vascular endothelial injury and dysfunction. In this respect, Böger et al. (25) reported that NO formation is decreased in untreated GH-deficient patients. Treatment of these patients with recombinant human GH normalized urinary nitrate and cyclic GMP excretion, possibly via IGF-1 stimulation of endothelial NO synthase.

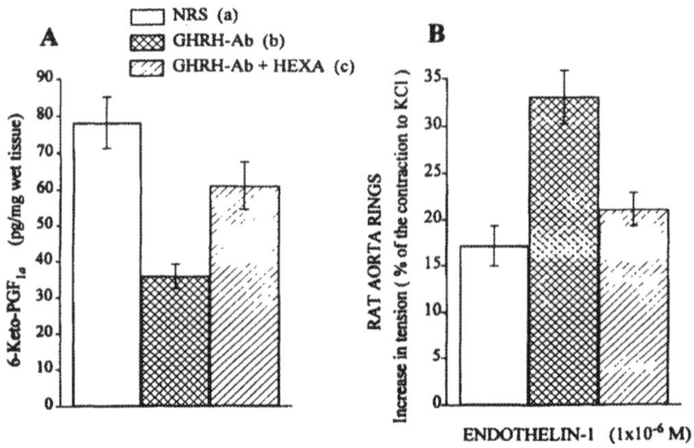

Figure 5. Unstimulated release of 6-Keto-PGF$_{1\alpha}$ in 20 min from isolated aortic rings (panel A) and vasopressor activity of endothelin-1 (panel B). Each column represents the mean values and vertical bars the SEM of 10 preparations. Statistical differences in panel A and B: b vs. a and c, $P < 0.01$; c vs. a, $P > 0.05$. For abbreviations see caption of Figure 1.

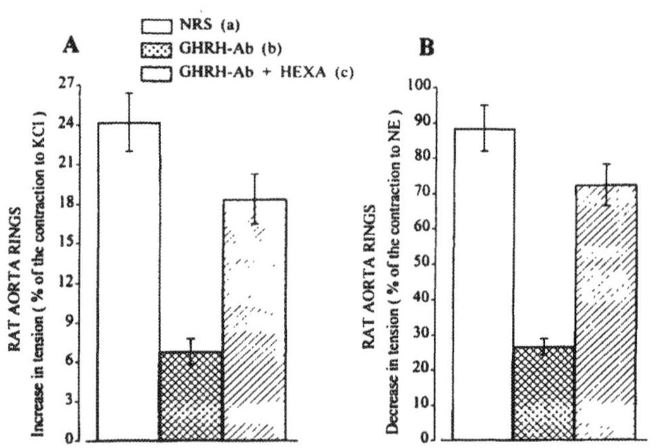

Figure 6. Vasopressor activity of N^G-monomethyl-L-arginine (L-NMMA) in isolated rat aortic rings (panel A) and relaxant activity of acetylcholine in norepinephrine-precontracted rat aortic rings (panel B). Each column represents the mean values and vertical bars the SEM of 10 preparations. Statistical differences in panel A and B: b vs. a and c, $P < 0.001$; c vs. a, $P > 0.05$. For abbreviations see caption of Figure 1.

HEXARELIN PROTECTS POST-ISCHEMIC VENTRICULAR DYSFUNCTION IN HEARTS OF AGED RATS

Aging has been shown to alter the spectrum of physiological and biochemical properties of the myocardium, including force production, excitation–contraction coupling, substrate use and mitochondrial oxidative capacity (26). However, new insights into myocardial-reperfusion injury indicate that aged rats, besides a reduction of the myocardial antioxidant defense mechanisms (27), are affected by alteration of calcium handling in cardiac cells

(28). In fact, abnormalities of regulation/modulation mechanisms normally involved in the restriction of calcium oscillation between sarcoplasmic reticulum and cytoplasm are associated with strong impairment of cardiac mechanics.

Myocardial ischemia, defined as an imbalance between fractional uptake of oxygen and the rate of cellular oxidation, may have several potential outcomes, especially in senescent hearts that are more prone to this pathologic event. Under these circumstances, when ischemia is brief, a transient post-ischemic ventricular dysfunction may occur, and this condition reflects many disturbances of cardiomyocytes and insufficient cellular antioxidant activity (29).

These findings and the awareness that GH secretion and its biological effects decline with aging in both experimental animals (30) and humans (31) prompted us to investigate the protective action of hexarelin in comparison with that of GH against post-ischemic myocardial dysfunction in hearts from *in vivo* treated senescent rats. Although strengthening of the somatotropic function would be instrumental in the anti-ischemic activity of hexarelin in aged rats, a direct action on the heart of the hexapeptide cannot be ruled out *a priori*. Favouring this view, previous findings of our laboratory have demonstrated that hexarelin given to young male rats was very effective in the cardiac ischemia–reperfusion model, despite the lack of an overt stimulation of the GH/IGF-1 axis (8).

Twenty-four-month-old male rats of Sprague–Dawley strain were used in these experiments. They were randomly assigned to three experimental groups and treated subcutaneously with: (a) 1 ml/kg saline; (b) biosynthetic human GH; (c) hexarelin. Hexarelin and GH were given to rats at the dose of 80 μg/kg and 400 μg/kg b.i.d., respectively, for 21 days. Animals were killed by cervical dislocation 14 h after the last injection, pituitaries were removed and used for determination of GH mRNA levels, whereas blood was collected for plasma determination of IGF-1 as reported above. The hearts were isolated and used for ischemia and reperfusion experiments. A moderate ischemia was induced by global reduction of the perfusion flow to 1 ml/min for a period of 20 min. A normal flow rate (15 ml/min) was then restored and reperfusion continued for 30 min.

In the present model of ischemia–reperfusion, hearts from old rats treated for the long term with hexarelin, achieved a strong protection: complete recovery of left ventricular function was present on reperfusion, and simultaneous blunting of creatine kinase (CK) leakage in the heart effluents bespoke the integrity of myocardial cell membranes and the preservation from the contractile impairment that follows oxygen readmission (Figures 7–9).

Under our experimental conditions, the beneficial effect disclosed by hexarelin in aged rats was not coupled to any apparent stimulation of the somatotropic function, because the levels of pituitary GH mRNA and plasma IGF-1 were unchanged. This would indicate, albeit inferentially, that the hexapeptide had a direct myocardial action divorced from that of GH. As we have reported above, Grilli et al. (17) and Howard et al. (32) recently reported that mRNA coding for a receptor related to GHS is expressed in peripheral organs of male rats, heart included.

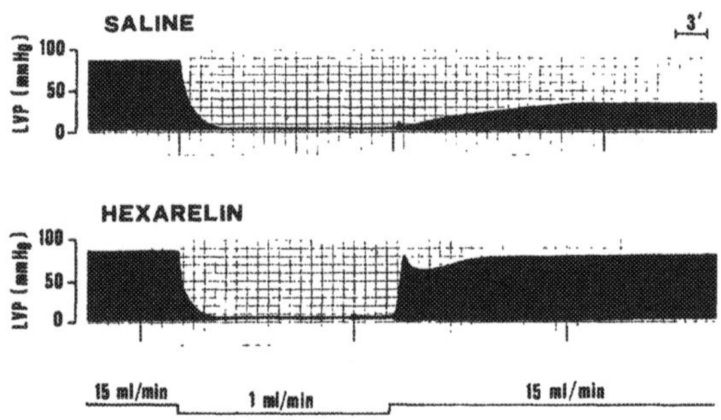

Figure 7. Left ventricular pressure (LVP) during post-ischemic reperfusion in heart preparations from saline- or hexarelin-treated old rats.

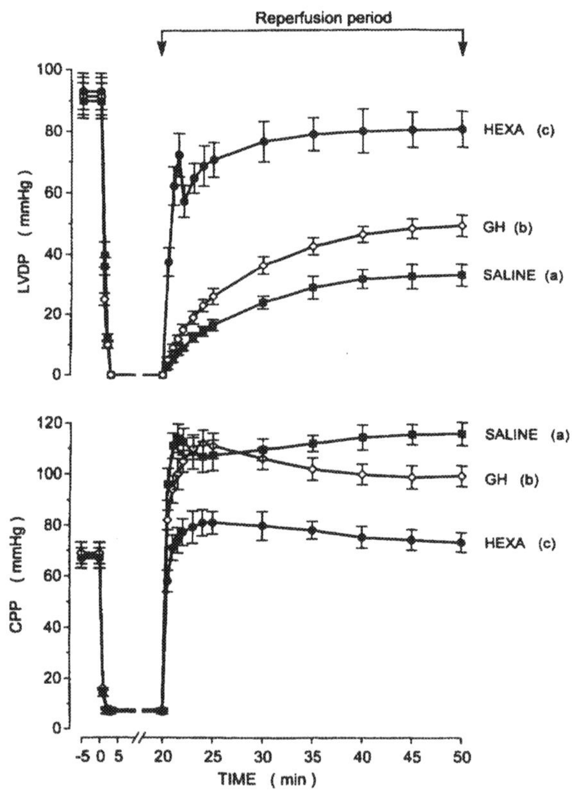

Figure 8. Left ventricular developed pressure (LVDP) and coronary perfusion pressure (CPP) in isovolumic left heart preparations submitted to low flow ischemia and reperfusion from old rats of the following experimental groups: (a) saline (controls, $n = 10$); (b) human growth hormone (GH, $n = 6$) and (c) hexarelin (HEXA, $n = 9$). Each point on the curves depicts mean values, and vertical bars, the SEM. The areas under the curve (AUCs) related to LVDP are: (a) 765 ± 46; (b) 1147 ± 88; (c) 2272 ± 66. Statistical differences: c vs. b and a, $P < 0.01$; b vs. a, $P < 0.05$. The AUCs related to CPP (increase in mm Hg over the pre-ischemic values) are: (a) 1284 ± 79; (b) 1008 ± 47 and (c) 235 ± 35. Statistical differences: c vs. b and a, $P < 0.01$; b vs. a, $P < 0.05$. AUCs was estimated according the trapezoid method: in ordinate, LVDP or CPP in mm Hg; in abscissa, time from 20 to 50 min.

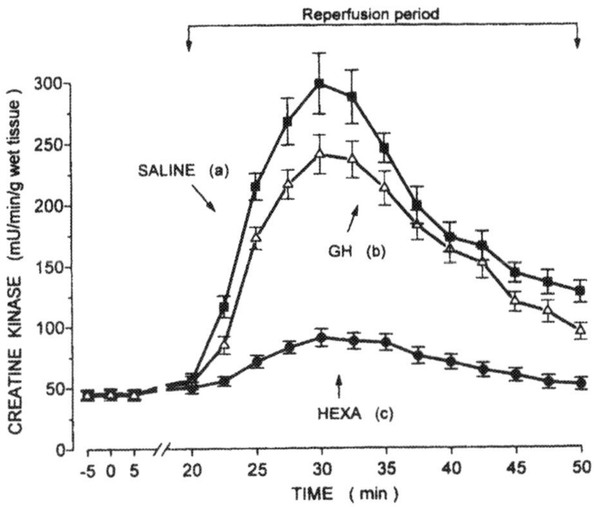

Figure 9. Creatine kinase (CK) release profile in ischemic and reperfusion conditions of old rat hearts. Caption as in Figure 8. Each point on the curves depicts mean values and vertical bars the SEM. The areas under the curve related to CK release during reperfusion are: (a) 4454 ± 352; (b) 3520 ± 278 and (c) 278 ± 56. Statistical differences: c vs. a and b, $P < 0.01$; b vs. a, $P < 0.05$.

We still ignore what kind of intracellular signal transduction is triggered by GH-secretagogue-receptor activation in peripheral organs, a point that deserves a thorough investigation. However, the striking hexarelin-induced inhibition of reperfusion damage in the isolated hearts would call for a restraint in the increase of cytosolic calcium that follows reperfusion. In this context, either the inhibitor of sarcoplasmic reticulum function, ryanodine, or the transsarcolemmal calcium channel blockers, diltiazem and verapamil, were shown capable of improving recovery of left ventricular developed pressure in rabbit hearts exposed to transient ischemia and hypoxia (33). However, because hexarelin, given directly to the heart through the perfusion system, does not depress myocardial contractility, the mechanism/s responsible for its anti-ischemic action may be different from those of calcium entry blockers. In fact, these compounds are known to protect the isolated rabbit hearts against abnormalities produced by transient hypoxia and low flow ischemia by a strong depression of myocardial contractility ultimately related to the inhibition of calcium entry into the myocardial cells. It is then possible that sustained administration of hexarelin in aged rats may have increased cardiomyocyte energy-stores to an extent compatible with the maintenance of basic cell organization that allows a normal recovery of the contractile function at the termination of the ischemic insult. In this vein, it is noteworthy that the amount of CK released during reperfusion from the heart of hexarelin-treated rats was significantly less than that of control preparations (Figure 9). This may indicate, in the former setting, a better preservation of the integrity of myocardial cell membranes, which is indispensable for a favourable osmotic control (prevention of free radicals accumulation and continuance of calcium homeostasis) and that maintenance of energy-rich phosphates had occurred. Our study, however, still lacks data on both the concentration of energy-storing nucleotides and on the grade of density of glycogen granules in ischemic hearts of control and hexarelin-treated rats. These aspects, together with the evaluation of ultrastructural changes of myocardial cells associated with ischemia, deserve further investigations to provide a clearer picture of the mode of action of hexarelin in post-ischemic ventricular dysfunction.

In this regard, unpublished observations of our laboratory showed that mechanical manifestations of the phenomenon called "calcium paradox" (34), (increase in resting tension and impaired contractility of the perfused hearts at the moment of calcium re-admission after transient exposure of the organ to a calcium free solution) are markedly inhibited in hearts prepared from young rats treated with hexarelin. These results can be again interpreted to mean that a better control of calcium influx in these hearts could be responsible for the sparing myocardial energy and improved recovery of depressed contractile forces.

Another interesting feature of our studies was the protective effect exerted by GH treatment in the heart preparations from senescent rats. The improvement of post-ischemic ventricular function was, however, modest and by no means comparable, under our experimental conditions, with that elicited by hexarelin, and it is probably attributable to a direct action of the hormone on the heart, where receptors for both GH and IGF-1 have been identified.

In conclusion, these findings clearly indicate that hexarelin, very likely through a mechanism divorced from its GH-releasing effect, strikingly reduces the reperfusion injury in isolated hearts from senescent rats. This action of hexarelin, which under our experimental conditions, overrides that exhibited by GH, opens new perspectives in the therapy of post-ischemic heart dysfunction in the elderly. This subject is of increasing interest because the aged population is continuously growing and is becoming one of the major targets of pharmacology; moreover, cardiac diseases are the first cause of mortality after the age of 65 years.

WORK IN PROGRESS

In line with the above findings, unpublished observations showed that hearts from hypophysectomized rats, exposed to global flow limitation and reperfusion have severe signs of ischemic and post-ischemic ventricular dysfunction, increased CK activity in the perfusates, arhythmias and constriction of the coronary vascular bed. In these hearts, the reduced rate of formation of 6-keto $PGF_{1\alpha}$ and the hyper-reactivity of the coronary vasculature to angiotensin-II suggest that the pituitary ablation impairs the endothelium-dependent relaxant function as in GH-deficient rats.

Administration of hexarelin (80 μg/kg once a day for 7 days) to hypophysectomized rats induced a significant and more prompt recovery of heart contractility with no disturbances of the electrical pacing. This was associated to a normalization of CK released in the perfusates, 6-keto $PGF_{1\alpha}$ generation and angiotensin II activity on coronary vessels.

These data again emphasize that hexarelin's beneficial effects are independent from GH release. In fact, under our experimental conditions, hexarelin, which *per se* was prohibited from acting directly at the pituitary levels, did not show any evidence of stimulation of the somatotropic axis. The protectant activity of hexarelin in hearts of hypophysectomized rats is striking and is probably exerted through cardiac and endothelial receptors activation and is divorced from the GH-releasing properties of the peptide.

REFERENCES

1. Markussis, V., Beshyah, S.A., Fischer, C., Sharp, P., Nicolaides, A.N. and Johnston, D.G. (1992) Detection of premature atherosclerosis by high-resolution ultrasonography in symptom-free hypopituitary adults. Lancet 340, 1188–1192.
2. Rosen, T. and Bengtsson, B.A. (1990) Premature mortality due to cardiovascular disease in hypopituitarism. Lancet 336, 285–288.
3. Shahi, M., Beshyah, S.A., Hackett, D., Sharp, P.S., Johnston, D.G. and Foale, R.A. (1992) Myocardial dysfunction in treated adult hypopituitarism: a possible explanation for increased cardiovascular mortality. Br. Heart J. 67, 92–96.
4. Monson, J.P. and Besser, G.M. (1997) The potential for growth hormone in the management of heart failure. Heart 77, 1–2.
5. Buerke, M., Murohara, T., Skurk, C., Nuss, C., Tomaselli, K. and Lefer, A.M. (1995) Cardioprotective effect of insulin-like growth factor I in myocardial ischemia followed by reperfusion. Proc. Natl. Acad. Sci. USA. 92, 8031–8035.
6. Buccellati, C., Rossoni, G., Bonazzi, A. et al. (1997) Nitric oxide modulation of transcellular biosynthesis of cys-leukotrienes in rabbit leukocyte-perfused heart. Br. J. Pharmacol. 120, 1128–1134.
7. De Gennaro Colonna, V., Rossoni, G., Bonacci, D. et al. (1996) Worsening of ischemic damage in hearts from rats with selective growth hormone deficiency. Eur. J. Pharmacol. 314, 333–338.
8. De Gennaro Colonna, V., Rossoni, G., Bernareggi, M., Müller, E.E. and Berti, F. (1997) Cardiac ischemia and impairment of vascular endothelium function in hearts from GH-deficient rats: protection by hexarelin. Eur. J. Pharmacol. 334, 201–207.
9. Rossoni, G., De Gennaro Colonna, V., Bernareggi, M., Polvani, G.L., Muller, E.E. and Berti, F. (1998) Protectant activity of hexarelin or growth hormone against post-ischemic ventricular dysfunction in hearts from aged rats. J. Cardiovasc. Pharmacol. (in press).
10. Deghenghi, R., Cananzi, M.M., Torsello, A., Battisti, C., Müller, E.E. and Locatelli, V. (1994) GH-releasing activity of hexarelin, a new growth hormone-releasing peptide, in infant and adult rats. Life Sci. 54, 1321–1328.
11. Saccà, L., Cittadini, A. and Fazio, S. (1994) Growth hormone and the heart. Endocr. Rev. 15, 555–573.
12. Cella, S.G., Locatelli, V., Broccia, M.L. et al. (1994) Long term changes of somatotropic function induced by deprivation of growth hormone-releasing hormone during the fetal life of the rat. J. Endocrinol. 140, 111–117.
13. Torsello, A., Luoni, M. and Grilli, R. (1997) Hexarelin stimulation of growth hormone release and mRNA levels in an infant and adult rat model of impaired GHRH function. Neuroendocrinology 65, 31–37.
14. Smith, R.G., Cheng, K. and Schoen, W.R. (1993) A nonpeptidyl growth hormone secretagogue. Science 260, 1640–1643.
15. Bowers, C.Y., Sartor, A.O., Reynolds, G.A. and Badger, T.M. (1991) On the actions of the growth hormone-releasing hexapeptide, GHRP. Endocrinology 128, 2027–2035.
16. Dickson, S.L., Leng, G. and Robinson, I.C. (1993) Systemic administration of growth hormone-releasing peptide activates hypothalamic arcuate neurons. Neuroscience 53, 303–306.
17. Grilli, R., Bresciani, E., Torsello, A. et al. (1997) Tissue-specific expression of GHS-receptor mRNA in the CNS and peripheral organs of the male rat. Proc. 79th Annual Meeting of the Endocrine Society, Minneapolis, 1997, p. 153.

18. Zapf, J. (1995) Physiological role of the insulin-like growth factor binding proteins. Eur. J. Endocrinol. 132, 645–650.
19. Donath, M.J., Jenni, R., Brunner, H.P. et al. (1996) Cardiovascular and metabolic effects of insulin-like growth factor-1 at rest and during exercise in humans. J. Clin. Endocrinol. Metab. 81, 4089–4094.
20. Freestone, N.S., Ribaric, S. and Mason, W.T. (1996) The effect of insulin-like growth factor-1 on adult rat cardiac contractility. Mol. Cell Biochem. 163, 223–229.
21. Duerr, R.L., Huang, S., Miraliakbar, H.R., Clark, R., Chien, K.R. and Ross, J. (1995) Insulin-like growth factor-1 enhances ventricular hypertrophy and function during the onset of experimental cardiac failure. J. Clin. Invest. 95, 619–627.
22. Berti, F., Rossoni, G., Magni, G. et al. (1988) Nonsteroidal anti-inflammatory drugs aggravate acute myocardial ischemia in the perfused rabbit heart: a role for prostacyclin. J. Cardiovasc. Pharmacol. 12, 438–444.
23. Berti, F., Rossoni, G., Omini, C. et al. (1987) Defibrotide, an antithrombotic substance which prevents myocardial contracture in ischemic rabbit heart. Eur. J. Pharmacol. 135, 375–382.
24. Berti, F., Rossoni, G., Della Bella, D. et al. (1993) Nitric oxide and prostacyclin influence coronary vasomotor tone in perfused rabbit heart and modulate endothelin-1 activity. J. Cardiovasc. Pharmacol. 22, 321–326.
25. Böger, R.H., Skamira, C., Bode-Böger, S.M., Brabant, G., von zur Mühlen, A., Frölich, J.C. (1996) Nitric oxide may mediate the hemodynamic effects of recombinant growth hormone in patients with acquired growth hormone deficiency. J. Clin. Invest. 98, 2706–2713.
26. Lakatta, E.G. and Yin, F.C. (1982) Myocardial aging: functional alterations and related cellular mechanisms. Am. J. Physiol. 242, H927–H941.
27. Ji, L.L., Dillon, D. and Wu, E. (1991) Myocardial aging: antioxidant enzyme systems and related biochemical properties. Am. J. Physiol. 261, R386–R392.
28. Mudumbi, R.V., Olson, R.D., Hubler, B.E., Montamat, S.C. and Vestal, R.E. (1995) Age-related effects in rabbit heart of N6-R-phenylisopropyladenosine, an adenosine A1 receptor agonist. Gerontology 50A, B351–B357.
29. Ferrari, R. (1995) Metabolic disturbances during myocardial ischemia and reperfusion. Am. J. Cardiol. 76, 17B–24B.
30. Sonntag, W.E., Steger, R.W., Forman, L.J. and Meites, J. (1980) Decreased pulsatile release of growth hormone in old male rats. Endocrinology 107, 1875–1879.
31. Rudman, D., Kutner, M.H., Rogers, C.M., Lubin, M.F., Fleming, G.A. and Bain, R.P. (1981) Impaired growth hormone secretion in the adult population. J. Clin. Invest. 67, 1361–1369.
32. Howard, A.D., Feighner, S.D., Cully, D.F., Arena, J.P., Liberator, P.A. and Rosenblum, C.I. (1996) A receptor in pituitary and hypothalamus that functions in growth hormone release. Science 273, 974–977.
33. Ruigrok, T.J.C., Boink, A.B.T.J., Slade, A., Zimmerman, A.N.E., Meijler, F.L. and Nayler, W.G. (1980) The effect of verapamil on the calcium paradox. Am. J. Pathol. 98, 769–790.
34. Cavero, I., Boudot, J.P. and Feuvray, D. (1983) Diltiazem protects the isolated rabbit heart from the mechanical and ultrastructural damage produced by transient hypoxia, low-flow ischemia and exposure to Ca^{++}-free medium. J. Pharmacol. Exp. Ther. 226, 258–268.

Growth Hormone Secretagogues
Edited by E. Ghigo, M. Boghen, F.F. Casanueva and C. Dieguez

Chapter 25

Potential Applications of Growth Hormone Secretagogues

ILAN SHIMON[1] and SHLOMO MELMED[2]

[1]*Institute of Endocrinology, Sheba Medical Center, Tel-Hashomer, Israel*
[2]*Department of Medicine, Cedars-Sinai Research Institute, Los Angeles, CA, U.S.A.*

INTRODUCTION

Since 1980 when Bowers et al. showed that short enkephalin analogs stimulate GH release (1), several synthetic hexapeptides with a similar GH-releasing ability, GH-secretagogues (GHS) (2,3), and nonpeptide GHS mimetics (4) have been developed and studied in animals and humans. Although a GHS-specific, G protein-coupled receptor has recently been cloned from the hypothalamus and pituitary (5), the endogenous GHS ligand(s) has not yet been identified.

The GHS exert their main effect in the arcuate nucleus at the hypothalamus where the GHS receptor is expressed, probably affecting GH-releasing hormone (GHRH)-containing neurons (6), and their *in vivo* GH-stimulation requires intact hypothalamo–pituitary function (7). Moreover, functional GHRH receptors are required for GHS function, as GHRH receptor mutation (*lit/lit* mouse) prevents the expected *in vitro* and *in vivo* GH responses to GHS (8). Thus, the hypothalamus is a major target for GHS *in vivo*, in addition to the direct effect of these peptides on the pituitary. We have recently shown that human fetal pituitary expresses GHS receptors, and a direct *in vitro* action of GHS on human somatotrophs also occurs (9). However, only additive effect of GHS with GHRH on GH secretion was shown in the human pituitary, confirming the central role of the hypothalamus for the synergistic interaction between these two modulators of GH.

Thus, based on the known mechanisms involved, GHS may play a significant role in regulating *in vivo* GH secretion in subjects with intact hypothalamo–pituitary axis and functioning anterior pituitary somatotrophs. Patients with hypothalamo–pituitary disconnection (7) or anterior pituitary dysfunction will not benefit from GHS treatment. The GHS and their nonpeptide mimetics provide a more physiological approach to GH replacement than daily GH injections. The recent development of nonpeptide secretagogues (MK-0677) with improved oral availability has identified these analogs as candidates for a once-daily oral drug capable of sustained stimulation of GH and IGF-I release (10,11). In this chapter we will discuss the potential clinical applications of GHS in adults. Use of GHS in childhood for growth disorders is beyond the scope of the chapter.

GHS AS REPLACEMENT THERAPY FOR ADULTS WITH ACQUIRED OR CONGENITAL GH DEFICIENCY

Patients with nonfunctioning pituitary macroadenomas resulting in sellar mass effect, and those with secreting tumors treated by transsphenoidal surgery or sellar irradiation, usually suffer from panhypopituitarism or deficiency of at least one anterior pituitary hormone. GH deficiency is very common, as somatotrophs comprise 40–50% of all pituitary cells and are located in the lateral wings of the gland. Thus, destruction of most but not all anterior pituitary mass may cause severe GH deficiency even if the residual GH-producing cells secrete some GH. Such patients may benefit from GH replacement therapy. Adult GH deficiency results in altered body composition with increased fat and decreased muscle volume and strength, lower psychosocial achievement and quality of life, and altered glucose and lipid metabolism. Some of these effects on body composition and metabolism are reversed by GH replacement (12,13). An alternative route for daily subcutaneous GH administration is continuous stimulation of the remaining small somatotroph cell population by a potent stimulator such as GHS, administered daily either intranasally or orally. Patients with GH deficiency due to pituitary or hypothalamic disease usually respond to a combination of GHRH and GHS (14), which stimulates GH and IGF-I significantly, indicating the potential for use of these peptides in treating GH deficient adults.

Another group of GH deficient adults are patients with congenital GH deficiency treated for short stature with recombinant GH during childhood until linear growth was completed, and treatment subsequently discontinued. These adults, when retested may continue to exhibit signs of GH deficiency, but currently most adults with childhood onset congenital GH deficiency are not yet treated with recombinant GH replacement after the age of 18 years. Injection inconvenience, economic limitations, GH-related adverse effects and controversy regarding the appropriate dose of GH for replacement have until very recently postponed its recommendation in most young adults with proven childhood-onset GH deficiency. These patients may potentially benefit from regular GHS treatment. Oral administration of MK-0677 to adults with congenital GH deficiency enhances GH secretion, thus GHS may have a role in the treatment of GH deficiency of childhood onset (15). However, most cases of hereditary GH deficiency result from GH gene deletion or lack of GHRH synthesis or secretion, and GH stimulation by GHS may prove to be modest in those patients.

GHS FOR SOMATOPAUSE

After puberty serum GH and IGF-I levels gradually decline during progression of the life span, and at the age of 70–80 years, 50% of all healthy subjects do not have significant circulating GH during the day. Serum IGF-I levels reflect these changes by declining progressively after the age of 40. The geriatric decline in endogenous GH production probably results from decreased hypothalamic GHRH secretion and increased hypothalamic somatostatin. Treatment with GHRH for two weeks restores the decreased GH and IGF-I levels in elderly men to normal ranges for young subjects (16).

Progressive changes in body composition with lean body mass shrinkage and reciprocal expansion of the adipose mass are well-documented with increasing age. In addition, loss of bone mass and progressive reduction of functional capacities of most organs occur. Thus, it has been postulated that geriatric decline in GH and IGF-I may contribute to age-related body composition and energy capacity changes. GH replacement treatment for one year in elderly men demonstrated beneficial effects on body composition, increasing lean body mass and decreasing adipose tissue (17). Adverse side effects associated with elevated IGF-I levels were common. It was therefore suggested, that GHS would stimulate a physiologic pattern of GH secretion. Oral administration of GHS in healthy elderly volunteers increased the amplitude of the GH pulses without changing the pulse number (18). Serum IGF-I levels also increased into the normal range for young adults. Moreover, GH response to GHS administration does not decline in late adulthood, compared to young healthy subjects (19). Thus, the impaired GH secretion in the healthy elderly population is a potentially reversible state, and GHS provide a more physiological approach to GH replacement in this frail elderly population.

GHS IN CHRONIC CATABOLIC STATES

Because of its anabolic effects reversing negative nitrogen balance, GH administration was studied in various catabolic states, including renal failure, burn injury, sepsis, during prolonged recovery after major surgery or trauma, and in AIDS patients with generalized wasting. During critical illness, GH secretion is reduced, resulting in decreased pulse amplitude and low circulating IGF-I levels (20,21). It was suggested that short-term GH therapy in these states, combined with appropriate medical and nutritional support would ameliorate ongoing catabolism, reverse this wasting syndrome and shorten the illness duration (22). For example, treatment of patients with AIDS-wasting with recombinant GH results in weight gain, increased lean body mass and functional capacity, without affecting clinical progression of AIDS (23). As these patients have an intact hypothalamo–pituitary axis not affected by the wasting syndrome, they are appropriate candidates for a more physiologic treatment pattern, i.e. GHS administration. In critically ill adults, including patients with respiratory failure, sepsis, shock, polytrauma, multiple organ failure, and necrotizing pancreatitis, GHRP-2 infusion increases both GH concentration (basal and burst amplitude levels) and IGF-I levels (21). Short-term oral treatment with MK-0677 to catabolic subjects reverses the diet-induced nitrogen wasting (24), thus emphasizing the therapeutic potential of GHS in the complex treatment support of catabolic patients with critical illness.

GHS FOR OSTEOPOROSIS

GH deficiency is associated with decreased bone mineral density (BMD) and a significantly increased fracture rate. GH administration to patients with GH deficiency results in increased BMD of the lumbar spine and femoral neck (25), and in acute increase of bone turnover markers (26). It is assumed that age-related GH deficiency may contribute to the osteoporosis seen in the elderly population, in addition to other important pathogenetic factors, including lack of estrogen in postmenopausal women. Thus, GH treatment may prevent the senile component of osteoporosis, which results mainly from defective bone formation. Cyclic GH treatment to postmenopausal women results in a significant increase of lumbar spine BMD. Long-term oral administration of a GHS preparation may be a reasonable and physiologic way to treat primary osteoporosis in the elderly population, and it should also be tested as an option in the treatment of postmenopausal osteoporosis.

GHS FOR OBESITY AND HEART FAILURE

Healthy obese men were treated for two months with MK-0677, resulting in a sustained increase in fat free mass and a transient increase in basal metabolic rate (27). Interestingly, in obesity the GH response to GHS is not as impaired as it is to GHRH. Further prospective studies are needed to elucidate the potential role of GHS in regulating body fat mass.

Recombinant human GH administered to patients with moderate-to-severe heart failure resulted in improved left ventricular function, exercise capacity, clinical symptoms and patients' quality of life (28). However, this treatment doubled the IGF-I serum concentrations, and the changes were partially reversed after GH was discontinued. Treatment of heart failure patients with GHS has not yet been reported, and this mode of treatment may prove effective in ameliorating cardiac symptoms without inducing acromegalic symptoms.

DIAGNOSTIC TESTS

Pituitary GH response to GHS administration can be used to assess pituitary functional reserve. A complete blockade of GH response to GHS in patients with pituitary stalk transection, suggests that this could be a sensitive test for the diagnosis of this specific condition (29). GH deficiency in adults, whether of adult or childhood onset, can also be evaluated using GHS. GHS are preferred over GHRH in various metabolic states, including obesity, non-insulin-dependent diabetes mellitus, and anorexia nervosa, because the GH response to GHRH is more impaired than it is to GHS in these states.

SUMMARY

The novel group of recently developed oral GHS, are potent stimulators of endogenous GH secretion. Their convenient potential once-daily administration, and the more physiologic pattern of GH replacement compared with daily GH injections, have opened the field of GH replacement to several new treatment options. These potential applications of GHS treatment should be evaluated in long-term controlled clinical studies to study their potential benefits and to monitor side effects. If proved effective and safe, the GHS may replace GH for either novel or well-established indications.

REFERENCES

1. Bowers, C.Y., Momany, F.A., Chang, D., Hong, A. and Chang, K. (1980) Structure–activity relationships of a synthetic pentapeptide that specifically releases GH *in vitro*. Endocrinology 106, 663–667.
2. Momany, F.A., Bowers, C.Y., Reynolds, G.A. and Newlander, K. (1984) Conformational energy studies and *in vitro* activity data on active GH-releasing peptides. Endocrinology 114, 1531–1536.
3. Bowers, C.Y., Momany, F.A., Reynolds, G.A. and Hong, A. (1984) On the *in vitro* and *in vivo* activity of a new synthetic hexapeptide that acts on the pituitary to specifically release growth hormone. Endocrinology 114, 1537–1545.
4. Smith, R.G., Cheng, K., Schoen, W.R. et al. (1993) A nonpeptidyl growth hormone secretagogue. Science 260, 1640–1643.
5. Howard, A.D., Feighner, S.D., Cully, D.F. et al. (1996) A receptor in pituitary and hypothalamus that functions in growth hormone release. Science 273, 974–977.
6. Smith, R.G., Van Der Poleg, L.H.T., Howard, A.D. et al. (1997) Peptidomimetic regulation of growth hormone secretion. Endoc. Rev. 18, 621–645.
7. Popovic, V., Damjanovic, S., Micic, D., Djurovic, M., Dieguez, C. and Casanueva, F.F. (1995) Blocked growth hormone-releasing peptide (GHRP-6)-induced GH secretion and absence of the synergic action of GHRP-6 plus GH-releasing hormone in patients with hypothalamo–pituitary disconnection: evidence that GHRP-6 main action is exerted at the hypothalamic level. J. Clin. Endocrinol. Metab. 80, 942–947.
8. Korbonits, M. and Grossman, A.B. (1995) Growth hormone-releasing peptide and its analogues. Novel stimuli to growth hormone release. Trends Endocrinol. Metab. 6, 43–49.
9. Shimon, I., Yan, X., Melmed, S. (1998) Human fetal pituitary expresses functional growth hormone-releasing peptide receptors. J. Clin. Endocrinol. Metab. 83, 174–178.
10. Jacks, T.M., Smith, R.G., Judith, F. et al. (1996) MK-0677, a potent, novel orally-active growth hormone (GH) secretagogue: GH, IGF-1 and other hormonal responses in beagles. Endocrinology 137, 5284–5289.
11. Hartman, M.L., Farello, G., Pezzoli, S.S. and Thorner, M.O. (1992) Oral administration of growth hormone (GH)-releasing peptide stimulates GH secretion in normal men. J. Clin. Endocrinol. Metab. 74, 1378–1384.
12. Salomon, F., Cuneo, R.C., Hesp, R. and Sonksen, P.H. (1989) The effects of treatment with recombinant human growth hormone on body composition and metabolism in adults with growth hormone deficiency. N. Engl. J. Med. 321, 1797–1803.
13. Cuneo, R.C., Judd, S., Wallace, J.D. et al. (1998) The Australian multicenter trial of growth hormone (GH) treatment in GH-deficient adults. J. Clin. Endocrinol. Metab. 83, 107–116.
14. Kendall-Taylor, P., Paxton, A. and Koppiker, N.P. (1996) Observations on the stimulation of growth hormone secretion in patients with growth hormone deficiency. Metabolism 45 (Suppl. 1), 127–128.
15. Chapman, I.M., Pescovitz, O.H., Murphy, G. et al. (1997) Oral administration of growth hormone (GH) releasing peptide-mimetic MK-677 stimulates the GH/insulin-like growth factor-I axis in selected GH-deficient adults. J. Clin. Endocrinol. Metab. 82, 3455–3463.
16. Corpas, E., Harman, S.M., Pineyro, M.A. et al. (1992) GH-releasing hormone (1-29) twice daily reverses the decreased GH and insulin-like growth factor-I levels in old men. J. Clin. Endocrinol. Metab. 75, 530–535.
17. Rudman, D., Feller, A.G., Nagraj, H.S. et al. (1990) Effects of human growth hormone in men over 60 years old. N. Engl. J. Med. 323, 1–6.
18. Chapman, I.M., Bach, M.A., Van Cauter, E. et al. (1996) Stimulation of the growth hormone (GH)-insulin-like growth factor-I axis by daily oral administration of a GH secretagogue (MK-0677) in healthy elderly subjects. J. Clin. Endocrinol. Metab. 81, 4249–4257.
19. Micic, D., Popovic, V., Kendereski, A., Macut, D., Casanueva, F.F. and Dieguez, C. (1995) Growth hormone secretion after the administration of GHRP-6 or GHRH combined with GHRP-6 does not decline in late adulthood. Clin. Endocrinol. (Oxf.) 42, 191–194.

20. Van den Berghe, G., de Zegher, F., Lauwers, P. and Veldhuis, J.D. (1994) Growth hormone secretion in critical illness: effect of dopamine. J. Clin. Endocrinol. Metab. 79, 1141–1146.
21. Van den Berghe, G., de Zegher, F., Veldhuis, J.D. et al. (1997) The somatotropic axis in critical illness: effect of continuous growth hormone (GH)-releasing hormone and GH-releasing peptide-2 infusion. J. Clin. Endocrinol. Metab. 82, 590–599.
22. Voerman, B.J., Strack van Schijndel, R.L.M., Groeneveld, A.B.J., de Boer, H., Nauta, J.P. and Thijs, L.G. (1995) Effects of human growth hormone in critically ill nonspecific patients: results from a prospective, randomized, placebo-controlled trial. Crit. Care Med. 23, 665–673.
23. Schambelan, M., Mulligan, K., Grunfeld, C. et al. (1996) Recombinant human growth hormone in patients with HIV-associated wasting. A randomized, placebo-controlled trial. Serostim Study Group. Ann. Intern. Med. 125, 873–882.
24. Murphy, M.G., Plunkett, L.M., Gertz, B.J. et al. (1998) MK-677, an orally active growth hormone secretagogue, reverses diet-induced catabolism. J. Clin. Endocrinol. Metab. 83, 320–325.
25. Johannsson, G., Rosen, T., Bosaeus, I., Sjostrom, L. and Bengtsson, B.A. (1996) Two years of growth hormone treatment increases bone mineral content and density in hypopituitary patients with adult-onset GH deficiency. J. Clin. Endocrinol. Metab. 81, 2865–2873.
26. Bianda, T., Glatz, Y., Bouillon, R., Froesch, E.R. and Schmid, C. (1998) Effects of short-term insulin-like growth factor-I (IGF-I) or growth hormone (GH) treatment on bone metabolism and on production of 1,25-dihydroxycholecalciferol in GH-deficient adults. J. Clin. Endocrinol. Metab. 98, 81–87.
27. Svensson, J., Lonn, L., Jansson, J.O. et al. (1998) Two-month treatment of obese subjects with the oral growth hormone (GH) secretagogue MK-677 increases GH secretion, fat-free mass, and energy expenditure. J. Clin. Endocrinol. Metab. 83, 362–369.
28. Fazio, S., Sabatini, D., Capaldo, B. et al. (1996) A preliminary study of growth hormone in the treatment of dilated cardiomyopathy. N. Engl. J. Med. 334, 809–814.
29. Pombo, M., Leal-Cerro, A., Barreiro, J. et al. (1996) Growth hormone releasing hexapeptide-6 (GHRP-6) test in the diagnosis of GH-deficiency. J. Pediat. Endocrinol. Metab. 9 (Suppl. 3), 333–338.

Index

www.ingramcontent.com/pod-product-compliance
Ingram Content Group UK Ltd.
Pitfield, Milton Keynes, MK11 3LW, UK
UKHW020238290926
15470UKWH00002B/598

9 780444 829337